Remote Sensing Change Detection

Environmental Monitoring Methods and Applications

EDITORS

Ross S. Lunetta
Christopher D. Elvidge

This book represents information obtained from authentic and highly regarded sources. Reprinted material is used with permission, and sources are indicated. A wide variety of references are listed. Every reasonable effort has been made to give reliable data and information, but the authors, publisher and copyright holder cannot assume responsibility for the validity of all materials or for the consequences of their use.

Published in the UK in 1999 by
Taylor & Francis Ltd., 1 Gunpowder Square, London, EC4A 3DF

British Library Cataloguing in Publication Data
A catalogue record for this book is available from the British Library.

Library of Congress Cataloguing-in-Publication Data

This edition not for sale in USA and Canada
ISBN 0-7484-0861-4 (cased)

Printed in the United States of America

DEDICATION

This book is dedicated to the memory of Mason J. Hewitt III, a loving father, patriot, distinguished colleague, and good friend who inspired through example. His memory lives on through his family, friends, and colleagues. He is sorely missed by all whose life he touched.

ACKNOWLEDGMENTS

The editors and authors would like to acknowledge their parents, who provided the unwavering encouragement, inspiration, and support required for us to achieve our personal goals and aspirations. They are the unsung heroes who gave their all without reservation to provide us with the freedom and means to pursue our dreams. Also, we would like to acknowledge those mentors and colleagues who through their contagious enthusiasm and support provided a fertile environment for our professional development. A special note of thanks is extended to our spouses and significant others who through their support, patience, and perseverance made this book possible. Special acknowledgment is extended to Dr. John G. Lyon for his guidance and numerous reviews in preparation of this book. Also, the editors would like to thank Mr. Adam Rifenberick, Project Manager, Sleeping Bear Press, for his exceptional production and coordination efforts.

ABOUT THE EDITORS

Ross S. Lunetta is currently an Environmental Scientist with the U.S. Environmental Protection Agency National Exposure Research Laboratory in Research Triangle Park, North Carolina. Academic work has included both bachelor's and master's degrees in biology from Northern Michigan University, post-master's work in aquatic ecology at Wayne State University and remote sensing at Eastern Michigan University. Current research interest is in the application of advanced remote sensor technologies for environmental monitoring and the study of ecosystem processes. Previous positions have included serving as a Senior Science Advisor, Remote Sensing Program Manager, and as the Technical Director for the development of the North American Landscape Characterization (NALC) interagency research program in support of the Landsat Pathfinder series. He was employed from 1983 to 1988 by the U.S. Army Corps of Engineers as an ecologist directing studies on the environmental impacts of federal water projects on the Laurentian Great Lakes. The views expressed in this book do not necessarily represent the views of the U.S. Environmental Protection Agency.

Christopher D. Elvidge is a Physical Scientist at the National Oceanic and Atmospheric Administration (NOAA) National Geophysical Data Center, Boulder, Colorado, and Associate Research Professor in Biological Sciences at the Desert Research Institute in Reno, Nevada. He received a Ph.D. in applied earth sciences from Stanford University in 1985. From 1985 to 1987 he was a Research Associate at NASA's Jet Propulsion Laboratory, Pasadena, California through a grant from the National Research Council. In 1988, Dr. Elvidge joined the faculty of the Desert Research Institute Biological Sciences Center. During 1991–1993 he was a visiting scientist at the U.S. Environmental Protection Agency Office of Research and Development's Global Change Research Program. At NOAA he is doing research on the remote sensing of fires, shoreline erosion, intertidal zone features, and is leading an effort to assemble the first global inventory of human settlements to be derived from satellite data.

CONTRIBUTORS

John D. Althausen
Department of Geography
University of South Carolina
Columbia, SC 29208

Bernard Armour
Atlantis Scientific Inc.
20 Colonnade Road
Nepean, Ontario
Canada, K2E 7M6

Getulio T. Batista
INPE, Caixa Postal 515
Avenida dos Astronautas
12227-010, Sho Jose Campos
Sao Paulo, Brazil

Kimberly E. Baugh
Cooperative Institute for Research in Environmental Sciences
University of Colorado
Boulder, CO 80303

Greg S. Biging
Center for the Assessment and Monitoring of Forest and Environmental Resources
Department of Environmental Science, Policy, and Management
University of California, Berkeley
Berkeley, CA 94720-3110

Vittorio Castelli
Image Information Systems
IBM T.J. Watson Research Center
P.O. Box 704
Yorktown Heights, NY 10598

Zhikang Chen
IMD/OEE
South Florida Water Management District
P.O. Box 24680
West Palm Beach, FL 33416-4680

Warren B. Cohen
Forestry Sciences Laboratory
Pacific Northwest Research Station
USDA Forest Service
Corvallis, OR 97331

David R. Colby
National Oceanic and Atmospheric Administration
Southeast Fisheries Science Center
Beaufort Laboratory
Beaufort, NC 28516-9722

Russell G. Congalton
Department of Natural Resources
University of New Hampshire
Durham, NH 03824

David J. Cowen
Department of Geography
University of South Carolina
Columbia, SC 29208

Christopher D. Elvidge
Solar-Terrestrial Physics Division
NOAA-NESDIS National Geophysical Data Center
325 Broadway
Boulder, CO 80303

Maria Fiorella
Department of Forest Science
Forestry Sciences Laboratory 020
Oregon State University
Corvallis, OR 97331

David P. Groeneveld
Resource Management
P.O. Box 3296
Telluride, CO 81435

Bert Guindon
Canada Centre for Remote Sensing
588 Booth Street
Ottawa, Canada K1A0Y7

Quinn J. Hart
Department of Land, Air, and Water Resources
University of California
Davis, CA 95616

Vinita Ruth Hobson
Cooperative Institute for Research in Environmental Sciences
University of Colorado
Boulder, CO 80303

Wally T. Jansen
WTJ Software Services
809 Lawrence Road
San Mateo, CA 94401

John R. Jensen
Department of Geography
University of South Carolina
Columbia, SC 29208

Jackie Kendall
NASA Goddard Space Flight Center
Greenbelt, MD 20770

Eric A. Kihn
Solar-Terrestrial Physics Division
NOAA-NESDIS National Geophysical Data Center
325 Broadway,
Boulder, CO 80303

Herbert W. Kroehl
Solar-Terrestrial Physics Division
NOAA-NESDIS National Geophysical Data Center
325 Broadway
Boulder, CO 80303

Chung-Sheng Li
IBM T.J. Watson Research Center
P.O. Box 704
Yorktown Heights, NY 10598

Ross S. Lunetta
U.S. Environmental Protection Agency
National Exposure Research Laboratory (MD-56)
Research Triangle Park, NC 27711

John G. Lyon
The Ohio State University
Department of Civil and Environmental Engineering and Geodetic Science
Columbus, OH 43210

Tomoaki Miura
Department of Soil, Water and Environmental Sciences
The University of Arizona
Tucson, AZ 85705

Sunil Narumalani
Department of Geography
University of South Carolina
Columbia, SC 29208

Bruce W. Nelson
INPA-Botanica
Avenida Andre Araujo 1756
69083 Manaus
Amazon, Brazil

Hiroshi Ohkura
National Research Institute for Earth Science and Disaster Prevention
Science and Technology Agency
3-1 Tennodai, Tsukuba
Ibaraki 305, Japan

Dee W. Pack
The Aerospace Corporation
Space and Environment Technology Center
Remote Sensing Department
P.O. Box 92957, M2/251
Los Angeles, CA 90009-2957

Jorge L.G. Pereira
INPE, Caixa Postal 515
Avenida dos Astronautas
12227-010, Sho Jose Campos
Sao Paulo, Brazil

Elaine Prins
NOAA-NESDIS
UW-Madison Cooperative Institute for Meteorological Satellite Studies
1225 West Dayton Street
Madison, WI 53706

Elijah W. Ramsey III
U.S. Geological Survey
Biological Resources Division
National Wetlands Research Center
700 Cajundome Blvd.
Lafayette, LA 70506

Jayita Ray
Department of City and Regional Planning
University of Texas at Arlington
Arlington, TX 76013

Dar A. Roberts
Department of Geography, EH3611
University of California
Santa Barbara, CA 93106

Genya Saito
Department of Environmental Management
National Institute of Agro-Environmental Sciences
1-1-3 Higashi, Tsukuba
Ibraki 305, Japan

Akiko Tanaka
Geophysics Department
Geological Survey of Japan
1-1-3 Higashi, Tsukuba
Ibaraki 305, Japan

John J. Turek
IBM T.J. Watson Research Center
P.O. Box 704
Yorktown Heights, NY 10598

Susan L. Ustin
Department of Land, Air, and Water Resources
University of California
Davis, CA 95616

Eric K. Waller
Department of Geography
University of California
Santa Barbara, CA 93106

Oliver Weatherbee
Department of Marine Sciences
University of Delaware
Newark, DE 19716

Ding Yuan
Resource 21, LLC
7257 South Tuscon Way
Englewood, CO 80112

TABLE OF CONTENTS

Chapter 12: Monitoring Trends in Wetland Vegetation Using Landsat MSS Time Series

Chapter 13: Radar Remote Sensing of Wetlands

Chapter 14: Radar Interferometry for Environmental Change Detection

CHAPTER 1

Applications, Project Formulation, and Analytical Approach

Ross S. Lunetta

1.0 INTRODUCTION

It has been said, "Change is continuous, but learning is optional." As humankind moves to the twenty-first century, environmental changes are predicted to accelerate, with unknown and potentially devastating consequences. Although, many scientific advances have occurred over the latter part of the twentieth century that have dramatically advanced our understanding of ecosystem processes, we are poorly positioned to predict with an acceptable degree of certainty what awaits humanity in the coming decades. The challenges are daunting: changing climate, sea level rise, changing hydrologic regimes, vegetation redistributions, and potential agricultural failures on a massive scale. Important scientific questions that will drive future environmental research include the following: How will we know if, when, and where predicted environmental impacts will occur? What ecosystems are most vulnerable to anthropogenic and natural stresses? Which ecosystems have the best potential for sustaining important ecological functions critical to maintain global ecosystem processes? Will it be possible to identify the onset of predicted changes early enough to act preemptively to either avoid or minimize the potential for catastrophic consequences? Can we successfully answer these questions and effectively formulate and implement plans to avert massive ecosystem failures, or are we destined to be passive participants in a game of environmental roulette that many believe is inevitable?

With the launch of the first Earth Resources Technology Satellite (ERTS-1) in 1972 (later renamed Landsat 1), the United States initiated the technological capability for the monitoring of environmental resources and the study of ecosystem processes from space. Since that time, the utility of space-based remote sensor platforms has been demonstrated and their potential applications for long term environmental monitoring recognized. Initially, the use of remote sensing data for large area applications was limited due to the lack of quantitative data processing techniques and insufficient computer technology required to process large data volumes. With the refinement and validation of mathematical data processing algorithms in the 1980s and major advancements in computer hardware technologies culminating in the 1990s, it is now technologically feasible to monitor large geographic regions of the planet on a periodic or regular basis. In addition, a new generation of satellites, scheduled for initial deployment at the turn of the century, will incorporate state-of-the-art 1990s advances in detector and electronic miniaturization technologies. This effort will represent the first coordinated attempt by the technologically advanced nations of the world to develop and deploy a comprehensive earth remote sensing satellite system capable of monitoring environmental changes over the entire planet, at the spatial scales and temporal frequencies required for both periodic and continual ecosystem monitoring.

The future challenge facing the scientific community of the early twenty-first century will be to utilize this myriad of remote sensing space-based technologies to answer the questions articulated above. Unlike prior scientific undertakings that were accomplished within narrow scientific disciplines, the issue of anthropogenic stressors and their subsequent impacts on ecosystem structure and function will require a truly multidisciplinary scientific approach. Armed with an arsenal of advanced technologies and an assemblage of analytical techniques, future scientific leadership will be of great importance to focus, coordinate, and link the multifaceted elements required to understand changing ecosystem processes. Space-based remote sensor technologies will provide the massive amounts of geospatial data required for ecosystem characterization for: the initialization, parameterization, and validation of earth process models; the detection of early-warning indicators of ecosystem change; and for documenting changes to ecosystem structure at multiple scales for the study of global, regional, and watershed processes to quantify anthropogenic impacts.

2.0 APPLICATIONS

Past applications of earth remote sensing satellite systems have largely focused on the mapping of environmental conditions or the characterization of ecosystem structure and function at one particular point in time, primarily to establish a baseline condition. Initially, the user community commonly utilized remote sensing data and derivative products in hard copy image or map formats. Although these products provided a powerful synoptic observation tool to better understand the spatial distribution of ecosystem resources and to study coarse-scale ecosystem processes, they did not provide quantitative information in a format amenable for integration with other types of data or for input to empirical or mechanistic ecosystem process models. However, with the widespread availability of analytical data processing algorithms and geographic information systems (GIS) in the 1990s, digital spatial products now predominate. The merging of remote sensing data coverages with other spatial ancillary data types is defined as the process of geospatial "data integration." This approach provides a particularly powerful tool for the science of ecosystem monitoring (Nellis et al., 1990).

Recent applications of digital remote sensor data largely focused on the development of ecosystem baseline conditions and investigation of coarse-scale vegetation dynamics. Past applications have included land cover characterization and classification (Townshend et al., 1987); coarse-scale vegetation dynamics (Justice et al., 1985; Justice and Hiernaux, 1986; Malingreau, 1986; Townshend and Justice, 1986); crop cover and acreage estimates (Carlson and Aspiazu, 1975; Lo et al., 1986; Wiegand et al., 1991); forest conversion monitoring and biomass estimates (Tucker et al., 1984; Nelson et al., 1987; Vogelmann, 1988); primary productivity measurements (Tucker and Sellers, 1986); post-fire forest recovery sequence monitoring (Fuller and Rouse, 1979); habitat condition mapping (Ormsby and Lunetta, 1987; Huber and Casler, 1990); crop monitoring (Teng, 1990), crop productivity by estimating leaf area index (Pollock and Kanemasu, 1979); and insect infestations mapping (McCulloch and Hunter, 1983; Nelson, 1983; Tucker et al., 1985).

Currently, researchers have been investigating the application of remote sensor data for the characterization of ecosystem structure, dynamics, and processes. These exploratory applications have included: deforestation and carbon budget dynamics and processes (Prins and Menzel,

1992; Keller et al., 1993; Lucas et al., 1993; Iverson et al., 1994); forest structure characterization (Peterson et al., 1986; Cohen and Spies, 1992) and seral stage determination (Cohen et al., 1995); documentation of tropical deforestation (Malingreau et al., 1989; Gilruth and Hutchinson, 1990; Malingreau and Tucker, 1990; Nelson and Horning, 1990; Mayaux and Lambin, 1995); reforestation processes (Fiorella and Ripple, 1993b; Sadler, 1995); forest defoliation monitoring (Muchoney and Haack, 1994); forest succession dynamics (Fiorella and Ripple, 1993a); forest wildfire monitoring (Kasischke et al., 1993; Kasischke and French, 1995); forest postfire revegetation monitoring (Jakubauskas et al., 1990; Ahern et al., 1991; Chuvieco and Martin, 1994; Marchetti et al., 1995); ecosystem diversity calculations (Stoms, 1992; Stoms and Estes, 1993; Wickham et al., 1995); analysis of landscape pattern and structure (Ritters et al., 1995); desertification processes (Mishra et al., 1994); habitat evaluation studies (Herr and Queen, 1993; Prasad et al., 1994; Lee and Marsh, 1995; Lunetta et al., 1997); wetland system dynamics (Jensen et al., 1993); leaf water stress measurements (Cohen, 1991); plant biochemical content measurements (Peterson and Hubbard, 1992); arid ecosystem vegetation changes (Price et al., 1992; Chavez and MacKinnon, 1994); agricultural crop growth parameters (Thenkabail et al., 1992; Thenkabail, et al., 1994); large-area/coarse-resolution vegetation monitoring (Justice et al., 1991; Eidenshink, 1992); land cover/use change detection (Foran, 1987; Quarmby and Cushine, 1989; Alwashe and Bokhari, 1993; Ram and Kolarkar, 1993; Green et al., 1994); and quantification of cumulative environmental effects (Green et al., 1993).

A major step forward in the application of remote sensing technologies for environmental monitoring and future change detection studies has been the initiation of consistent land cover/use (LCLU) mapping and database development efforts at the global, continental, and national scales. At the national scale, efforts began with construction of the biweekly conterminous United States composite data sets using the National Oceanic and Atmospheric Administration (NOAA) Advanced Very High Resolution Radiometer (AVHRR) polar orbiting satellite, and companion normalized vegetation index (NDVI) composites developed by the EROS Data Center beginning in 1990 (Eidenshink, 1992). Subsequent to the development of biweekly composites, the first coarse-scale land cover mapping over the conterminous United States was accomplished (Loveland et al., 1991). Currently, there is an international effort underway to map land cover for most Earth continents by the year 2002. The North American Landscape Characterization (NALC) program began in 1992 to compile time series Landsat Multispectral Scanner (MSS) "triplicate" data sets for the 1970s, 1980s, and 1990s. This was done for the majority of the North American land mass, for the purpose of generating LCLU change detection products (Lunetta et al., 1993). In 1993 the interagency Multi-Resolution Land Characteristics (MRLC) project purchased multitemporal Landsat Thematic Mapper (TM) data for the continental United States to provide a consistent data set for land cover/change categorization (Bara and Shaw, 1995). As part of the MRLC program a consistent LCLU database is scheduled for completion by the year 2000. These efforts mark an important milestone for the application of remote sensing for environmental studies because they serve to provide baseline data sets for future remote sensing based change detection analyses.

3.0 PROJECT FORMULATION

Most remote sensing change detection applications utilize the standard project formulation approach outlined in this section. The project formulation elements include problem definition, development of product specifications, data requirements analysis, determination of data availability, calculation of data acquisition costs, and data analysis cost estimates (Figure 1.1). Sub-

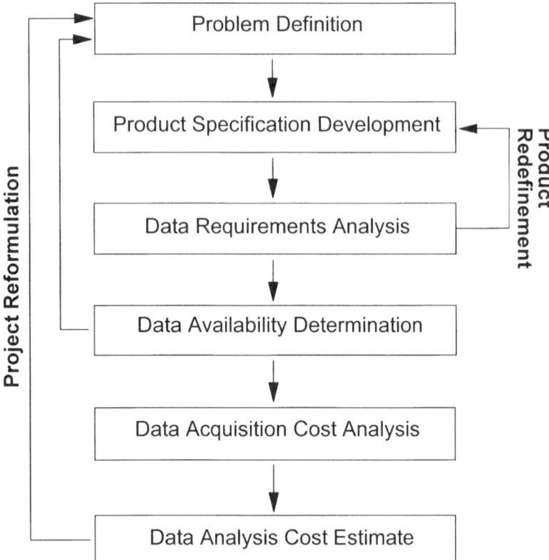

Figure 1.1. Sequential steps outlining the project formulation process for remote sensing change detection projects. Based on data availability and data analysis costs, project reformulation is frequently required.

sequent to the determination of data availability and the calculation of project costs, project objectives and products, may necessitate reformulation. If modifications are necessary, the redefined requirements must feed back into the project formulation process. The process of project reformulation may need to be repeated several times before a suitable project plan is achieved.

Equally as important is the development of the data analysis approach and data processing flow diagram, although most remote sensing change detection analysis typically involves all steps outlined in Figure 1.2. The analytical methods employed to accomplish each individual element will vary considerably based on the specific goals and objectives defined using the project formulation process above. The final result is the development of a data processing flow diagram that sequentially details all analysis steps and data processing time lines (optional). The data processing flow diagram is particularly important to facilitate the implementation of complex data analysis projects, especially those involving numerous investigators and/ or technicians.

3.1 Problem Definition

The first step in developing a change detection analysis project is to precisely define the client's needs or science objectives. This can simply be stated in one of two ways:

- what does the client need to know, or
- what is the specific science question(s) to be answered.

Detailed documentation of client needs and science questions is essential information for the development of the data analysis approach. The greater the level of detail, the higher the probability of a successful outcome with the lowest possible expenditure of resources. If the goal is to answer a science question or to conduct a research study, problem definition should be

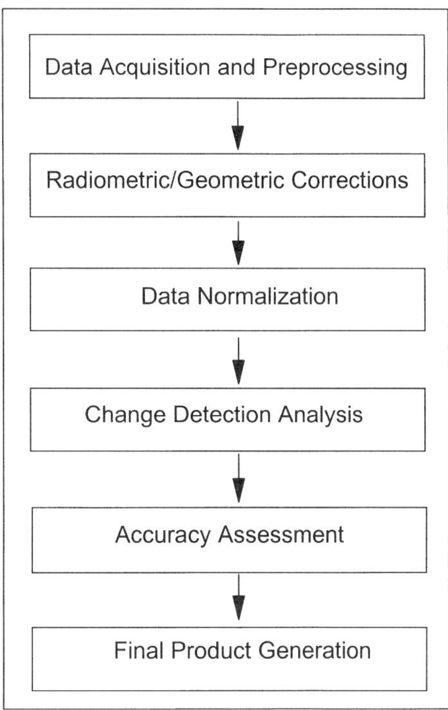

Figure 1.2. General data processing elements for a remote sensing change detection application project or scientific study. Analytical methods and techniques for individual elements are selected to best meet project objectives.

accomplished using the process of hypothesis formulation to support a subsequent experimental design for hypothesis testing. Failure to clearly define the problem in sufficient detail will inevitably result in either a product that will not meet client needs or not provide a satisfactory scientific outcome, and has a high probability of being cost-excessive. Efforts spent in the initial problem formulation stage to clearly detail and document scientific objectives are essential for the steps to follow.

3.2 Product Specifications

The development of product specifications is most effectively accomplished in concert with the problem definition process outlined above. This approach fosters the logical progression of client thought processes to provide a complete understanding of project goals, objectives, and/or output deliverables. Product deliverables can be in the form of hard copy (plotter or film) and spatial database (raster or vector) products. When developing specifications for spatial database products it is important to include requirements for both thematic and geometric accuracy and precision. Thematic accuracy requirements are typically expressed in terms of percent accurate at a specified confidence interval. Individual class accuracies are typically reported by the proportion correctly categorized with a corresponding percent of omission and commission errors by class. In addition, the error (or confusion) matrices facilitate the identification of confusion associated with errors of omission and commission among the various thematic categories (Congalton, 1991). Geometric accuracy requirements are expressed in terms of root mean square (RMS) error and are typically based on the degree of absolute correspondence to an earth coordinate system or reference image data set. Values are commonly reported as a plus/minus pixel value (e.g., ± 0.5 pixel).

3.3 Data Requirements

There are three principal criteria used to establish data requirements to support project goals and objectives. They include the following:

- spectral data resolution
- spatial data resolution and
- temporal data resolution.

The simplest approach is to address and establish requirements for each of the above criterion separately, then consider them in total to identify the remote sensor data acquisition system(s) capable of meeting all specified source data requirements.

Spectral data resolution includes two elements, the identification of the specific portion or location along the electromagnetic spectrum to be detected, and requirements (if any) detailing the range in wavelengths to be sampled (band width). These requirements can only be established based on a thorough understanding of the interaction of the ecosystem characteristics of interest with incoming solar radiation or artificially propagated electromagnetic energy. A common approach used for vegetation discrimination is to select wavelengths that correspond to pigment absorbance minima and reflectance maxima to provide unique spectral signatures for targets of interest. Band width becomes an issue for a given material or target when adjacent wavelength locations along the electromagnetic spectrum will either mute the primary signal or suppress a complementary signal of interest.

Spatial data resolution dictates the scale to which the data may be useful for ecosystem analysis. The concept of a minimum mapping unit (MMU) is an important consideration when establishing spatial data requirements. The MMU is defined as the highest resolution ground element that can be consistently and repetitively mapped over the area of interest. A general rule of thumb is that the spatial resolution of the remotely sensed data should be an order of magnitude greater than the required MMU. When establishing MMUs for ecological studies, an important consideration is the scale at which physical and biological processes of interest are occurring. Ideally, all remotely sensed data should support a MMU that corresponds to the ecosystem processes of interest.

Temporal data resolution for the study of ecosystem condition is based largely on the annual phenological cycles of vegetation that include green-up, growth and maturation cycles, and senescence. The importance of phenology for the characterization of land-cover vegetation types and condition has long been recognized. Kanemasu (1974) recorded seasonal canopy reflectance of wheat using a portable spectral radiometer. The application of the NDVI for the study of vegetation on a regional to global scale was demonstrated using AVHRR meteorological data by Justice et al. (1985) and further demonstrated by other studies (Justice and Hiernaux, 1986; Malingreau, 1986; Townshend and Justice, 1986; Townshend et al., 1987; Justice et al., 1989; Justice et al., 1991). The NDVI continues to be the vegetation index of choice for the characterization of vegetation phenology using remote sensor data. It should be noted that the NDVI is particularly sensitive to variations in precipitation during vegetation growth cycles (Di et al., 1994).

3.4 Data Availability

For change detection studies, data are usually selected from existing archives and new data acquisitions are sometimes ordered to provide data to represent the most current condition. It is essential when studying ecosystems that the phenological conditions across all time periods be carefully matched. The better the phenological correspondence between data acquisitions, the greater the potential for distinguishing land cover change, with the highest possible degree of accuracy. Data sets exhibiting similar phenological conditions minimize the interscene variability associated with phenological cycles. Therefore, the resultant spectral change between different times, has a greater correspondence to land cover changes versus phenological differences.

There are several approaches that can be applied to match data acquisition dates to phenological condition. These techniques include the use of: (a) simple annual calendar, (b) vegetation calendar, and (c) historical phenological data methods. The simple annual calendar method is applied by selecting a specific annual time frame to establish time "windows" that best correspond to delineating the target(s) of interest. Using established vegetation development calendars (i.e., crop calendars), time "windows" can be established based on specific phenological conditions. However, because phenological cycles vary on an annual basis due to climatic conditions, time "windows" must be set over a sufficiently broad range to account for annual variations. This shortcoming has not been significant to date, because data archives have not been sufficiently populated to accommodate narrow time "windows." However, with the deployment of the National Aeronautics and Space Administration's (NASA's) new EOS generation sensors these methods may not be optimal for future applications.

The use of historical phenology data provides a technique to establish very narrow time "windows" for targeting data acquisitions. Biweekly AVHRR composite data sets have been processed to provide corresponding NDVI data coverages back to 1990 (Eidenshink, 1992). The NDVI data can be used as an analytical tool to characterize the annual patterns of vegetation phenology at a 1.1×1.1 km pixel resolution. A better understanding of vegetation phenology patterns will be possible with the use of new satellite sensors such as NASA's Moderate-Resolution Imaging Spectrometer (MODIS). This technique is particularly powerful for large area studies that cross numerous ecoregion boundaries.

3.5 Data Acquisition Costs

Remote sensing data acquisition costs vary dramatically on a per unit basis. Predominate factors that regulate costs include (a) data collection platform, (b) spectral data resolution, and (c) spatial data resolution. Costs associated with radiometric and geometric corrections using satellite ephemerous data contribute only a minor additional cost. Whereas, similar corrections to aircraft collected data have traditionally represented a major cost investment. Generally, the older the data the lower the per unit cost of acquisition. Most Landsat data 10 years or older, held in the national data archive at the EROS Data Center, are available for only the cost of duplication and distribution

Although the precise cost for remote sensing data acquisitions have fluctuated over the past 20 years, some relative cost relationships have remained consistent. Coarse (spectral and spatial) NOAA-AVHRR satellite data are many orders of magnitude less expensive than moderate (spectral and spatial) resolution Landsat and SPOT satellite imagery. Correspondingly, high resolution aircraft data are approximately one order of magnitude greater in cost compared to Landsat and SPOT data. For aerial photography, archival imagery is approximately one order

Table 1.1. Comparison of Sensor Spatial and Temporal Resolution, Coverage Area, and Approximate Cost per Unit Area. Cost Estimates are for General Comparisons Only.

Sensor	Spatial Resolution	Temporal Resolution	Coverage (km^2)	Acquisition Costs (km^2)
NOAA-AVHRR	1km × 1km	2/day	9,000,000	$0.0001
Landsat MSS	79m × 79m	16 day	26,000	$0.02
Landsat TM	30m × 30m	16 day	26,000	$0.20
Archive Color-IR Photography (1:40,000)	0.3m–0.9m	Variable	83	$0.50
SPOT (XS)	20m × 20m	26 day (nadir), ±4 day (off-nadir)	3,650	$0.75
SPOT (Pan.)	10m × 10m	26 day (nadir), ±4 day (off-nadir)	3,650	$0.75
Color-IR (new) Photography (1:40,000)	0.3m–0.9m	Variable	83	$5.00–$6.00
Aircraft MS and Color-IR Photography (1:40,000)	1m × 1m to 20m × 20m 0.3m–0.9m	Variable	Variable	$5.00–$10.00

of magnitude less expensive than the cost of new data acquisition. Table 1.1 contains approximate data costs on a per unit area for many commonly used remote sensing data collection systems. Note that these comparisons are based on historical price structures and current costs may vary considerably.

3.6 Data Analysis Costs

Actual data analysis costs are quite variable and beyond the scope of this chapter. However, one important cost requirement that is often neglected or substantially underestimated should be noted. These are the costs associated with the performance of a rigorous accuracy assessment. It is not uncommon for accuracy assessment costs to represent 25 to 50 percent of the total data analysis budget.

4.0 ANALYTICAL APPROACH

The analytical or data processing steps involve data acquisition and preprocessing, radiometric and geometric corrections, data normalization, change detection analysis, accuracy assessment, and final product generation (Figure 1.2).

4.1 Data Acquisition and Preprocessing

This first step of the change detection process involves data acquisition and a number of initial preprocessing steps prior to data analysis. These typically include creating data mosaics, sub-setting, and masking, and the preparation of required ancillary data sets. When acquiring satellite imagery it is always necessary to preview either a film product or digital browse file prior to final purchase to evaluate cloud cover conditions and data quality. For most change detection projects, multiple satellite scenes or aircraft flight lines are required to provide complete area coverage.

An important step prior to data processing is assembling a complete data mosaic of the study area. The goal is to minimize cloud cover and maximize the area coverage corresponding to the highest quality data. In overlap areas between scenes or flight lines, options are presented to select data from multiple scenes or flight lines. This provides a means to avoid clouds and poor quality data. However, it is important to also create a pixel identity coverage documenting the path/row scene identity for every pixel contained in the mosaic.

Subsequent to mosaic development the study area must be subset to conform to the outer boundaries of the area of interest. Study boundaries are frequently established with the aid of ancillary data coverages such as hydrologic or geopolitical boundary coverages. The mosaic may again be subset corresponding to individual scene source identities using the pixel identity coverage as a guide. At this point the masking operations are typically performed. Masking provides a convenient means of eliminating data within the study area boundaries that will not be analyzed. For example, in a vegetation change detection project, open-water areas are frequently eliminated. Masking can be accomplished either through preliminary spectral data processing or by employing the use of ancillary data coverages (e.g., water-land boundary data). The net effect is to reduce the date dimensionality associated with the scene to optimize the performance of classification algorithms.

4.2 Geometric and Radiometric Corrections

Accurate geometric fidelity is particularly important for change detection analysis. Overlay analysis of post-classification thematic coverages, or the simultaneous analysis of multispectral data necessitates accurate spatial orientation of the input datasets. Because analysis is performed on a pixel-by-pixel basis, any misregistration greater than one pixel will provide an anomalous result for that pixel. To overcome this problem, the RMS error between any two dates should not exceed 0.5 pixel. This is typically accomplished by performing an image-to-image registration. Usually, one date is selected for absolute registration to ground coordinates, followed by the image-to-image registration(s) to the geocoded image. In locations with a high degree of topographic relief, the employment of a digital elevation model (DEM) data may be required to achieve the necessary degree of geometric agreement.

When performing multitemporal image data analyses, radiometric corrections are required because of the following atmospheric effects:

- modifications of the spectral and spatial distribution of the radiation incident on the earth surface,
- attenuation of the radiation reflected by the surface, and
- addition of path radiance.

Accordingly, multitemporal analysis approach necessitates corrections for the variations in atmospheric conditions which affect the transmittance of surface reflected radiance. In two TM images collected 16 days apart, a resultant 6° change in solar elevation was sufficient to prevent accurate radiometric comparisons of steeply sloping areas (Thomson, 1992). As a first approximation, radiometric normalization can be applied to minimize the above atmospheric effects (Maracci and Aifadopoulou, 1990).

Additionally, variations across identical sensors also contribute to radiometric discontinuity. For example, although the Landsat 1-4 MSS and Landsat 4-5 TM sensors are essentially identical among platforms, their spectral characteristics vary slightly. In a study conducted by Markham and Baker (1983), the spectral variation between the Landsat MSS sensor's spectral response ranged from 3.0 to 10.0 percent in the red band (0.6–0.7 μm) and 3.0 to 11.0 percent in the near-IR band (0.7–0.8 μm). Because the Landsat MSS and TM sensors have no onboard calibration systems, there are uncorrected differences between data collected by different sensors (e.g., Landsats 1-5). Drift in the radiometric performance also occurs over time (Elvidge et al., 1995). The spectral variations between identical sensors can also be minimized by using a first approximation radiometric normalization correction.

4.3 Data Normalization

Variations in solar illumination conditions, atmospheric scattering, atmospheric absorption and detector performance results in differences in radiance values unrelated to the reflectance of the land surface. Given sufficient time and resources, it would be possible to model or calibrate each of these effects and generate corrected multispectral data sets for use in the analysis of land cover change. However, this effort is beyond the financial and technical means of most project applications (Elvidge et al., 1995). Radiometric data normalization represents a first-order data transformation approach used to reduce the variability between multitemporal data sets acquired over the same geographic area. The process substantially reduces or normalizes the interscene variability resulting from differing atmospheric conditions, radiation incidence angle, and detector disparity. Relative radiometric normalization uses one image as a reference and adjusts the radiometric properties of the subject image(s) to match the reference (Hall et al., 1991). Among the most commonly used radiometric normalization techniques are (a) pseudo-variat features (Schott et al., 1988), (b) dark-pixel subtraction (Gonima, 1993); and (c) relative radiometric normalization (Elvidge et al., 1995).

4.4 Change Detection Analysis

The first difference detection analysis using satellite imagery was accomplished using the "blink" principle which was originally used to detect of changes in the position and luminosity of stars (Menzel, 1970). Masry et al. (1975) applied this approach to detect differences between multitemporal Landsat MSS images. This difference visualization technique was subsequently refined using high resolution film recording devices to visually display change locations using a color additive or subtraction processes. Although effective, the difference visualization technique provides only a visual reference of change locations; no resultant spatial database products. With the current demand for GIS compatible data coverages for integration with other data, visual change detection techniques have limited application for ecosystem monitoring. Change detection analysis approaches can be broadly divided into either post-classification change methods or pre-classification spectral change detection.

Post-Classification Approach

The post-classification approach involves the analysis of differences between two independent categorization products. Originally used for Landsat change detection in the late 1970s, this approach was then considered to be the most reliable and was used as a standard for quantitative evaluation of emerging image differencing techniques (Weismiller et al., 1977). Recently the approach has been successfully applied for regional deforestation studies in the Amazon by Skole and Tucker (1993). Application approaches can include either visually imaged interpretation (pattern recognition) or computer data categorization (spectral analysis), or a combination of both. Subsequent to multitemporal categorization, a comparison of the categorizations is performed, typically using vector or raster format GIS-based analysis. An important advantage of post-classification change detection is that data normalization is not required because the two dates are classified separately (Singh, 1989).

Factors that limit the application of post-classification change detection techniques can include (a) cost, (b) consistency, and (c) error propagation. Resource expenditures tend to escalate with the employment of visual interpretation or when comparisons are accomplished using vector format GIS. Inevitably, vector polygon "sliver" areas must be reconciled using an editing process. Consistency issues associated with visual-based interpretations include difference or limitations of interpreters (Westmorelans and Stow, 1992). Categorization errors associated with post-classification data products are best characterized by a multiplicative (Stow et al., 1980) error propagation model approach (Lunetta et al., 1991). Hence, this approach can produce a large number of erroneous change indications since an error on either date gives a false indication of change (Singh, 1989).

Pre-Classification Approach

The purpose of this section is to provide an introduction and brief overview of the various pre-classification analysis approaches for digital (computer assisted) remote sensing change detection. These include composite analysis, image differencing, principal components analysis, change vector and spectral analysis methods. Individual methods are detailed and applications examples provided in the sections that follow.

Composite Analysis

Among the earliest semi-automated computer assisted approaches used to generate LCLU change maps from satellite data was the composite analysis (CA) or direct multidate classification technique. CA was described as a spectral-temporal change classification (STCC) technique by Weismiller et al. (1977). This technique is based upon a single analysis of a multidate data set using standard pattern recognition and spectral classification techniques to identify land cover change areas. By using data sets collected under similar conditions (e.g., vegetation phenology and solar zenith angle) but from different years, change areas are significantly different statistically (Muchoney and Haack, 1994). Initially developed for use with Landsat MSS data sets, analysis was performed by first merging two four-band data sets into a composite eight-band data set followed by the application of a maximum-likelihood classification algorithm. Landsat TM data analyses are typically performed using a subset of bands (e.g., one each, Visible/Near-IR, and shortwave-IR) or a transformation of the original data into vegetation indices prior to composite analysis, to first reduce data dimensionality. Although STCC requires only a single classification, the combined interpretation of LCLU versus change classes can be complex. This basic technique has also been applied to Landsat MSS data sets acquired over different seasons during the same year to improve the accuracy of LCLU categories for a heterogenous landscape (Walsh, 1979).

Image Differencing

Recent research has focused on the development and evaluation of various analytical methods to accomplish semiautomated change detection using various image differencing techniques. The advantages associated with image differencing include (a) low-cost, and (b) potential for massive data processing volume. The major disadvantages associated with image differencing include (a) required optimization of change/no-change threshold level, and (b) subsequent interpretation of image difference products. Numerous analytical methods discussed below have been applied and evaluated for image differencing for ecosystem monitoring.

Image Subtraction and Thresholding. Univariate image differencing is performed by first calculating a subtraction product for two digital complementary data sets, followed by the application of a threshold to distinguish significant spectral differences as areas of land cover change. This data transformation was previously the most fundamental and widely used approach to accentuate spectral change, hence land cover change (Nelson, 1983). Thresholds are typically set based on a standard deviation value; the lower the standard deviation, the greater the inclusion and potential for errors of commission. Optimally, thresholds should be based on the accuracy of categorizing pixels as change or no-change (Fung and LeDrew, 1988). Grey level thresholding can be performed interactively with a monitor and operator-controlled software capable of level slicing. Threshold values may also be derived from the histogram of the change image (Singh, 1989). Near-infrared (greenness) differences best detect crop-type changes and changes between vegetated and nonvegetated features. Changes in the visible part of the spectrum (brightness) detect changes due to rural-to-urban land conversion. The shortwave infrared (SWIR) detects change in moisture content (wetness) (Fung, 1990).

Data Transformation, Subtraction, and Thresholding. Numerous data transformation techniques have been applied to first reduce data dimensionality prior to data subtraction. The most important of these include (a) band ratioing, (b) NDVI, and (c) the tasseled-cap transformation. Band ratioing, which involves the calculation of a simple ratio data set, has long been used to reduce the data dimensionality associated with multispectral data (Howarth and Wickware, 1981). Based on the bands used for ratioing, a positive or negative subtraction result can provide valuable insights as to the nature of changes. To this end, the NDVI and tasseled-cap vegetation indices have been widely applied, although, other vegetation indices may provide the best results for specific applications. Other experimental vegetation indexes that have been proposed by Everitt (1992) include: the difference vegetation index (DVI); perpendicular vegetation index (PVI); ratio vegetation index (RVI); soil adjusted vegetation index (SAVI); soil adjusted ratio vegetation index (SARVI); and transformed soil adjusted vegetation index (TSAVI).

The NDVI is the most widely used of all vegetation indices because it requires data from only the red and near-infrared portions of the electromagnetic spectrum, and it can be applied to virtually all remote sensor, multispectral data types. Vogelmann (1990) determined that NDVI data transformation is particularly well suited for monitoring broadleaf forest conditions. The NDVI was determined to provide the best results for detecting changes in Longleaf Pine-Wiregrass ecosystems (Houhoulis and Michner, 1996). Of significance for many applications, the NDVI has been demonstrated to be the least affected by topographic features compared to numerous other techniques (Lyon et al., 1998).

The tasseled-cap transformation (rotation) first developed by Kauth and Thomas (1976) for agricultural crop monitoring using Landsat MSS data can also be applied to Landsat TM data resulting in three vector outputs, corresponding to brightness, greenness, and wetness. The advantage of this method is the enhanced data interpretability by emphasizing the structures in the spectral data which arise as a result of particular characteristics of scene classes specifically related to vegetation. However, because the tasseled-cap data structure is influenced by sensor calibration and sensor response, the transformations are sensor dependent (Crist, 1985).

Principal Components Analysis

Principal components analysis (PCA) is a powerful data transformation technique for information extraction in remote sensing for the analysis of multispectral or multidimensional data (Lillesand and Kiefer, 1979). The PCA transformation is a redundancy reduction technique that leads to the description of multidimensional data in which the axis variables are uncorrelated, with the first variable or component (PC1) containing most variance and succeeding components containing decreasing proportions of data scatter. The monitoring of land cover change with multitemporal Landsat MSS data using PCA analysis was first reported by Byrne et al. (1980). Due to its decorrelation nature, it has been shown to be of value in enhancing regions of localized change in Landsat multitemporal data (Richards, 1984). PCA transformations have been applied to reduce multitemporal data to fewer dimensions, with as much as 99 percent of the original information preserved along the first four axes (Lo et al., 1986).

Experiments have demonstrated that when applied in a multitemporal data analysis, that the PC1 and PC2 tend to represent the unchanged land cover, whereas the PC3 and later principal components contain the changed land cover information (Byrne et al., 1980; Richards, 1984). These higher-order principal components are related to changes in brightness and greenness when analysis is performed using multitemporal Landsat MSS data (Ingebritsen and Lyon, 1985). Jiaju (1988) demonstrated the advantages of the PCA approach to separate crop types over others in maintaining classification accuracies without performing data pre-processing calibrations to multitemporal Landsat scenes.

Change Vector Analysis

Vector analysis of multitemporal Landsat data was first introduced by Engvall et al. (1977) using the temporal trend of mean vectors to categorize agricultural crops across the wheat growing regions of the United States. This temporal trend procedure was referred to as the "Delta Classifier" and was originally used to validate results obtained from maximum likelihood categorization. The change vector can be used to compare the differences in the time-trajectory of a biophysical indicator such as the NDVI or tasseled-cap for successive time periods. In establishing the time-trajectory, the indicator is composited for each pixel in a multitemporal image sequence. The change vector is simply the vector difference between successive time-trajectories, represented as a vector in multidimensional measurement space. The length of the change vector indicates the magnitude of change, while its direction indicates the nature of change (Lambin and Strahler, 1994a).

Vector analysis provides a powerful tool to analyze successive years of remotely sensed indicators (e.g., NDVI) derived from high temporal resolution data, such as AVHRR. When the time trajectory of these indicators over a particular pixel departs from that expected for that pixel, a land cover change can be detected (Lambin and Strahler, 1994b). Change vector analysis is being developed for application to the land cover change product to be produced using NASA's MODIS. MODIS is scheduled for space deployment in 1998–2000 on EOS-AM and -PM platforms. The advantage of high temporal resolution vector change analysis is the ability to monitor subtle changes within individual land cover class (Lambin and Strahler, 1994a).

Spectral Mixture Analysis

Spectral mixture analysis (SMA) was first introduced by Gillespie et al. (1990) for the interpretation of high spectral resolution (HSR) Advanced Visible/Infrared Image Spectrometer (AVIRIS) images and was later expanded by Adams et al. (1995) for use with broadband Landsat TM and MSS data. This approach is based on the premise that a multispectral image elements are composed of multiple spectral signatures or "endmembers" that contribute to the overall image reflectance. Typically a linear mixing model is assumed; thus, the total image reflectance can be calculated by the proportion of the individual endmember area weighted contributions. Endmembers can be derived from the image data (image endmembers) or from laboratory spectra (reference endmembers). Image endmembers are most commonly used because they are easy to obtain and can be readily interpreted and analyzed for their location of origin. This method requires that data first be converted to reflectance values and that the data be normalized or otherwise corrected for atmospheric effects prior to analysis.

The SMA approach is particularly attractive for use with HSR data because the high degree of data dimensionality associated with these data affords the maximum separation of endmembers. HSR data such as AVIRIS have the potential to resolve unique spectral signatures with greater accuracy than broadband sensors. Attributable to the narrow band-width sensitivity, they are capable of detecting spectral detail unique to individual endmembers. In theory, HSR spectral library signatures can be exported for use in any location where the endmembers are known to exist. However, due to the natural variability of biotic ecosystem components, most applications utilize image endmembers. Also, HSR data can be corrected for atmospheric effects using radiative transfer models that utilize data collected by the sensor as direct variable inputs to the modeling process. It should be noted that the simple normalization correction techniques that can be applied to broadband sensors are insufficient for use with HSR data. The major advantage of the SMA approach for change detection analysis is the ability to perform spectral unmixing which can be used to identify subtle land cover changes (e.g., forest thinning).

5.0 ACCURACY ASSESSMENT

The accuracy associated with post-classification products can be accomplished using one of two approaches. If the accuracy of categorization products from different times have been quantified, the accuracy of the change product can be approximated by multiplying the accuracies of each individual classification (Stow et al., 1980). If the accuracies of the original land cover data are unknown, a statistically rigorous accuracy assessment of the change map product is required using an independent source of validation data. Similarly, for all change detection products generated by the simultaneous analysis of multiple date multispectral data, a rigorous accuracy assessment of the change map is required. Issues associated with sample size, statistical design, spatial autocorrelation, and assessment technique must be considered in accuracy assessment design. Accuracy assessment analysis is based on the use of an error matrix or contingency table (Congalton, 1991). See Chapter 15 for a detailed discussion of accuracy assessment approaches for land cover change detection products.

6.0 SUMMARY

The definition of project objectives, development of a data processing flow diagram, and the generation of conceptual models detailing vegetation phenological patterns are required first steps leading to the implementation of a change detection project to characterize ecosystem changes or to support the study of ecosystem processes over time. Although the general elements for remote sensing change detection projects are similar for most project applications, specific data analysis steps vary considerably. Thus, data processing steps must be specifically formulated to meet specific project objectives and deliverables. There are numerous change detection techniques and analytical methods that can be employed to achieve required objectives. No single approach or assemblage of methods will work equally well for most change detection projects. Instead, the investigator must carefully craft technical approaches and analytical techniques to meet specific objectives. A statistically valid assessment of the accuracy associated with change detection maps and database products is an essential element for all change detection analyses. Subsequent chapters in this book will present many state-of-the-art methodologies and numerous examples illustrating the application of remote sensing change detection for environmental applications.

REFERENCES

Adams, J.B., D.E. Sabol, V. Kapos, R. Almeida Filho, D.A. Roberts, M.O. Smith, and A.R. Gillespie. Classification of multispectral images based on fractions of endmembers: Application to land-cover change in the Brazilian Amazon. *Remote Sens. Environ.*, 52, 137–154, 1995.

Ahern, F.J., T. Erdle, D.A. Maclean, and I.D. Kneppeck. A quantitative relationship between forest growth rates and Thematic Mapper reflectance measurements. *Int. J. Remote Sensing*, 12(3), 387–400, 1991.

Alwashe, M.A. and A.Y. Bokhari. Monitoring vegetation changes in Al Madinah, Saudi Arabia, using Thematic Mapper data. *Int. J. Remote Sensing*, 14(2), 191–197, 1993.

Bara, T.J. and D.M. Shaw. *Multi-Resolution Land Characteristics Consortium: Documentation Notebook*. Internal Report, U.S. EPA, Research Triangle Park, NC, May, 1995.

Byrne, G.F., P.F. Crapper, and K.K. Mayo. Monitoring land-cover change by principal component analysis of multitemporal Landsat data. *Remote Sens. Environ.*, 10, 175–184, 1980.

Carlson, R.E. and C. Aspiazu. Cropland estimates from temporal, multispectral ERTS-1 data. *Remote Sens. Environ.*, 4, 237–243, 1975.

Chavez, P.S. and D.J. MacKinnon. Automatic detection of vegetation changes in the southwestern United States using remotely sensed images. *Photogrammetric Engineering and Remote Sensing*, 60(5), 571–583, 1994.

Chuvieco, E. and M.P. Martin. A simple method for fire growth mapping using AVHRR channel 3 data. *Int. J. Remote Sensing*, 15(16), 3141–3146, 1994.

Cohen, W.B. Response of vegetation indices to change in three measures of leaf water stress. *Photogrammetric Engineering and Remote Sensing*, 57(2), 195–202, 1991.

Cohen, W.B. and T.A. Spies. Estimating structural attributes of douglas-fir/western hemlock forest stands from Landsat and SPOT imagery. *Remote Sens. Environ.*, 41, 1–17, 1992.

Cohen, W.B., T.A. Spies, and M. Fiorella. Estimating the age and structure of forests in a multi-ownership landscape of western Oregon, U.S.A. *Int. J. Remote Sensing*, 16(4), 721–746, 1995.

Congalton, R.G. A review of assessing the accuracy of classifications of remotely sensed data. *Remote Sens. Environ.*, 37, 35–46, 1991.

Crist, E.P. A TM tasseled cap equivalent transformation for reflectance factor data. *Remote Sens. Environ.*, 17, 301–306, 1985.

Di, L., D.C. Rundquist, and H. Luoheng. Modeling relationships between NDVI and precipitation during vegetation growth cycles. *Int. J. Remote Sensing*, 15(10), 2121–2136, 1994.

Eidenshink, J.C. The 1990 conterminous U.S. AVHRR data set. *Photogrammetric Engineering and Remote Sensing*, 58(6), 809–813, 1992.

Engvall, J.L., J.D. Tubbs, and Q.A. Holmes. Pattern recognition of Landsat data based upon temporal trend analysis. *Remote Sens. Environ.*, 6, 303–314, 1977.

Fiorella, M. and W.J. Ripple. Determining successional stage of temperate conifer forests with Landsat satellite data. *Photogrammetric Engineering and Remote Sensing*, 59(2), 239–246, 1993a.

Fiorella, M. and W.J. Ripple. Analysis of conifer forest regeneration using Landsat Thematic Mapper data. *Photogrammetric Engineering and Remote Sensing*, 59(9), 1383–1388, 1993b.

Foran, B.D. Detection of yearly cover change with Landsat MSS on pastoral landscapes in central Australia. *Remote Sens. Environ.*, 23, 333–350, 1987.

Fuller, S.P. and W.R. Rouse. Spectral reflectance changes accompanying a post-fire recovery sequence in a subarctic spruce lichen woodland. *Remote Sens. Environ.*, 8, 11–23, 1979.

Fung, T. An assessment of TM imagery for land-cover change detection. *IEEE Transactions of Geoscience and Remote Sensing*, 28(4), 681–692, 1990.

Fung, T. and E. LeDrew. The determination of optimal threshold levels for change detection using various accuracy indices. *Photogrammetric Engineering and Remote Sensing*, 54(10), 1449–1454, 1988.

Gillespie, A.R., M.O. Smith, J.B. Adams, S.C. Willis, A.F. Fischer, and D.E. Sabol. Interpretation of residual images: A spectral mixture analysis of AVIRIS images, Owens Valley, California. *Proceedings 2nd Airborne Visible/Infrared Imaging Spectrometer (AVIRIS) Workshop*, R. Green, Ed., Pasadena, CA, June 4–5, JPL Publication No. 90-54, 243–270, 1990.

Gilruth, P.T. and C.F. Hutchinson. Assessing deforestation in the guinea highlands of west Africa using remote sensing. *Photogrammetric Engineering and Remote Sensing*, 56(10), 1375–1382, 1990.

Gonima, L. Simple algorithm for the atmospheric correction of reflectance images. *Int. J. Remote Sensing*, 14(6), 1179–1187, 1993.

Green, K.S., D.K. Kempka, and L. Lackey. Using remote sensing to detect and monitor land-cover and land-use change. *Photogrammetric Engineering and Remote Sensing*. 60(3), 331–337, 1994.

Green, K., S. Bernath, L. Lackey, M. Brunengo, and S. Smith. Analyzing the commutative effects of forest practices: Where do we start? *Geo. Info. Systems*, February, 1993, pp. 31–41.

Hall, F.G., D.E. Strebel, J.E. Nickerson, and S.J. Goetz. Radiometric rectification: Toward a common radiometric response among multidate, multisensor images. *Remote Sens. Environ.*, 35, 11–27, 1991.

Herr, A.M. and L.P. Queen. Crane habitat evaluation using GIS and remote sensing. *Photogrammetric Engineering and Remote Sensing*. 59(10), 1531–1538, 1993.

Houhoulis, P.F. and W.K. Michener. Detection of vegetation changes associated with tropical storm Alberto in a longleaf pine-wiregrass ecosystem. *Proceedings APSRS Annual Convention*, Baltimore, MD, April 22–25, 1996, pp. 99–108.

Howarth, P.J. and G.M. Wickware. Procedures for change detection using Landsat digital data. *Int. J. Remote Sensing*, 2(3), 277–291, 1981.

Huber, T.P. and K.E. Casler. Initial analysis of Landsat TM data for elk habitat mapping. *Int. J. Remote Sensing*, 11(5), 907–912, 1990.

Ingebritsen, S.E. and R.J.P. Lyon. Principal components analysis of multispectral image pairs. *Int. J. Remote Sensing*, 6(5), 687–696, 1985.

Iverson, L.R., E.A. Cook, and R.L. Graham. Regional forest cover estimates via remote sensing: The calibration center concept. *Landscape Ecology*, 9(3), 159–174, 1994.

Jakubauskas, M.E., K.P. Lulla, and P.W. Mausel. Assessment of vegetation change in a fire altered forest landscape. *Photogrammetric Engineering and Remote Sensing*, 56(3) 371–377, 1990.

Jensen, J.R., S. Narumalani, O. Weatherbee, and H.E. Mackey, Jr. Measurement of seasonal and yearly cattail and waterlily changes using multidate SPOT Panchromatic data. *Photogrammetric Engineering and Remote Sensing*, 59(4), 519–525, 1993.

Jensen, J.R., E.W. Ramsey, H.E. Mackey, Jr., E.J. Christensen, and R.R. Sharitz. Inland wetland change detection using aircraft MSS data. *Photogrammetric Engineering and Remote Sensing*, 53(5), 521–529, 1987.

Justice, C.O. and P.H.Y. Hiernaux. Monitoring the grasslands of Shel using NOAA AVHRR data: Niger 1983. *Int. J. Remote Sensing*, 7(11), 1475–1497, 1986.

Justice, C.O., J.R.G. Townshend, B.N. Holben, and C.J. Tucker. Analysis of the phenology of global vegetation using meteorological satellite data. *Int. J. Remote Sensing*, 6(8), 1271–1318, 1985.

Justice, C.O., J.R.G. Townshend, and V.L. Kalb. Representation of vegetation by continental data sets derived from NOAA-AVHRR data. *Int. J. Remote Sensing*, 12(5), 999–1021, 1991.

Kanemasu, E.T. Seasonal canopy reflectance patterns of wheat, sorghum, and soybean. *Remote Sens. Environ.*, 3, 43–47, 1974.

Kasischke, E.S., N.H.F. French, P. Harrell, N.L. Christensen Jr., S.L. Ustin, and D. Barry. Monitoring of wildfires in boreal forests using large area AVHRR NDVI composite image data. *Remote Sens. Environ.*, 45, 61–71, 1993.

Kasischke, E.S. and N.F. French. Locating and estimating the areal extent of wildfires in Alaska boreal forests using multiple-season AVHRR NDVI composite data. *Remote Sens. Environ.*, 51, 263–275, 1995.

Kauth, R.J. and G.S. Thomas. The tasselled-cap: A graphic description of spectral-temporal development of agricultural crops as seen by Landsat. *Proceedings of the 2nd International Symposium on Machine Processing of Remotely Sensed Data*, Purdue University, West Lafayette, IN, 1976.

Keller, M., E. Veldkamp, A.M. Weitz, and W.A. Reiners. Effects of pasture age on soil trace-gas emissions from a deforested area in Costa Rica. *Science*, 365, 244–246, 1993.

Lambin, E.F. and A.H Strahler. Change-vector analysis in multispectral space: A tool to detect and categorize land-cover change processes using high temporal-resolution satellite data. *Remote Sens. Environ.*, 48, 231–244, 1994a.

Lambin, E.F. and A.H Strahler. Indicators of land-cover change for change-vector analysis in multitemporal space at coarse spatial scales. *Int. J. Remote Sensing*, 15(10), 2099–2119, 1994b.

Lee, C.T. and S.E. Marsh. The use of archival Landsat MSS and ancillary data in a GIS environment to map historical change in an urban riparian habitat. *Photogrammetric Engineering and Remote Sensing*, 61(8), 999–1008, 1995.

Lillesand, T.M. and R.W. Keifer. *Remote Sensing and Image Interpretation*. Second Edition, John Wiley & Sons, 1979.

Lo, T.H.C., F.L. Scarpace, and T.M. Lillesand. Use of multitemporal spectral profiles in agricultural land-cover classification. *Photogrammetric Engineering and Remote Sensing*, 52(4), 535–544, 1986.

Loveland, T.R., J.W. Merchant, D.O. Ohlen, and J.F. Brown. Development of a land-cover characteristics data base for the conterminous U.S. *Photogrammetric Engineering and Remote Sensing*, 57(11), 1453–1463, 1991.

Lucas, R.M., M. Honzak, G.M. Foody, P.J. Curran, and C. Corves. Characterizing tropical secondary forests using multi-temporal landsat sensor imagery. *Int. J. Remote Sensing*, 14(16), 3061–3067, 1993.

Lunetta, R.S., B.L. Cosentino, D.R. Montgomery, E.M. Beamer, and T.J. Beechie. GIS-based evaluation of salmon habitat in the Pacific northwest. *Photogrammetric Engineering and Remote Sensing*, 63(10), 1219–1229, 1997.

Lunetta, R.S., R.G. Congalton, L.K. Fenstermaker, J.R. Jensen, K.C. McGwire, and L.R. Tinney. Remote sensing and geographic information system data integration: Error sources and research issues. *Photogrammetric Engineering and Remote Sensing*, 57(6), 677–687, 1991.

Lunetta, R.S., J.G. Lyon, J.A. Sturdevant, J.L. Dwyer, C.D. Elvidge, L.K. Fenstermaker, D. Yuan, J.R. Hoffer, and R. Werrackoon. *North American Landscape Characterization: Research Plan*. U.S. EPA Report No. 600/R-93/135, 1993.

Lyon, J.G., D. Yuan, R.L. Lunetta, and C.D. Elvidge. A change detection experiment using vegetation indices. *Photogrammetric Engineering and Remote Sensing*, 64(2), 143–150, 1998.

Mahlke, J.E. and M. Jakubauskas. A multitemporal approach to characterizing Oklahoma reservoir wetlands. *Proceedings ASPRS Annual Convention*, Baltimore, MD, April 22–25, 1996, pp. 238–244.

Malingreau, J.P. Global vegetation dynamics: satellite observations over Asia. *Int. J. Remote Sensing*, 7(9), 1121–1146, 1986.

Malingreau, J.P. and C.J. Tucker. Ranching in the Amazon Basin: Large-scale changes observed by AVHRR. *Int. J. Remote Sensing*, 11(2), 187–189, 1990.

Malingreau, J.P., C.J. Tucker, and N. Laporte. AVHRR for monitoring tropical deforestation. *Int. J. Remote Sensing*, 10(4/5),855–867, 1989.

Maracci, G. and D. Aifadopoulou. Multi-temporal remote sensing study of spectral signatures of crops in the Thessaloniki test site. *Int. J. Remote Sensing*, 11(9), 1609–1615, 1990.

Marchetti, M., C. Ricotta, and F. Volpe. A quantitative approach to the mapping of post-fire regrowth in Mediterranean vegetation with Landsat TM data. *Int. J. Remote Sensing*, 16(13), 2487–2494, 1995.

Markham, B.L. and J.L. Baker. Spectral characteristics of the Landsat-4 MSS sensors. *Photogrammetric Engineering and Remote Sensing*, 49(6), 811–833, 1983.

Masry, S.E., B.G. Crawley, and W.H. Hilborn. Difference detection. *Photogrammetric Engineering and Remote Sensing*, 41(9), 1145–1148, 1975.

Maxwell, S.K. and R.M. Hoffer. Mapping agricultural crops with multidate Landsat data. *ASPRS Annual Convention Proceedings*, Baltimore, MD, April 22–25, 1996, pp. 433–443.

Mayaux, P. and E.F. Lambin. Estimation of tropical forest area from coarse spatial resolution data: A two-step correction function for proportional errors due to spatial aggregation. *Remote Sens. Environ.*, 53, 1–15, 1995.

McCulloch, L. and D.M. Hunter. Identification and monitoring of Australian plague locust habitats from Landsat. *Remote Sens. Environ.*, 13, 95–102, 1983.

Menzel, D.H. *Survey of the Universe*. Prentice-Hall, Englewood Cliffs, NJ, 1970.

Mishra, J.K., M.D. Joshi, and R. Devi. Study of decertification processes in Aravalli environment using remote sensing techniques. *Int. J. Remote Sensing*, 15(1), 87–94, 1994.

Muchoney, D.M. and B.N. Haack. Change detection for monitoring forest defoliation. *Photogrammetric Engineering and Remote Sensing*, 60(10), 1243–1251, 1994.

Nellis, M.D., K. Lulla, and J. Jensen. Interfacing geographic information systems and remote sensing for rural land-use analysis. *Photogrammetric Engineering and Remote Sensing*, 56(3), 329–331, 1990.

Nelson, R.F. Detecting forest canopy change due to insect activity using Landsat MSS. *Photogrammetric Engineering and Remote Sensing*, 49(9), 1303–1314, 1983.

Nelson, R. and N. Horning. AVHRR-LAC estimates of forest area in Madagascar. *Int. J. Remote Sensing*, 14(8), 1463–1475, 1990.

Nelson, R., N. Horning, and T.A. Stone. Determining the rate of forest conversions in Mato Grosso, Brazil using Landsat MSS and AVHRR data. *Int. J. Remote Sensing*, 8(12), 1767–1784, 1987.

Ormsby, J.P. and R.S. Lunetta. Whitetailed food availability maps from Landsat Thematic Mapper data. *Photogrammetric Engineering and Remote Sensing*, 53(11), 1585–89, 1987.

Peterson, D.L. and G.S. Hubbard. Scientific issues and potential remote-sensing requirements for plant biochemical content. *J. of Imaging Science and Tech.*, 36(5), 446–456, 1992.

Peterson, D.L., W.E. Westman, N.J. Stephenson, V.G. Ambrosia, J.A. Brass, and M.A. Spanner. Analysis of forest structure using Thematic Mapper simulator data. *IEEE Transactions on Geoscience and Remote Sensing*, GE-24(1), 113–121, 1986.

Pollock, R.B. and E.T. Kanemasu. Estimating leaf area index of wheat with Landsat data. *Remote Sens. Environ.*, 3, 307–312, 1979.

Prasad, S.N., S.P. Goyal, P.S. Roy, and S. Sing. Changes in wild ass (*Equus hemionus khur*) habitat conditions in Little Rann of Kutch, Gujarat from a remote sensing perspective. *Int. J. Remote Sensing*, 15(16), 3155–3164, 1994.

Price, K.P., D.A. Pyke, and L. Mendes. Shrub dieback in a semiarid ecosystem: The integration of remote sensing and geographic information systems for detecting vegetation change. *Photogrammetric Engineering and Remote Sensing*, 58(4), 455–463, 1992.

Prins, E.M. and W.P. Menzel. Geostationary satellite detection of biomass burning in South America. *Int. J. Remote Sensing*, 13(15), 2783–2799, 1992.

Quarmby, N.A. and J.L. Cushine. Monitoring urban land cover change at the urban fringe from SPOT HRV imagery in south-east England. *Int. J. Remote Sensing*, 10(6), 953–963, 1989.

Ram, B. and A.S. Kolarkar. Remote sensing application in monitoring land-use changes in arid Rajasthan. *Int. J. Remote Sensing*, 14(17), 3191–3200, 1993.

Richards, J.A. Thematic mapping from multitemporal image data using the principal components transformation. *Remote Sens. Environ.*, 16, 25–46, 1984.

Richardson, A. and J. Everitt. Using spectral vegetation indices to estimate rangeland productivity. *Geocarto International*, 1, 63–77, 1992.

Ritters, K.H., R.V. O'Neill, C.T. Hunsaker, J.D. Wickham, D.H. Yankee, S.P. Timmins, K.B. Jones, and B.L. Jackson. A factor analysis of landscape pattern and structure metrics. *Landscape Ecology*, 10, 23–39, 1995.

Sadler, S.A. Spatial characteristics of forest clearing and vegetation regrowth as detected by Landsat Thematic Mapper imagery. *Photogrammetric Engineering and Remote Sensing*, 61(9), 1145–1151, 1995.

Singh, A. Digital change detection techniques using remotely-sensed data. *Int. J. Remote Sensing*, 10(6), 989–1003, 1989.

Skole, D. and C. Tucker. Tropical deforestation and habitat fragmentation in the Amazon: Satellite data from 1978 to 1988. *Science*, 260, 1905–1910, 1993.

Stoms, D.M. Effects of habitat map generalization in biodiversity assessment. *Photogrammetric Engineering and Remote Sensing*, 58(11), 1587–1591, 1992.

Stoms, D.M. and J.E. Estes. A remote sensing research agenda for mapping and monitoring biodiversity. *Int. J. Remote Sensing*, 14(10), 1839–1860, 1993.

Stow, D.A., L.R. Tinney, and J.E. Estes. Deriving land use/land cover change statistics from Landsat: A study of prime agricultural land. *Proceedings of the 14th International Symposium on Remote Sensing of the Environment*, Ann Arbor, MI, 1980, pp. 1227–1237.

Teng, W.L. AVHRR monitoring of U.S. crops during the 1988 drought. *Photogrammetric Engineering and Remote Sensing*, 56(8), 1143–1146, 1990.

Thenkabail, P.S., A.D. Ward, J.G. Lyon, and P. Van Deventer. Landsat Thematic Mapper indices for evaluating management and growth characteristics of soybeans and corn. *Am. Society of Agricultural Engineers*, 35(5), 1441–1448, 1992.

Thenkabail, P.S., A.D. Ward, J.G. Lyon, and C.J. Merry. Thematic Mapper vegetation indices for determining soybean and corn growth parameters. *Photogrammetric Engineering and Remote Sensing*, 60(4), 437–442, 1994.

Thomson, A.G. A multi-temporal comparison of two similar Landsat Thematic Mapper images of upland North Wales, U.K.. *Int. J. Remote Sensing*, 13(5), 947–955, 1992.

Townshend, J.R.G. and C.O. Justice. Analysis of the dynamics of African vegetation using the normalized difference vegetation index. *Int. J. Remote Sensing*, 7(11), 1435–1445, 1986.

Townshend, J.R.G., C.O. Justice, and V. Kalb. Characterization and classification of South American land cover types using satellite data. *Int. J. Remote Sensing*, 8(8), 1189–1207, 1987.

Tucker, C.J., J.U. Hielkema, and J. Roffey. The potential of satellite remote sensing of ecological conditions for survey and forecasting desert-locust activity. *Int. J. Remote Sensing*, 6(1), 127–138, 1985.

Tucker, C.J., B.N. Holben, and T.E. Goff. Intensive forest clearing in Rondonia, Brazil, as detected by satellite remote sensing. *Remote Sens. Environ.*, 15, 255–261, 1984.

Tucker, C.J., and P.J. Sellers. Satellite remote sensing of primary productivity. *Int. J. Remote Sensing*, 7(11), 1395–1416, 1986.

Vogelmann, J.E. Detection of forest change in the Green Mountains of Vermont using Multispectral Scanner data. *Int. J. Remote Sensing*, 9(7), 1187–1200, 1988.

Walsh, M.B. *Landsat Eight Band two Season Land Cover Classification Compared with Aerial Photographic Interpretation of the Toledo, Ohio Area*. U.S. Army Corps of Engineers, Report No. 139100-1-F, 1979.

Weismiller, R.A., S.J. Kristof, D.K. Scholz, P.E. Anuta, and S.A. Momin. Change detection in coastal zone environments. *Photogrammetric Engineering and Remote Sensing*, 43(12), 1533–1539, 1977.

Wickham, J. D., T.G. Wade, K.B. Jones, K.H. Ritters, and R.V. O'Neill. Diversity of ecological communities of the United States. *Vegetatio*, 119, 91–100, 1995.

Wiegand, C.L., A.J. Richardson, D.E. Escobar, and A.H. Gerbermann. Vegetation indices in crop assessments. *Remote Sens. Environ.*, 35, 105–119, 1991.

CHAPTER 2

Survey of Multispectral Methods for Land Cover Change Analysis

Ding Yuan, Christopher D. Elvidge, and Ross S. Lunetta

1.0 INTRODUCTION

1.1 Land Cover Change

More than two thousand years ago, the Greek sage Heraclitus claimed that "It is impossible to step into the same river twice." About 1,600 years ago, the Chinese Taoist, Gehong, noted that "The seas change into mulberry fields, while the mulberry fields change into seas." We do not know exactly how Gehong observed or detected those magnificent landscape changes, nor do we know exactly if Heraclitus meant changes in the contents of the river or changes its morphology. What we do know is that people—the most intelligent residents of the Earth—have been aware of changes in their living environment for thousands of years. Some of those changes are big and significant, as noted by Gehong; and some are small, hard to discern, but can be logically ascertained—like a river of only minutes apart—as observed by Heraclitus.

During Heraclitus and Gehong's time, nature was probably still the major factor resulting in land cover change. Today, however, mankind is the most important force in altering the land surface. Through human activities to provide food, building materials, building sites, firewood, and clothing, man has been continually altering land cover. Since human population numbers are expected to double in 40 to 50 years, we expect to witness widespread land cover change as people around the world seek their livelihoods from the land surface.

Land cover changes can be divided into conversions from one land cover type into another and transformations within a given land cover type, as follows:

- Conversion between land cover types (between-class changes). This kind of change occurs when a land surface is sufficiently disturbed that it is no longer classified as the same land cover type. For instance, the change from agricultural area to urban area, and change from forest to nonforest areas are of this kind. Most remote sensing change detection studies focus on this type of land cover change.
- Changes within a land cover type (within-class change). These are transformations within one land cover type. For instance, the conversion of a forest from 80 percent to 50 percent tree canopy coverage. This type of change will not produce new land cover classifications. Both 80 percent and 50 percent canopy coverage are classified as forest. Multispectral remote sensing change detection can frequently detect these changes, providing evidence of changes in habitat quality or carbon stocks.

1.2 Digital Remote Sensing of Land Cover Change

Digital change detection methods have been broadly divided into either pre-classification spectral change detection or post-classification change detection methods (Nelson, 1983; Pilon et al., 1988; and Singh, 1989). In post-classification change detection two images from different dates are independently classified and labeled. The area of change is then extracted through the direct comparison of the classification results (e.g., Colwell and Weber, 1981; Howarth and Wickware, 1981). The advantage of post-classification change detection is that it bypasses the difficulties in change detection associated with the analysis of images acquired at different times of year or by different sensors. The main disadvantage of the post-classification approach is the high dependency of the land cover change results on the individual classification accuracies.

Spectral change detection techniques rely on the principle that land cover changes result in persistent changes in the spectral signature of the affected land surface. These techniques involve the transformation of two original images to a new single-band or multi-band image in which the areas of spectral change are highlighted. The spectral change data must be further processed by other analytic methods, such as a classifier, to produce a labeled land cover change product. Most of the spectral change detection techniques are based on some style of image differencing or image ratioing (Weismiller et al., 1977; Toll et al., 1980). It has been shown that image equalization in the data pre-processing stage usually improves the results of change detection (Hall et al., 1991). Techniques like band-to-band regressions and principal components analysis have been used to simultaneously perform the image-to-image equalization and the detection of change areas (Ingram et al., 1981).

Since all spectral change detection methods are based on pixel-wise operations or scene-wise plus pixel-wise operations, accuracy in image registration and coregistration is more critical for these methods than for other methods. The greatest challenge to the successful application of the spectral change detection methods is the discrimination of "change" and "no-change" pixels from the continuous spread of data. When looking at a histogram from a single band there are no sharp boundaries separating the value resulting from areas with change from those with no-change (Figure 2.1).

2.0 SCENE SELECTION AND PREPROCESSING: OPTIMIZING FOR DIGITAL LAND COVER CHANGE DETECTION

The basic requirement for remote sensing land cover change detection is the availability of two dates of imagery upon which the same area of land can be observed. Depending on the characteristics of the two data sets, change detection can be either an easy or difficult task. To a large extent, these complicating factors can be addressed by restricting the analysis to a single sensor series and the selection of low cloud cover imagery with careful attention to matching the dates from year to year. Below is a description of some of the complicating factors which tend to make change detection more difficult, along with a brief description of the scene selection and preprocessing steps which can be used to address these complications for ease in land cover change analyses.

- Spatial resolution and spectral bandpass differences between images acquired with two sensors complicates the direct comparison or the digital analysis of the data to detect change; for instance, combining Landsat Multispectral Scanner (MSS – four spectral bands and ~80 meter pixels) with Landsat Thematic Mapper data (TM – seven spectral bands and ~30 meter pixels). With the

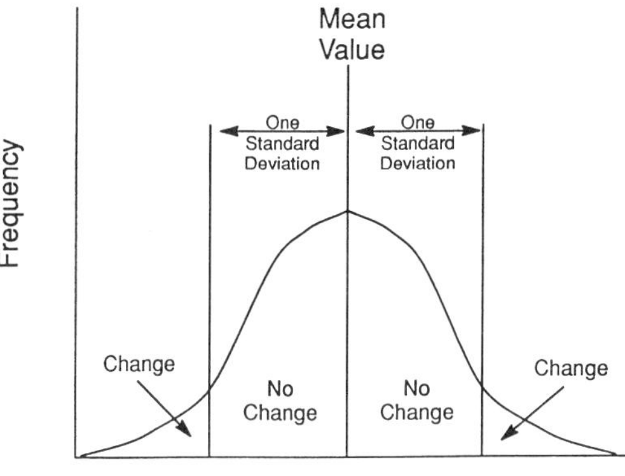

Figure 2.1. Histogram curve showing mean and a one-standard deviation threshold used to extract areas of change.

spectral bandpass differences, there may not be a direct comparability of the detectable land cover classes between the two dates of imagery. Land cover classes which are distinct when observed with one sensor may be indistinguishable in data acquired by a sensor with broader or fewer numbers of spectral bands. If there are substantial spatial resolution differences between the two input images, ground features may be visible in one data set and undetectable in the other. These complications can be minimized by using data from a single sensor series.

- If clouds are present on one or both dates of imagery it is impossible to detect land cover change between the two dates of imagery. To minimize the effects of clouds, low cloud cover should be one of the primary factors directing the selection of scenes for a project. Alternatively, it is possible to composite multiple cloud-impacted scenes to assemble cloud-reduced image products. The compositing to reduce cloud cover can be quite effective if the composited scenes are closely matched radiometrically and phenologically.
- Phenological variations in vegetation result in large changes in the reflectance patterns of the land surface. If images of leaf-on and leaf-off conditions are compared digitally, large areas can appear to have been "deforested." This problem can be minimized by selecting scenes from the same time of year for each of the three time periods. Spectral differences in vegetation between wet years and dry years can be quite pronounced, even if image dates are closely matched. In this event it may be necessary to select imagery from alternate years, which are more closely matched in their precipitation pattern. For areas with high cloud cover and sparse satellite data archives, it may not be possible to select scenes that are temporally matched. In this case, the post-classification route to land cover change analysis is indicated.
- Differences in the radiometric performance between sensors or changes in the performance of an individual sensor over time can complicate change detection by inducing spectral differences that are unrelated to land cover changes. This style of effect is best addressed using data from an onboard calibration system to provide gains and offsets for conversion of the digital data to radiance units (watts/cm^2/sr/um). Unfortunately, not every sensor is equipped with onboard calibration systems. For example, Landsat MSS data were collected over a 20-year period by five different MSS sensors (Landsat 1-5), none of which had onboard calibration systems. Although each of the MSS sensors were calibrated before launch, it is known that their response drifted over time (Markham and Baker, 1983). The simplest way to address sensor-related radiometric differences is through scene-to-scene radiometric normalization.
- Variations in solar irradiance, solar zenith angle, and solar azimuth will affect scene brightness levels and the location of shadows, all of which may impact a land cover change analysis. These

effects can be minimized by selecting scenes from different years at or near the same Julian date in order to match the solar conditions as far as possible. In general, high sun angles (low amount of shadowing) are better than low sun angles for the detection of land cover change.
- Scene-to-scene variations in atmospheric effects (scattering and absorption) can affect scene characteristics sufficiently that they must be considered in evaluating change detection methods. Data sets can be visually screened to avoid scenes where within-scene atmospheric variations are obvious. By assuming that the atmospheric effects on the selected scenes are uniform across the entire scene area, it is possible to compensate for scene-to-scene variations in atmospheric effects through atmospheric correction or radiometric normalization. In some cases even the best "low-cloud" imagery has obvious atmospheric haze or cirrus cloud cover which are not evenly distributed across the scene, confounding simple methods to prepare the data for digital change detection.
- Spatial misregistration of images will tend to reduce the accuracy of any digital change detection effort. Generally, coregistration accuracies should be on the order of half a pixel or less, creating data sets which can be analyzed for change with minimal errors due to misregistration.

3.0 SPECTRAL CHANGE IDENTIFICATION METHODS

The largest number of change detection techniques are in the spectral change identification category. Except for the raw image subtraction method, all of these methods involve some type of transformation designed to normalize or bypass radiometric differences between the raw images. Techniques in this category utilize data from two images to generate a new single-band or multi-band unclassified image representing spectral change. Most spectral change identification techniques are based on the concepts of spectral distance between pixels (differencing techniques) or pixel similarity (ratios, inner product, or correlation). All of them use pixel-wise operations or scene-wise plus pixel-wise operations. Accurate image coregistration is critical for the methods in this group. A typical final product is an image indicating locations where change occurred, labeled to indicate the nature of the landcover change (e.g., forest converted into agricultural land). For certain applications, only information on change versus no-change is needed. The change/no-change information can then be used as a field guide for future work. In other cases, the degree of change may be the feature of interest, such as in vegetation index differencing.

Once a change image is generated, additional analysis is required to separate the change and no-change pixels and to produce a labeled land cover product. The most widely used analytic technique for identifying the change pixels is the histogram thresholding method (see Figure 2.1). Pixels showing significant change are found in the tails of the histogram distribution, while pixels showing no significant change tend to be grouped around the mean (Stauffer and McKinney, 1978; Singh, 1986). The change/no-change thresholds are commonly determined by the mean plus a number of standard deviations in corresponding bands. Stauffer and McKinney (1978) and Nelson (1983) selected several different thresholds on the basis of the number of standard deviations from the mean. Ingram et al. (1981) derived thresholds from a t-test in comparing differencing of averaged images. Weismiller et al. (1977) and Woodwell et al. (1983) have advocated other threshold computing methods for raw image differencing. Once the threshold is determined, the change and no-change pixels can be distinguished directly by their change values in the change image. Alternatively, the change image can be classified to produce a map of change based on spectral classes (Anuta and Bauer, 1973; Weismiller et al., 1977).

3.1 Spectral Change Identification Methods with Pixel-Wise Operations

Spectral change detection techniques based on pixel-wise operations are among the simplest change detection techniques (Singh, 1989). In this category, coregistered pixels from two dates are processed together to produce an output image in which areas of spectral change can be distinguished from background areas where little or no change has occurred.

Raw Image Differencing

In raw imaging differencing, coregistered images from two dates are subtracted, pixel by pixel, to produce a new image that represents the digital change between the two dates. Mathematically, this method produces a new change image with the same number of bands as the input images.

Raw image differencing is easy to implement under various computational environments and has been widely used in a variety of geographical environments. Weismiller et al. (1977) found that change detection based on raw image differencing worked well in the Texas coastal zone environment despite many small areas of change not being identified accurately. Miller et al. (1978) applied Landsat image differencing successfully to the mapping of changes in tropical forest cover in northern Thailand. Williams and Stauffer (1978) utilized the technique to monitor gypsy moth defoliation in the forests of Pennsylvania. They observed that the color composite created by superimposing three of the four bands in the change image depicts the area of change in unique color tones which are easier to interpret than a one-band black and white change image. Singh (1986) used the technique for monitoring changes due to shifting cultivation in a tropical forest environment.

Riordan (1980) criticized the raw image differencing method and found it ineffective. One criticism is that the method is sensitive to image misregistration, to the existence of mixed pixels, and to radiometric differences between the input images. It is unable to differentiate spectral differences resulting from different original spectral values. For instance, a pixel that has changed from 150 to 180 cannot be distinguished from another pixel that has changed from 10 to 40 (*180–150* = *30* = *40–10*). Thus, the raw image differencing has the potential for generating misleading data.

Change Vector Analysis

The change vector of a pixel is defined as the vector difference between the multi-band digital vectors of the pixel on two different dates. When a forest stand undergoes a change, its spectral appearance changes accordingly. The vector describing the direction and magnitude of change from the first to the second date is a spectral change vector. Malila (1980) introduced the concept of change vector analysis. In this method, two associated one-band images are computed. The first contains the magnitude of the pixel change vectors. The second contains the direction of the change vectors. The decision that a change has occurred is made if the magnitude of the change vector exceeds a specified threshold criterion. Once a pixel is identified as having sustained a spectral change, the direction image should be examined to determine the type of change. The direction of the vector contains the information about the type of change; i.e., clear-cut or regrowth (Malila, 1980; Colwell and Weber, 1981; Virag and Colwell, 1987). Similar to the raw image differencing method, the change vector analysis method is sensitive to misregistration, to the existence of mixed pixels, and to the radiometric differences between the input images. Thresholding of the change magnitude image and accurate interpretation of the change direction apparently requires some practice to master.

Inner Product Analysis

In a procedure called inner product analysis, the spectral values of a pixel in an image are again considered as multispectral vectors (Inamura et al., 1982). The difference between the two multispectral vectors is measured as the cosine of the angle between them ($cos(a)$). The concept behind this approach is that if two multispectral vectors coincide with each other, their inner product would be equal to one (1). If some change took place between two dates of the concerned pixel, the inner product would be somewhere between negative one (–1) and one (1) (Yasuoka et al., 1988). A one-band image is generated to record the inner products.

Let x and y be defined as the two spectral vectors derived from a pixel occurring in two dates of imagery. The inner product of the two spectral vectors is:

$$< x, y \geq \sum_{k=1}^{b} x_k y_k$$

The difference of the surface reflectance is evaluated by:

$$d = \frac{< x, y >}{\sqrt{< x, x >< y, y >}} = \frac{\sum_{k=1}^{b} x_k y_k}{\sqrt{\sum_{k=1}^{b} x_k^2 \sum_{k=1}^{b} y_k^2}}$$

Since $-1 < d < 1$, the actually recorded inner product value is:

$$c = a_1 d + a_0$$

Where a_0 and a_1 are two constants such that c is scaled to an appropriate non-negative interval.

Inamura et al. (1982) and Yasuoka et al. (1988) used the inner product method for land cover change detection. Similar to other pixel-wise techniques, this method is sensitive to misregistration and mixed pixels, and sensitive to the change in absolute radiance and the fluctuation in the sensor gain and offsets (Yasuoka et al., 1988).

Correlation Analysis (Spectral Signature Similarity)

Conceptually, the correlation method is similar to the inner product method. The difference between the correlation method and the inner product method is that correlation takes into account the means of the multispectral vectors. This is helpful in reducing effects due to absolute values of the two multispectral vectors. It has the potential of reducing scene-to-scene radiometric influence on the analysis induced by differences in total solar irradiance, sun angles, atmosphere effects, and sensors.

The correlation coefficient or spectral similarity between two spectral vectors x and y for a given pixel is given by:

$$r = \frac{\sum_{k=1}^{b} (x_k - \bar{x})(y_k - \bar{y})}{\sqrt{\sum_{k=1}^{b} (x_k - \bar{x})^2 \sum_{k=1}^{b} (y_k - \bar{y})^2}}$$

The correlation coefficient or spectral signature similarity, r, takes the value $-1.0 < r < 1.0$ and $|r| = 1$ only when the two radiance vectors have a linear relation. The actual recorded correlation coefficient in the change image is given by, $c = a_1 r + a_0$, where a_0 and a_1 are two constants such that c can be scaled to an appropriate non-negative and recordable interval.

Coiner (1980) used the correlation method to monitor changes in vegetation cover induced by desertification processes. Yasuoka et al. (1988) applied this method for detecting land cover changes. They compared the effectiveness of this method with that of the inner product method. Their study found the inner product method to be sensitive to scene-to-scene radiometric differences. The correlation method was found to substantially reduce these radiometric effects.

Image Ratioing

Ratioing is considered to be a rapid means of identifying areas of change (Howarth and Wickware, 1981; Howarth and Boasson, 1983; Nelson, 1983; Todd, 1977; Wilson et al., 1976). In ratioing, two coregistered images from different dates with one or more bands in an image are ratioed, band by band. The concept is that all areas without significant spectral change will yield similar ratio values. In areas of change, the ratio value would be either higher or lower than the values in the no-change areas.

Todd (1977) used the ratio of Landsat MSS near-infrared band data from two dates to determine urban change in Atlanta, Georgia. Only ratios on the low side of the mean were considered as change. His study correctly identified 91.4 percent of all true land cover change. Nelson (1983) used this technique along with image differencing and vegetation index differencing techniques to delineate gypsy moth defoliation in Pennsylvania.

Similar to the raw image differencing, image ratioing is easy to implement under various computational environments. Its output may be used as the input for many other analytical methods. Similar to other techniques based on pixel-wise operations, image ratioing is sensitive to misregistration and the existence of mixed pixels. It is also sensitive to the offsets and gains of the two images if two sensors are involved. After the ratio image has been obtained, it needs extensive human interpretation efforts or other type of analysis to identify land cover change. Ratioing has been criticized due to the non-normal distribution on which it is based (Robinson, 1979). Some type of normality transformation of the original image data may be useful in improving the interpretation.

Vegetation Index Differencing

The development of vegetation indices from red and near-infrared multispectral values is based on the differential absorption and reflectance of solar energy by green vegetation (Derring and Haas, 1980). Numerous vegetation indices have been formulated to utilize this difference. Most formulas fall into one of two basic categories: those that use ratios or those that use differences (Perry and Lautenschlager, 1984). In change detection, the computed vegetation indices for two dates are subtracted to generate a band of vegetation index differences.

Nelson (1982, 1983) tested the method quantitatively in the study of gypsy moth defoliation in Pennsylvania. His results indicated that of the three methods tested, i.e., image differencing, ratioing, and vegetation index difference, the latter most accurately delineated forest canopy change. Angelici et al. (1977) used the difference of NIR/red band ratio data and the thresholding technique to delineate the changed area. Banner and Lynham (1981) used a vegetation index difference transformation and thresholding to delineate forest clear-cuts. They found that the vegetation index difference method was less accurate for delineating forest clear-cuts than a supervised classification approach.

Since the vegetation indices emphasize the spectral response differences of different ground features and tends to normalize the differences in irradiance, the vegetation index differencing has been reported a better choice for forest canopy change detection (Lyon et al., 1998). Vegetation index differencing techniques are sensitive to misregistration and the existence of mixed pixels.

3.2 Spectral Change Identification Methods with Scene-Wise Operations

Not all the changes recorded in the multispectral image pairs are of interest to the analyst. This includes changes due to variations of the sensors, the illumination conditions at the times when the images were taken, and the atmospheric scattering conditions at the times of exposure of the images. If two images were obtained by two completely different types of sensors, say, MSS and Thematic Mapper (TM) or Systeme Probatoire d'Observation de la Terre (SPOT) images, the changes due to types of sensors are also adverse for land cover change detection. All these conditions can be viewed as scene-to-scene radiometric differences, which can be addressed and largely removed prior to spectral change analysis. Based on these considerations, scene-wise radiometric correction or normalization has been developed.

Normalized Image Differencing

Normalized image differencing is basically a variation of the raw image differencing method with considerations given to the radiometric differences between the two dates of imagery. In this method, two raw images are normalized, band by band, to yield images with comparable means and standard deviations. The two normalized images are then subtracted, yielding a spectral change image. There are a number of ways to normalize an image. A very common normalization procedure uses the mean and the standard deviation.

The normalization of the subject image can be done using the following formula:

$$u_k = a_k + b_k(x_k - \bar{x}_k)$$

Where x_k is the kth band value for the given pixel, x_k bar is the mean of x_k for the given image, u_k is the kth band output value, a_k and b_k are the appropriate scaling factors. It can be shown that:

$$a_k = \bar{u}_k \quad b_k = \frac{s_{u_k}}{s_{x_k}}$$

Where u_k bar is the mean of u_k, S_{xk} and S_{uk} are the standard deviations for x_k and u_k, respectively.

Ingram et al. (1981) used the normalized image differencing technique for change detection in urban areas. They found that the normalization method improved the results for change detection in urbanized areas. Yasuoka et al. (1988) used the normalized image differencing technique for land cover change detection.

The normalized image differencing technique provides better results than raw image differencing. Scene-dependent effects are reduced. For normalization, selection of the transformation parameters is also important. Since this method is also based on pixel-wise operations, misregistration can be sensitive. In normalization, the band mean and band standard deviation need to be computed before transformation; thus, slightly more computer time may be needed to perform this procedure.

Radiometrically Normalized Image Differencing

Relative radiometric normalization techniques are based on the linear comparison of the statistical characteristics of two images or of selected spectral control subsets from two images. One image is defined as the "reference." The "subject" image is then adjusted to match the radiometric condition of the reference. Empirical methods for relative radiometric normalization have a similar linear transformation form: $u_k = a_k x_k + b_k$, where y_k is the kth band of the subject image y, x_k is the kth band of reference image x, a_k and b_k are the gain and offset used to achieve normalization of band k in image y.

Regression-based approaches to relative radiometric normalization have been recognized for many years (Jenson, 1983). Regression techniques are based on the observation that within a given spectral band there is an overall linear relationship between the digital number (DN) values for two images acquired of the same ground area. In image pairs where such linearity exists, regression analysis can be used to derive a gain and offset for radiometrically normalizing the subject image to match the reference image. In the Simple Regression (SR) normalization, the full contents of scenes are analyzed using a linear regression (Jensen, 1983). Successful application of the SR normalization requires image pairs that are devoid of statistical outliers which are present in only one of the images (e.g., clouds). In addition, the SR normalization should only be applied to images in which the vegetation is in a comparable growth stage (phenology).

The SR method works well where there are no major clouds or other phenomena present in one date but absent in the other. Other procedures have been developed to select subsets of the image contents, believed to be spectrally stable, for use in radiometric normalization procedures. Schott et al. (1988) developed a Pseudo-Invariant Feature (PIF) method applies a threshold to the ratio of the near-infrared/red spectral bands to locate a set of land surface pixels in each image with low green vegetation cover. In the PIF method, the gains and offsets that are applied to the subject image are determined based on the linear shift required to make the mean and standard deviation of the subject pixel set match the reference image pixel set. Hall et al. (1991) devised an approach which uses dark and bright pixels sets selected independently from each image based on a greenness–brightness transformation (Kauth and Thomas, 1976). The dark set typically consists of deepwater pixels and the bright set contains land surface pixels containing bright materials (e.g., soils or concrete) and little green vegetation. The mean values for the dark and bright pixel sets in each image are used to define a gain and offset for each spectral band, which is then applied to the full subject scene.

Elvidge et al. (1995) developed an Automatic Scattergram-Controlled Regression (ASCR) for use in the digital detection of land cover change using Landsat MSS data. Instead of using the whole image to derive gains and offsets as in the SR method, the ASCR approach uses pixels from a region of "no-change" identified using scattergrams. By identifying a "no-change" pixel set, it is possible to avoid the use of pixels containing cloud, cloud shadow, and land cover change in the regression analysis. The ASCR procedure assumes that the majority of pixels in a scene have the same land cover and phenological (vegetation growth) stage for both image dates. The "no-change" pixels are automatically selected based on scattergrams of the near-infrared data from subject versus the reference images. The near-infrared data were used because at these wavelengths the spectral clusters for water and land are clearly separated, and a distinct axis of "no-change" can be observed. Thresholds set an envelope above and below the "no-change" axis, to define the "no-change" set of pixels, avoiding inclusion of clouds, shadows, and areas of significant land cover change from the analysis. These pixels were then used to perform the regression to derive the normalization gain and offset for each spectral band. Comparative studies of these normalization techniques have been reported by Yuan and Elvidge (1996).

An advantage of these procedures is that the original radiometric condition of the reference image is retained, obviating the computational effort required to convert each image to units of radiance or reflectance. Once the two images are radiometrically matched, they are subtracted from each other on a band by band basis, revealing the locations where spectral changes have occurred.

Albedo Differencing

Albedo, or reflectance, is defined as the ratio of the amount of electromagnetic radiance reflected by a body to the amount of incident upon it. In the albedo differencing method, albedo calculated for each pixel is used to create an albedo image, whose digital number values are proportional to the albedo. Differencing coregistered image pairs creates quantitative images of the albedo changes which have occurred across a terrain (Robinove et al., 1981).

There are several published methods for computing Landsat band albedo (Otterman and Fraser, 1976; Struve et al., 1977; Robinove et al., 1981). The first two methods include use of either a computerized atmospheric model or an atmospheric correction that requires meteorological data. Robinove's method utilizes only Landsat data for the atmospheric correction. Therefore, Robinove's method is recommended for processing historical remote sensing data.

Multidate Principal Component Analysis

Principal component analysis (PCA) is one of the most popular multivariate analysis techniques for data reduction. In multitemporal studies, two b-band image scenes of the same area, which are recorded on different dates, can be superimposed and treated as a single $2b$-band image. Principal component analysis of this data set should result in the gross difference associated with overall surface albedo and radiometric differences appearing in the major component images and statistically minor changes associated with local changes in land cover appearing in the minor component images (Byrne et al., 1980; Richardson and Milne, 1983).

To composite the new image, denote

$$z = (z_1, \ldots, z_b, z_{b+1}, \ldots, z_{2b}) = (x_1, \ldots, x_b, y_1, \ldots, yb)$$

the composite image, and Z^* the standardization of Z. For PCA without standardization, one considers the variance-covariance matrix of Z. For PCA with standardization, one considers the correlation matrix $V = COV(Z,Z)R = CORR(Z^*,Z^*)$ of Z. For PCA with standardization, one considers the correlation matrix $V = COV(Z,Z)R = CORR(Z^*,Z^*)$ of Z^*.

Consider the nonstandardized situation. Let the eigenvalue decomposition of V be $V = P^T APR = Q^T BQ$, where P is an orthogonal matrix, and A is non-negative diagonal matrices with nondecreasing diagonal elements (eigenvalues of V). Then, the PCA transformation is given by

$$U = PZ + C_1 \quad \textit{for} \quad \textit{nonstandardized} \quad \textit{data}$$

$$W = QZ^* + C_2 \quad \textit{for} \quad \textit{standardized} \quad \textit{data}$$

for non-standardization data, where c is a constant vector to make U non-negative.

Byrne et al. (1980) and Richardson and Milne (1983) studied the effectiveness of principal component analysis for identifying land cover changes and mapping brush fires and subsequent vegetation regeneration, respectively. Richards (1984) pointed out what PCA could do for change detection was to show how different globally one subscene is from another. Therefore, this method is better applied to a study area where a large proportion of the area belongs to areas of no-change. Both Bryne et al. (1980) and Richards (1984) demonstrated that higher order principal components (PCs—e.g., PC 3 and PC 4) were able to identify land cover changes. Following along these lines, Ingebritsen and Lyon (1985) performed multidate principal component analysis, also finding land cover changes isolated in the minor PCs (e.g., PC 3 and PC 4) and viewed these PCs as "change in brightness" and "change in greenness."

Multidate PCA is good for global comparison of image scenes. It takes into account global effects, and is insensitive to image registration. However, Toll et al. (1980) reported that when used for urban change detection, principal component transformation produced poor change detection results compared with simple image differencing of band-2 or -4 data. Singh and Harrison (1985) have reported that the PCA using standardized variables (variance-covariance matrix used) yielded significantly different results from those using nonstandardized variables (correlation matrix used).

Principal Component Comparison

Comparison of individual image PCA results is another way to use principal component analysis for change detection. Principal component analysis is performed on the image data of each date. The derived PCs are again analyzed by other change detection methods such as differencing and regression methods. The principal component analysis can be performed on original or standardized data. The former uses the covariance matrices, while the latter uses the correlation matrices. If the original (nonstandardized) data is used, PC1 usually has equal loading on all spectral bands and PC2 represents the difference between visible and infrared bands. If standardized data is used, PC2 usually has equal loading on all spectral bands and PC1 represents the difference between visible and infrared bands.

Singh and Harrison (1985) compared the use of nonstandardized (the use of covariance matrix) versus standardized (the use of correlation matrix) data for rotation. Nonstandardized principal components were justified because of the possible differences in radiometric resolution between the spectral bands. They found that the nonstandardized PC1 usually has positive loading of eigenvectors on all of the spectral bands. The nonstandardized PC2 was the difference between the visible and infrared bands of MSS data. The first two PCs account for over 95 percent of the total variance of a single date's MSS data.

In contrast to the findings of Singh and Harrison (1985), Fung and LeDrew (1987) found that the standardized PCs were clearly distinct from nonstandardized PCs. The standardized PC1s were found to represent a contrast between the visible and infrared bands. They usually have positive loading on the visible bands (bands 1 and 2) and negative loading on the infrared bands (bands 3 and 4). The negative loading on the infrared bands suggest that the PC1s are negative greenness measures. The PC2s are positively loaded on all the bands, which is interpreted as a brightness measure. They reported that the standardized principal components were better in describing the underlying data dimension of MSS data in their study.

Although PCA is good for global comparison of multitemporal image data and insensitive to image registrations, its interpretation often requires experience and knowledge of the particular data used. It may need further visual or analytic analysis, such as the differencing or the ratioing method, to extract change information. It should be noted that the results produced by standardized and nonstandardized principal component analysis often require different interpretations.

4.0 POST-CLASSIFICATION COMPARISON METHODS

In post-classification change detection, land cover change is detected as a change in land cover label between two image dates. These methods were among the first change detection techniques applied to digital image data (Weismiller et al., 1977; and Rubec and Thie, 1987). The post-classification category consists of two types of change detection methods: (1) methods based on two independent spectral or spectral-spatial classifications (can be both supervised or unsupervised), and (2) methods based on two independent true land cover class classifications (must be supervised).

For land cover change output, products reported at the scene or block level image registration and coregistration are not critical. For output products at the pixel level, the coregistration accuracy requirements are the same as for the spectral change identification techniques. As long as the scenes are classified and labeled independently, there is no requirement for any type of scene-to-scene radiometric normalization.

The change detection accuracy of the techniques in this category is generally dependent on the accuracy of the adapted classifiers. A crude estimate of accuracy for most of the change detection methods in this category is approximately the product of the accuracy of the two image classifiers employed. For instance, if two image classifiers are used, and one has 80 percent accuracy, and the other has 70 percent accuracy, then the accuracy of change detection based on their classifications is about 56 percent. Therefore, it is often reported that post-classification change detection methods have relatively low accuracy.

5.0 DIRECT MULTIDATE CLASSIFICATION METHODS

Direct multidate classification methods classify two images obtained on two dates using one classifier simultaneously (Weismiller et al., 1977). The two images of two dates are either parallel merged (mosaicked) or superimposed (one over the other). In the first case, the classification produces two classifications for two images separately as in the post-classification comparison situation. The results are compared for detecting changes in land cover labels. In the second case, two N-band images are superimposed to form a $2N$-band image. The classification of this superimposed image produces a set of stable land cover classes (locations where land cover did not change between dates) and a set of change classes, which when properly labeled, yield a land cover change image.

Depending on whether ground truth information is used, the direct multidate classification may be subdivided into two categories: (1) methods based on one spectral or spectral-spatial (unsupervised) classification, and (2) methods based on ground truth (supervised) classifications. As a direct method of change detection, it may produce a class-to-class change map. Change statistics can be obtained directly from measuring the areas of corresponding change classes in the change map. This method may be performed using transferred data such as principal components (Estes et al., 1982).

Weismiller et al. (1977) used a direct multidate classification method for detecting changes in a Texas coastal zone environment. They found that this method showed the best agreement with the ground post-classification comparison results. These methods require a single, but very complex classification which requires many classes. Thus, a large computer core is needed for this method. Another problem is that the temporal and the spectral features have equal status in the combined data set (Schowengerdt, 1983). Thus, spectral changes within one multispectral image cannot be easily separated from temporal changes between images in the classification. Spectral change due to atmospheric factors may not be easy to separate from the vegetation temporal change. Some variable normalization or transformation before classification may be helpful.

6.0 HYBRID METHODS

The spectral change identification methods and the classification based methods can be combined in various ways to minimize errors in land cover change analysis. For instance, radiometrically normalized image differencing may be used to identify the area of significant spectral change, then apply a post-classification comparison method within areas where spectral change was detected to obtain class-to-class change information. This approach can substantially reduce the errors of commission (misidentifying no-change and change pixels).

Pilon et al. (1988) developed a hybrid land cover change detection method and applied it to the detection of changes in a northwestern Nigeria semiarid environment. The procedure performs post-classification change detection in the areas where spectral changes were detected by other methods. Pilon et al. (1988) report this hybrid approach satisfactorily enhanced the results obtained by spectral post-classification comparison and ground post-classification comparison, reducing errors of commission.

7.0 EXAMPLES

7.1 Post-Classification Land Cover Change Analysis

The North American Landscape Characterization (NALC) Landsat MSS triplicates have been selected from the available archive of data held by the USGS EROS Data Center. A considerable effort was made to select scenes with low cloud cover and at or near the same date for all three time periods in order to have data sets optimized for digital change detection (Lunetta et al., 1993). However, in the extreme southern portions of the project area it was frequently not possible to find low cloud cover scenes at the same time of year. An example of data availability for Path 21, Row 47 in southern Mexico is provided in Table 2.1.

Visual inspection of imagery revealed that there are substantial phenological changes between the four dates of imagery, indicating that only post-classification change detection could be used to derive land cover change products in Path 21, Row 47. An outline of a post-classification land cover change analysis for two dates of imagery is shown in Figure 2.2. The procedure involves georeferencing one date to a map base, coregistration of the second scene to the georeferenced image, clustering and labeling of the land cover classes in both scenes, and comparison of pixel labels to identify areas of land cover change.

Table 2.1. NALC Datasets Identified by Path/Row, Date, and Scene Identification Numbers for Southern Mexico.

Path/Row	Date	Scene ID#
21/47	92/04/29	LE84357515342x0
21/47	84/11/25	LM85026915590x0
22/57	74/02/15	LM8157215552500
23/47	78/09/17	LM83019615545x0

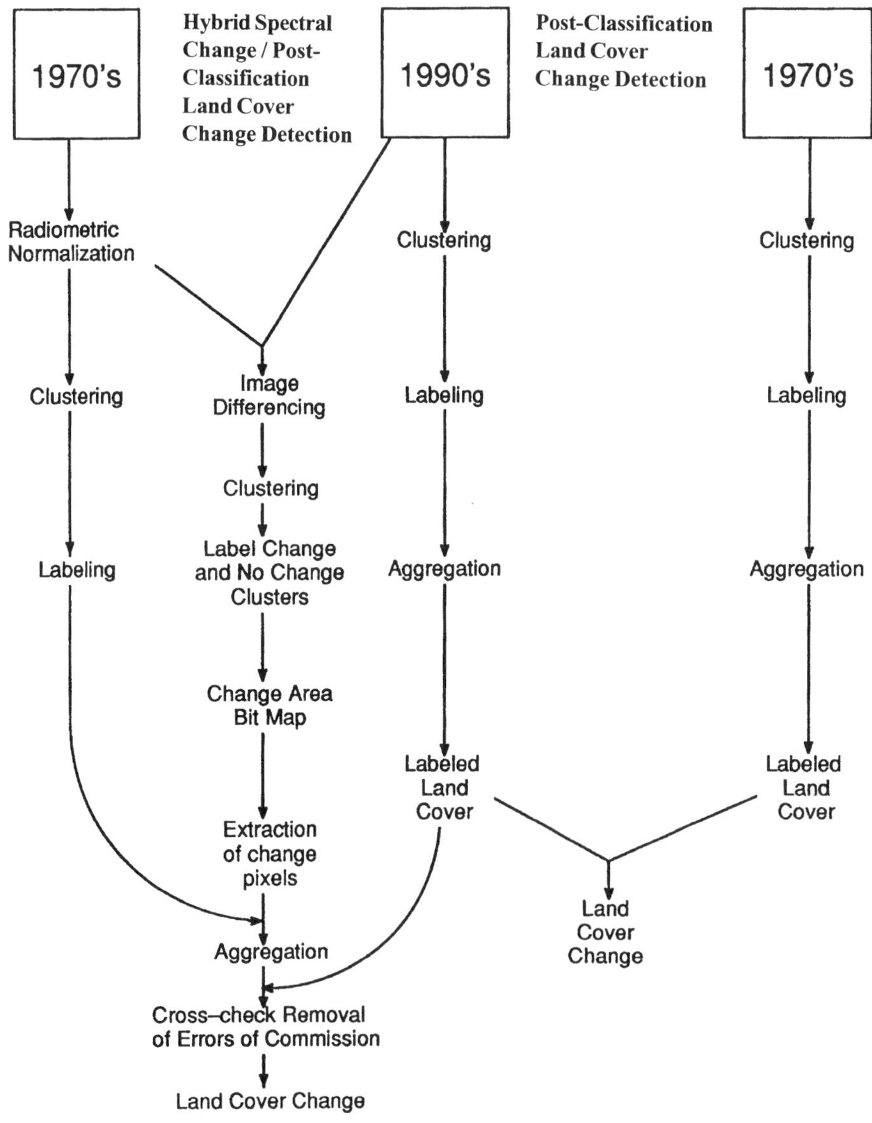

Figure 2.2. Outline of a post-classification land cover change procedure.

Table 2.2. NALC Datasets Identified by Path/Row, Date, and Scene Identification Numbers for the Oregon Study Area.

Path/Row	Date	Scene ID#
45/29	92/08/03	LM85307718130x0
45/29	86/08/19	LM85090118114x0
49/29	72/09/02	LM8104118265500
48/29	72/07/27	LM8100418210500

7.2 Land Cover Change Analysis Using Spectral Change Identification

For the majority of the NALC triplicates, it has been possible to find data in the archive having low cloud cover and closely matched dates for use in the triplicates. An example of this type of triplicate is provided in Table 2.2 for Path 45, Row 29 in Oregon.

For the NALC project, a hybrid spectral change identification/post-classification land cover change analysis procedure has been developed (Figure 2.2). As in the classic post-classification method, images are first georeferenced and coregistered. Then both images are clustered and the land cover classes are labeled by an image analyst.

The spectral change identification method involves the radiometric normalization of the older imagery to match the overall radiometric condition of the 1990's imagery (see color Plate 1) using the ASCR method described by Elvidge et al. (1995) and Yuan and Elvidge (1996). This is followed by image differencing to create a four-band "change" image, which is further analyzed to identify areas of spectral change (see color Plate 2).

The most common method for separating change and no-change pixels in the change images generated through image differencing and other methods is the use of statistical thresholds. For instance, all values exceeding a standard deviation from the mean may be extracted as the change areas (Figure 2.1). Research by Singh (1989) indicated that the errors of commission and omission are highly sensitive to the changes in the thresholding. Also, it is unlikely that the optimal threshold would be the same above and below the mean for each spectral band. There are no clear guidelines for making this separation between change and no-change pixels using thresholds (Singh, 1989). Our solution is to cluster the difference images and to have an image analyst label the clusters as "change" and "no-change." The "no-change" clusters are centered on the mean values of each band difference image and contain the majority of the scene pixels. The "no-change" clusters are generally devoid of defined spatial features and are widely distributed throughout the difference image. In contrast, the "change" clusters tend to occupy sharply defined boundaries, to have substantial departure from the mean values, and exhibit color in three-band color composites made from the band difference images. After labeling the "change" classes, a bit-map is produced. As a final step the land cover labels are compared for all pixels under the change bit map.

8.0 CONCLUSION

Digital change detection methods have been broadly divided into either spectral change identification methods or post-classification methods (Nelson, 1983; Pilon et al., 1988; Singh, 1989). In post-classification change detection, two images from different dates are independently classified and labeled. The area of change is then extracted through the direct comparison of the classification results (e.g., Colwell and Weber, 1981; Howarth and Wickware, 1981). The advantage of post-classification change detection is that it bypasses the difficulties in change detection associated with the analysis of images acquired at different times of year or by different sensors. The major disadvantage of the post-classification approach is the high sensitivity to the individual classification accuracies.

Spectral change identification techniques involve the transformation of two original images to a new single-band or multi-band image in which the areas of land cover change are detected. The change-enhanced data can be further processed by other analytic methods, such as a classifier, to produce a labeled change detection output product. Most of the enhancement techniques are based on the concepts of image differencing or image ratioing (Weismiller et al., 1977; Toll et al., 1980). Differencing of vegetation indices has been a popular form of enhancement change detection (e.g., Nelson, 1983). It has been shown that radiometric normalization in the data pre-processing stage usually improves the results of change detection (Hall et al., 1991). Techniques like band-to-band regressions and principal components analysis have been used to simultaneously perform the image-to-image equalization and the detection of change areas (Ingram et al., 1981).

The advantage of the spectral change identification techniques is that they are based on the detection of physical changes between image dates. This avoids the errors introduced in post-classification change detection where inaccuracies in the land cover classification between dates are propagated into the land cover change analysis. The greatest challenge to the successful application of the spectral change identification methods is the discrimination of "change" and "no-change" pixels from the continuous spread of data. When looking at a histogram from a single band there are no sharp boundaries separating the value resulting from areas with change from those with no-change.

Through the NALC project, we have combined these two approaches to land cover change analysis procedures, using spectral change identification to define the areas where post-classification land cover change analysis can be applied. Radiometric normalization is applied to one date of imagery in order to match the radiometric condition of the two dates of data prior to image subtraction. The band difference or "change" image is clustered and then aggregated to "change" and "no-change" spectral classes by an image analyst. This hybrid approach eliminates the identification of land cover change in areas where no significant spectral change has occurred between the two dates of imagery.

ACKNOWLEDGMENTS

This chapter had been partially funded by the U.S. EPA North American Landscape Characterization Program through Cooperative Research Agreement #CR816826. Final completion of the chapter was partially supported by the NASA Commercial Remote Sensing Program Office, under Contract Number NAS13-650 at the John C. Stennis Space Center.

REFERENCES

Angelici, G.L., N.A. Brynt, and S.Z. Friendman. Techniques for land use change detection using Landsat imagery. *Proceedings of the American Society of Photogrammetry*, 1977, pp. 217–228.

Anuta, P.E. and M. Bauer. An analysis of temporal data for crop species classification and urban change detection. *Laboratory for Applications of Remote Sensing*, Information Note 110873, Purdue University, West Lafayette, IN, 1973.

Banner, A.V. and T. Lynham. Multitemporal analysis of Landsat data for forest cut over mapping - a trial of two procedures. *Proceedings of the 7th Canadian Symposium on Remote Sensing,* Winnipeg (Canadian Remote Sensing Society), 1981, pp. 233–240.

Byrne, G.F., P.F. Crapper, and K.K. Mayo. Monitoring land cover change by principal component analysis of multitemporal Landsat data. *Remote Sens. Environ.*, 10, 175–184, 1980.

Coiner, J.C. Using Landsat to monitor change in vegetation cover induced by desertification processes. *Proceedings of the 14th International Symposium of Remote Sensing of Environment*, San Juan, Costa Rica (Environmental Research Institute of Michigan), 1980, pp. 1341–1351.

Colwell, J.E. and F.P. Weber. Forest change detection. *Proceedings of the 15th International Symposium on Remote Sensing of Environment*, Ann Arbor, MI (Environmental Research Institute of Michigan). 1981, pp. 65–99.

Derring, D.W. and R.H. Haas. Using Landsat digital data for estimating green biomass. *NASA Technical Memorandum*, No. 80727, 1980.

Elvidge, C.D., D. Yuan, D.W. Ridgeway, and R S. Lunnetta. Relative radiometric normalization of Landsat Multispectral Scanner (MSS) data using automatic scattergram-controlled regression. *Photogrammetric Engineering and Remote Sensing*, 61(10), 1255–1260, 1995.

Estes, J.E., D. Stow, and J.R. Jensen. Monitoring land use and land cover changes. In: *Remote Sensing for Resources Management*, C. J. Johannsen and J. L. Sanders, Eds., Iowa: Soil Conservation Society of America, 1982, pp. 100–110.

Fung, T. and E. LeDrew. Application of principal components analysis to change detection. *Photogrammetric Engineering and Remote Sensing*, 53(12), 1649–1658, 1987.

Hall, F.G., D.E. Strebel, J.E. Nickeson, and S.J. Goetz. Radiometric rectification: toward a common radiometric response among multidate, multisensor images. *Remote Sens. Environ.*, 35, 11–27, 1991.

Howarth, J.P. and E. Boasson. Landsat digital enhancements for change detection in urban environment. *Remote Sens. Environ.*, 13, 149–160, 1983.

Howarth, J.P. and G.M. Wickware. Procedure for change detection using Landsat digital data. *Int. J. Remote Sensing,* 2, 277–291, 1981.

Inamura, M., H. Toyota, and S. Fujimura. Exterior algebraic processing for remotely sensed multispectral and multitemporal images. *IEEE Transaction on Geosciences and Remote Sensing*, 20(1), 112–118, 1982.

Ingebritsen, S.E. and R.J.P. Lyon. Principal component analysis of multitemporal image pairs. *Int. J. Remote Sensing*, 6(5), 687–696, 1985.

Ingram, K., E. Knapp, and J.W. Robinson. Change detection technique development for improved urbanized area delineation. Technical Memorandum CSC/TM-81/6087, *Computer Sciences Corporation*, Silver Springs, MD, 1981.

Jensen, J.R. Urban change detection mapping using Landsat Digital Data. *The American Cartographer*, 8, 127–147, 1983.

Kauth, R.J. and G.S. Thomas. The tasseled cap—a graphic description of the spectral temporal development of agricultural crops as seen by Landsat. *Proceedings of the 2nd Annual Symposium on Machine Processing of Remotely Sensed Data*, Purdue University, 1976, pp. 4B41–4B49.

Lunetta, R.S., J.G. Lyon, J.A. Sturdevant, J.L. Dwyer, C.D. Elvidge, L.K. Fenstermaker, D. Yuan, S.R. Hoffer, and R. Weerackoon. *North American Landscape Characterization (NALC) Research Plan.* USEPA, EPA/600/R-93/135, 1993.

Lyon, J.G., D. Yuan, R.S. Lunetta, and C.D. Elvidge. A change detection experiment using vegetation indices, photogrammetric. *Engineering and Remote Sensing,* 64(2), 143–150, 1998.

Malila, W.A. Change vector analysis: an approach for detecting forest changes with Landsat. *Proceedings of the 6th Annual Symposium on Machine Processing of Remotely Sensed Data*, Purdue University, West Lafayette, IN, 1980, pp. 326–335.

Markham, B.L. and J.L. Baker. Spectral characteristics of the Landsat MSS sensor. *Photogrammetric Engineering and Remote Sensing,* 49(6), 811–833, 1983.

Miller, L.D., K. Nualchawee, and C. Tom. Analysis of the dynamics of shifting cultivation in the tropic forest of northern Thailand using landscape modeling and classification of Landsat imagery. *NASA Goddard Space Flight Center*, Technical Memorandum No. 79545, Greenbelt, MD, 1978.

Nelson, R.F. Detecting forest canopy change using Landsat, *NASA Goddard Space Flight Center*, Technical Memorandum No. 83918, Greenbelt, MD, 1982.

Nelson, R.F. Detecting forest canopy change due to insect activity using Landsat MSS. *Photogrammetric Engineering and Remote Sensing*, 49, 1303–1314, 1983.

Otterman, J. and R.S. Fraser. Earth-atmosphere system and surface reflectivities in arid region from Landsat MSS data. *Remote Sens. Environ.*, 5, 247–266, 1976.

Perry, C.R. and L.F. Lautenschlager. Functional equivalence of spectral vegetation indices. *Remote Sens. Environ.*, 14, 169–182, 1984.

Pilon, P.G., P.J. Howarth, and R.A. Bullock. An enhanced classification approach to change detection in semi-arid environments. *Photogrammetric Engineering and Remote Sensing*, 54(12), 1709–1716, 1988.

Richardson, A.J. and A.K. Milne. Mapping fire burns and vegetation regeneration using principal components analysis. *Proceedings of IGARSS '83*, San Francisco, CA, 1983, pp. 51–56.

Richards, J.A. Thematic mapping from multitemporal image data using the principal component transformation. *Remote Sens. Environ.*, 16, 35–46, 1984.

Riordan, C.J. Non-urban to urban land cover change detection using Landsat data. *Summary Report of the Colorado Agricultural Research Experiment Station*, Fort Collins, CO, 1980.

Robinove, C.J., P.S. Chavez, Jr., and D. Gehring. Arid land monitoring using Landsat Albedo differencing images. *Remote Sens. Environ.*, 11, 133–156, 1981.

Robinson, C.J. A critical review of the change detection and urban classification literature. Technical Memorandum CSC/TM-79/6235, *Computer Sciences Corporation*, Silver Spring, MD, 1979.

Rubec, C.D. and J. Thie. Land use monitoring with Landsat digital data in southwestern Manitoba. *Proceedings of the fifth Canadian Symposium on Remote Sensing*, Victoria, BC, 1987, pp. 136–150.

Schott, J.R., C. Salvaggio, and W. Volchok. Radiometric scene normalization using pseudoinvariant features, *Remote Sens. Environ.*, 26, 1–6, 1988.

Schowengerdt, R.A. *Techniques of Image Processing and Classification in Remote Sensing*, Academic Press, New York, 1983.

Singh, A. Change detection in the tropical forest environmental of northern India using Landsat. In: *Remote Sensing and Tropical Land Management*, M.J. Eden and J.T. Parry, Eds. John Wiley & Sons, London, 1986, pp. 237–254.

Singh, A. Digital change detection techniques using remotely-sensed data. *Int. J. Remote Sens*ing, 10(6), 989–100, 1989.

Singh, A. and A. Harrison. Standardized principal components, *Int. J. Remote Sensing*, 6(6), 883–896, 1985.

Stauffer, M.L. and R.L. McKinney. Landsat image differencing as an automated land cover change detection technique. *Computer Sciences Corporation*, Technical Memorandum CSC/TM-78/6215, Silver Spring, MD, 1978.

Struve, H., W.E. Grabau, and H.W. West. Acquisition of terrain information using Landsat multispectral data: correction of Landsat multispectral data for extrinsic effects. *U.S. Army Engineer Waterways and Experiment Station*, Technical Report No. M-77-2, 1977.

Todd, W.J. Urban and regional land use change detected by using Landsat data, *Journal of Research*, U.S. Geological Survey, 5, 527–534, 1977.

Toll, D.L., J.A. Royal, and J.B. Davis. Urban area up-date procedures using Landsat date. *Proceedings of the Fall Technical Meeting of the American Society of Photogrammetry*, Niagara Falls, Canada, 1980, pp. RS-E1-17.

Virag, L.A. and J.E. Colwell. An improved procedure for analysis of change in thematic mapper image-pairs. *Twenty-First International Symposium on Remote Sensing of Environment*, Ann Arbor, MI, Oct. 26–30, 1987.

Weismiller, R.A., S.J. Kristoof, D.K. Scholz, P.E. Anuta, and S.A. Momen. Change detection in coastal zone environments. *Photogrammetric Engineering and Remote Sensing,* 43, 1533–1539, 1977.

Williams, D.L. and M.L. Stauffer. Monitoring gypsy moth defoliation by applying change detection techniques to Landsat imagery. *Proceedings of the Symposium on Remote Sensing for Vegetation Damage Assessment*, Falls Church, VA (American Society for Photogrammetry), 1978, pp. 221–229.

Wilson, J.R., C. Blackman, and G.W. Spann. Land use change detection using Landsat data. *Proceedings of the 5th Annual Remote Sensing of Earth Resources Conference*, University of Tennessee, Tullhama, TN, 1976, pp. 79–91.

Woodwell, G.M., J.E. Hobbie, R.A. Houghton, J.M. Melillo, B.J. Peterson, G.R. Shaver, T.A. Stone, B. Moore, and A.B. Park. Deforestation measured by Landsat: Step towards a method. *U.S. Department of Energy*, Report No. DOE/EV/10468-1, Washington, DC, 1983.

Yasuoka, Y. et al. Detection of land-cover change from remotely sensed images using spectral signature similarity. *Proceedings of the 9th Asian Conference on Remote Sensing*, November, 1988, pp. G-3-1–G-3-6.

Yuan, D. and C.D. Elvidge. Comparison of relative radiometric normalization techniques. *ISPRS Photogrammetry and Remote Sensing,* 51, 117–126, 1996.

CHAPTER 3

North American Landscape Characterization: "Triplicate" Data Sets and Data Fusion Products*

Ross S. Lunetta, John G. Lyon, Bert Guindon, and Christopher D. Elvidge

1.0 INTRODUCTION

The North American Landscape Characterization (NALC) project is a component of the National Aeronautics and Space Administration (NASA) Landsat Pathfinder program of experiments to study global change issues. The NALC project is one of three Landsat Pathfinder projects approved by the Landsat Pathfinder Science Working Group along with the Humid Tropical Forest Inventory Project and the Global Land Cover Test Site Project. The NALC project was sponsored by the U.S. Environmental Protection Agency's (EPA) Office of Research and Development (ORD), the U.S. Global Change Research Program (USGCRP), and the U.S. Geological Survey, EROS Data Center. In addition, substantial collaboration for NALC efforts in the Laurentian Great Lakes was performed by the Canada Centre for Remote Sensing (CCRS), Ottawa, Canada. Landsat Pathfinder coordination and technical review was provided by the NASA's Ames Research Center at Moffett Field, California and Goddard Space Flight Center at Greenbelt, Maryland.

This chapter provides a descriptive overview of the NALC program, provides examples of NALC land cover categorization products, presents provisional carbon inventory data fusion products, and describes a two-step model for accuracy assessment.

The primary goal of the NALC project was to assemble and distribute standard Landsat Multi-Spectral Scanner (MSS) image data sets to the scientific community, and to provisionally apply these data on a selective basis to study land cover and land cover change in support of global change research issues. Land cover data products at a 3.2 hectare spatial resolution have been produced for priority locations in the Mexican humid tropics (Lyon et al., 1998). Other projects have been completed using NALC data, including a mosaic of the Canadian and United States Laurentian Great Lakes watershed (Guindon, 1995). The program was originally designed to provide source data for land cover categorization and change detection analysis

* Reprinted with permission by the American Society for Photogrammetry and Remote Sensing. North American Landscape Characterization: Dataset Development and Data Fusion Issues. Photogrammetric Engineering and Remote Sensing, R.S. Lunetta, J.G. Lyon, B. Guindon, and C.D. Elvidge, Vol. 64(8), pp. 821–829, 1998.

across North America. Initially, these data have been used to supply information on the spatial distribution of forest cover types required for regional and national inventories of terrestrial carbon stocks and trace greenhouse gas (i.e., CH_4, N_2O) emissions, in support of the Intergovernmental Panel on Climate Change (IPCC) science objectives (Houghton et al., 1992) and the climate-change action plan target of returning emissions of greenhouse gasses in the United States to 1990 emission levels by the year 2000 (Clinton and Gore, 1993).

The main objective of the NALC project was to produce standardized remote sensing data sets or "triplicates" that consisted of three or more registered Landsat MSS images corresponding to the 1990s, 1980s, and 1970s time periods. On average, a NALC triplicate consists of one scene from the 1990s and 1980s and two from the 1970s for each path/row.

2.0 BACKGROUND

The principal clients of the NALC project were the USGCRP, and in particular, the EPA's GCRP. An important objective of the USGCRP was to document global change through an integrated, comprehensive, long-term program of observing and analyzing change on a global scale (CEES, 1993). The NALC project will provide documentation of land cover change for large portions of North America by establishing a database of Landsat MSS images acquired over 20 years. The principal science goal was to monitor forest land cover, including forest distributions across the North American continent, land cover conversion (i.e., forests to alternate land cover types), and forest regrowth. NALC data can support the creation of land cover database products by exploiting the extensive Landsat MSS data archive.

Many researchers believe that to conserve significant quantities of carbon (C), deforestation must be controlled and management strategies need to be implemented to improve forest system net primary productivity (NPP) (Houghton, 1992; Clinton and Gore, 1993; Haynes et al., 1994). Current estimates indicate that the world's forests contain up to 80 percent of the aboveground and 40 percent of belowground C stocks. Uncertainties in balancing the contribution of forest landscapes to the global C budget will continue until source data are available that can accurately characterize the spatial distribution and condition of forest resources (Dixon et al., 1994).

National C pools have traditionally been determined by multiplying the area (ha) for each major land cover type (e.g., forest, woodland/scrub, wetland, mangrove, etc.) by its respective C density (MgC/ha) and summing the resultant quantities (PgC) for entire countries. Ideally, all aspects of NPP including changes in above- and belowground phytomass should be measured. In practice, this rarely occurs due in large part to the lack of reliable land cover and land cover change datasets (Cairns et al., 1997). Aboveground forest C content can be crudely estimated based on general forest type land cover distributions or quantified with a higher degree of certainty by resolving individual forest community types and densities. The following relationship is proposed as a potential quantification approach:

$$\text{forest community C content} = \sum_{j=1}^{n} fc_i \,(s_i \bullet a_i \bullet d_i \bullet k_i) \tag{1}$$

where c = carbon content (PgC), n = number of forest community types, fc = forest community type, s = forest seral stage coefficient, a = stand area (ha), d = stand density (percent canopy cover), and k = forest community specific carbon content density for late seral and 100 percent canopy cover (MgC/ha).

Also, the long-term monitoring of cleared forests is important to quantify the rate of C storage or sequestration (Dixon et al., 1994), and the emissions of other radiative important trace gases (RITG). For example, forest to pasture conversions in the humid tropics of Costa Rica resulted in a threefold increase of N_2O flux relative to the original forest soil. The rate of flux peaked during the first 10 years of conversion, then dropped below original forest flux rates (Keller et al., 1993). This example indicates the value of general land cover and land cover change data when used in combination with field sampling measurements to further our understanding of regional scale processes.

Numerous attempts have been made to use satellite data to monitor forest stand condition and change in the humid tropics of Brazil or the Pan-Amazon (Nelson and Holben, 1986; Nelson et al., 1987; Skole and Tucker, 1993). Results indicate that Advanced Very High Resolution Radiometer (AVHRR) - Local Area Coverage (LAC) data can provide reconnaissance level data of forests at a regional to continental scale, when used in conjunction with higher resolution data (e.g., Landsat MSS) (Nelson and Horning, 1990). However, Landsat MSS and TM data provide the finer spatial resolution necessary to accurately characterize forest change (Nelson et al., 1987). Landsat Thematic Mapper (TM) and SPOT Multi-Spectral (XS) data provide comparable estimates of forest extent and distribution (Skole and Tucker, 1993).

3.0 APPROACH

NALC triplicate data sets have been assembled for a large portion of the North American continent land mass. The NALC study area was defined as the area north of the Panama-Colombia border including the Caribbean and the Hawaiian Islands. Due to funding restraints and insufficient Landsat data holdings in U.S. archives for Canada, the NALC project did not assemble triplicate data sets for all of Canada. Canadian archives contain an extensive collection of Landsat MSS data from the 1990s, 1980s, and 1970s that could be used in support of a NALC/Canada effort.

3.1 Satellite Data Acquisitions

Landsat MSS data were selected as the source data for NALC triplicate development. This was for several reasons, including the 20-year MSS digital data archive and the relatively lower data acquisition and processing costs as compared to Landsat TM or SPOT XS data. Data used to assemble the NALC triplicates included Landsat MSS data for three epochs, 1990–92, 1985–87, and 1972–74. These epoch periods were selected to coincide with archival data maxima and provide roughly equal temporal spacing. The objective was to select high quality scenes for each epoch period; however, numerous deviations occurred due to the lack of available high quality archive data.

A total of 801 new (i.e., non-archival) Landsat MSS path/row scenes were acquired for the time period of 1991 plus or minus one year. The acquisition of these 1990s scenes was a major program accomplishment. This purchase was significant because the existing archive had very few 1990s scenes and additional acquisitions were not scheduled before the Landsat MSS was scheduled to be decommissioned in 1992. New data purchases for the 1990s were made for Mexico (119 scenes), Central America (32 scenes), the Caribbean islands (48 scenes), Alaska (180 scenes), continental United States (414 scenes), and the Hawaiian Islands (8 scenes).

3.2 New Data Acquisition and Archival Scene Selection

Vegetation Phenology

The first step in the process was the establishment of time "windows" for Landsat MSS data acquisitions. Time "windows" were defined using temporal variables based on plant phenology to provide temporal continuity essential for performing accurate land cover change analysis. The goal was to minimize spectral differences associated with vegetation phenology in order to optimize accuracy.

AVHRR NDVI biweekly composite data sets, or "greenness" images (Eidenshink et al., 1991), were used to establish suitable vegetation phenology time windows for each Landsat path/row location. The suitability of times were based on the onset, peak, and duration of "greenness." An optimal 90-day window was defined for use in scheduling of new data acquisitions and to search for suitable archival data.

Temporal Continuity

The Landsat MSS archive was evaluated to determine the availability of existing high-quality and cloud-free MSS images for 1980s and 1970s that met established acquisition "windows." An EROS Data Center document that describes the historical cloud cover conditions for each world reference system 2 (WRS-2) path/row was used to further refine the phenology-based acquisition windows to optimize for new data acquisitions. Subsequent to data acquisitions in 1991 and 1992, approximately 3,000 Landsat MSS NALC images were screened to obtain an initial list of scenes that fell within the established acquisition windows and exhibited a cloud cover of less than 30 percent. Microfiche of all candidate scenes were then viewed, and scene selections made for the three epoch periods. When an image was not available within a given three-year window that met the phenology and cloud cover requirements, images were selected outside of the window to coincide with the established phenology windows. All triplicate scene selections have been published in the NALC Research Plan (Lunetta et al., 1993).

3.3 Triplicate Data Sets

The standard data used in support of NALC featured Landsat MSS triplicates that have been georeferenced, coregistered, and registered to earth coordinates (rectified). Standard triplicate data sets included coregistered digital elevation model (DEM) data in addition to the image data (Lunetta et al., 1993). All data sets were clipped to the Landsat WRS-2 image sampling frame. In the case of the 1970s imagery, which conforms to the WRS-1 sampling frame, multiple scenes were acquired and clipped to the WRS-2 boundaries and processed as separate images to provide complete coverage for each scene in each decade.

3.4 Database Development

For each triplicate set, the 1980s image was rectified, and then used as a template to coregister 1970s and 1990s images. Image control points were selected from the 1980s images, and corresponding map control points were obtained from maps or a library of ground control points (GCP) for use in developing the model for geometric transformations (geocoding) model. A single-step cubic convolution algorithm was used to rectify and resample the image into a Universal Transverse Mercator (UTM) projection with a 60 x 60 meter pixel size. Quantitative assessment was undertaken to ensure that residual geometric correction was within acceptable limits established for both within and outside the United States to correspond with available map scales.

Images were preprocessed to compensate for variations in the radiometric response of the individual detectors prior to geometric registration to correct scan-line dependent noise. An interim systematic correction was applied, using the satellite ephemeris data and platform navigation model, to generate a UTM image that was oriented with north at the top. Automated cross-correlation procedures were then implemented to extract control points from triplicate images to compute coefficients for image-to-image registration. When an accurate transformation was developed, the grid for the interim systematic correction was convolved with the image-to-image transformation grid to produce coefficients that facilitated a single-step registration of the 1970s and 1990s products with the previously map-registered 1980s image. The result was an image registration procedure that only involved one step of resampling (Lunetta et al., 1993). The last step involved creation of a pixel identity image to accompany each of the 1970s images. Each pixel value in the identity image identifies the specific 1970s scene used to build the WRS-2 path/row equivalent.

The final database development task involved the mosaicking and projection transformation of the DEM data. The DEM data were derived from the Defense Mapping Agency (now National Image and Mapping Agency or NIMA) digital terrain and elevation data (DTED) which were digitized from standard 1:250,000-scale topographic maps. Complete coverage existed for the United States and Mexico. These data, often referred to as three arc-second DTED, were provided as gridded DEM corresponding to 1- by 1-degree data blocks. For the NALC project, these blocks were mosaicked for the path/row of interest, and then reprojected and resampled to 60- by 60-meter pixels in the UTM projection.

3.5 Land Cover Categorization

The NALC land cover categorization system was designed to provide information for documenting land cover change and for the calculation of standing carbon stocks (Table 3.1). The original NALC classification system was based primarily on vegetation structure and was hierarchical. This hierarchical system provided the capability to translate (convert) between other major land cover categorization systems (e.g., Anderson et al., 1976; Brown et al., 1979; Cowardin et al., 1979). This multiple layer approach to categorization system development was consistent with the ongoing effort by the United Nations Environmental Program to develop a Global Vegetation Categorization System (United Nations Workshop, 1993).

The initial effort to categorize NALC imagery involved a three-step process. First, unsupervised spectral clustering and statistical analysis of the raw Landsat MSS data were performed to generate training sets. Second, a maximum likelihood classifier was applied to individual MSS scenes to produce a spectrally categorized data set using the training sets developed from clustering and statistical analysis. Third, the image categorization step involved labeling the spectral categories as to land cover types corresponding to the NALC land cover categorization system. Labeling was accomplished through the use of National High Altitude Photography (NHAP) leaf-on photography and subsequent field identification of spectral clusters with a low interpretation confidence.

Table 3.1. NALC Pathfinder Categorization System.

LEVEL 0	LEVEL 1	LEVEL 2
Land	1.0 Barren or Developed Land	1.1 Exposed Land
		1.2 Developed Land
	2.0 Woody	2.1 Forest
		2.2 Scrub/Shrub
	3.0 Herbaceous	3.1 Herbaceous
	4.0 Arid	4.1 Arid Vegetation
		4.2 Riparian
	5.0 Snow/Ice	5.1 Snow/Ice
Water	6.0 Water & Submerged Land	6.1 Ocean
		6.2 Coastal
		6.3 Near-Shore
		6.4 Inland
Other	7.0 Other	7.1 Cloud
		7.2 Shadow
		7.3 Missing
		7.4 Indeterminable

3.6 Accuracy Assessment

An integral component of the NALC project was the quality control of data and subsequent validation of land cover change products. Data and products should be assessed in both spatial and thematic accuracies (Lunetta et al., 1991). The data quality objectives were:

Absolute Geometric Accuracy

The data quality objective for image-to-map registration RMS error was better than ±1.0 pixel Root-Mean-Square (RMS) error within the United States (1:24,000-scale base maps), and ±1.5 pixels outside the United States (1:50,000-scale base maps).

Relative Geometric Accuracy

Image-to-image registration of better than ±0.5 pixel was the established as a data quality objective. A high level of importance was placed on the image-to-image registration so as to accommodate image differencing for change detection analysis. This accuracy objective was established so as to provide an absolute registration accuracy of ±1.0 pixel in the resultant change image product. This accuracy was calculated using the following formula.

$$e = 2k \quad (2)$$

Where e = change image relative misregistration, and k = original image misregistration.

Thematic

Categorized data products were designed to be 80 percent accurate by class at a 75 percent confidence interval. The target of 80 percent accuracy was established based on results reported in the literature (Jensen et al., 1984; Ormsby and Lunetta, 1987; Hopkins et al., 1988). Reporting procedures for thematic accuracies included an error matrix with the percent commission error by category, percent omission error by category, total percent correct, number of points sampled, thematic accuracy at 75 percent confidence interval, and Kappa statistic (Lunetta et al., 1991).

3.7 Database Availability and Distribution

An information management system (IMS) that consisted of a relational database and descriptive metadata was developed to facilitate the indexing, archiving, and distribution of data sets. The NALC and other Pathfinder data sets are managed by the Earth Observing System Data and Information System (EOSDIS) IMS. The EOSDIS IMS provides the data information and management system for all Landsat Pathfinder Project data including source data and derivative products. Distribution is provided by the Distributed Active Archive Center (DAAC) at the USGS EROS Data Center.

The following elements are the capabilities provided by an IMS: graphic display of the extent of data coverage, geographic query, interrogation of metadata attributes, data richness or status tracking, overlays of basic cartographic elements (such as political boundaries and hydrography), viewing of selected browse products, linkages between value added products and their source data, and the ability to execute product orders.

The EDC DAAC is distributing the NALC triplicate products via file transfer protocol (ftp), on magnetic tape media, and CD-ROM. These contain the data descriptor record (DDR) files in ASCII format and the image data in band sequential format. The DDRs contain information describing the image projection parameters and UTM coordinates. The order in which files are written to tape is constant, except the number of files depends on the number of scenes used for each of the 1970s and 1990s components. The image files, in turn, vary in size (number of lines, samples, and bands). A DDR file is written before each image file. All files (images and DTED) have the same dimensions; the MSS images are 8-bit data, and the DTED are 16-bit data. Data are provided either at no cost or at the marginal cost of filling the user request.

4.0 NALC PROJECT RESULTS

4.1 Triplicate Images

Figure 3.1 contains a subset of the NALC data for path 21, row 49, located in the State of Chiapas, Mexico. Portions of three georegistered Landsat MSS "triplicate images" and DEM data are shown. The DTED data used to construct the Mexico DEMs were acquired from the Instituto Nacional de Estadistica, Geografia e Informatica (INEGI), Aguascalientes, AGS, Mexico. As a collaborator in the NALC/Mexico effort, INEGI provided the Mexico DTED data as their contribution to the project. These data have been provided by the INEGI for use in scientific investigations and projects conducted by government agencies, universities, and non-profit research institutions. For-profit organizational use of the Mexico DEM data requires prior approval from the INEGI Director General. Data may be acquired at the cost of duplication and distribution and can be queried and accessed through the EOSDIS/V.0 IMS.

An excellent example of a significant change in land cover between the 1970s data acquisitions is illustrated in the above path 21/row 29 1970s image. Although the reservoir was not present in the 02/15/74 image, in a companion image collected on 05/12/75 (not shown), the reservoir was present. This inconsistency in land cover between the 1970s dates necessitated that the images be held as separates in the triplicate image set.

As an example NALC product, a Great Lakes watershed pilot study was conducted by the CCRS. The CCRS Pilot differs from non-Canadian NALC project activities in that the Great Lakes watershed pilot developed a seamless 200-meter MSS data product (see color Plate 3) (Guindon, 1995 and 1996) in order to demonstrate the feasibility of (a) integrating NALC and Canadian operational MSS scene products and (b) creating value-added image products from the NALC archive.

4.2 Categorization Products

Provisional NALC Level II land cover products are presented in color Plates 4 and 5. The images are a subset of the path 15 and row 33 and include the Washington, DC and Baltimore, MD metropolitan areas. General trends in land cover change associated with the process of urbanization are apparent. Over the short time period of 1987–90, there has been a significant increase in developed and herbaceous lands at the expense of forested land cover. The high degree of image-to-image coregistration between these data sets yielded land cover products ideal for post-categorization change detection analysis using a raster-based Geographic Information System (GIS). Note that the land cover categorization depicted in these products is slightly different from that presented in Table 3.1.

4.3 Carbon Inventory Products

Land cover products developed from NALC data alone are insufficient for global change research applications because the measurement of forest canopy density is an important variable required for content calculations (Lunetta et al., 1993). However, land cover data developed from NALC data can be combined with other raster data sets to produce interpolated data products that provide superior spatial and thematic resolution for GIS-based calculation of standing C stocks. In support of this effort, prototype carbon inventory products were developed. Color Plate 6a presents a product created by integrating a high spatial resolution NALC land cover categorization product with the coarse resolution Forest Type Group data set for the

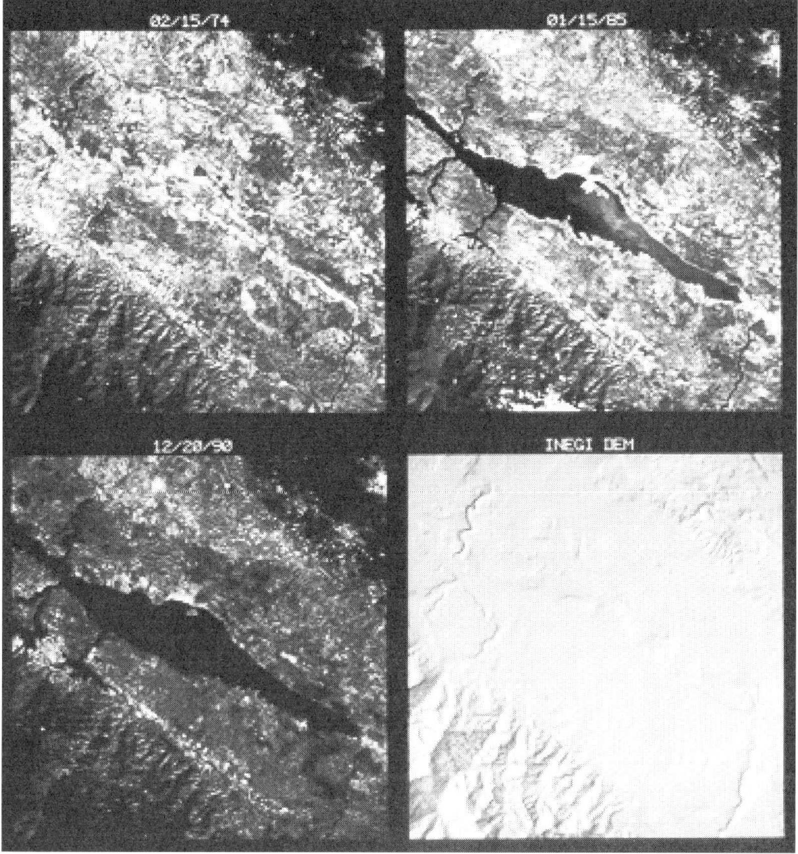

Figure 3.1. NALC Landsat MSS "triplicate images" and digital elevation model data (DEM) for WRS-2 Path 21, Row 29 in the State of Chiapas, Mexico.

continental United States. Color Plate 6b illustrates the integration of a NALC categorized product with the USFS Forest Type Group and Forest Density data. These coarse resolution data sets were derived from AVHRR data by the U.S. Forest Service (USFS) (Zhu and Evans, 1994).

5.0 DISCUSSION

The "higher order" interpolation products presented here integrate the strengths associated with each data type; the relatively high spatial resolution and categorical accuracy associated with the NALC forest land cover complements the forestry attributes contained in the USFS data. This is consistent with the successful use of other types of ancillary data (e.g, DEMs, city lights, census block, etc.) to improve categorization accuracies (Bara and Shaw, 1995). The "higher order" NALC product provides information detailing the specific forest community type and density data required for calculation of forest C content. A working hypothesis is that this product provides a data set of superior quality and accuracy for the calculation of standing C stocks throughout the North American continent. The USFS is currently working on expanding the Forest Type Group and Density data to include Mexico. Once completed, this product will be available to complement the high spatial resolution NALC-based land cover data products currently under development in the continental United States and Mexico.

The propagation of error associated with these "higher order" C inventory products must be quantified to evaluate their full potential application for global change C studies. What is the accuracy of these integration products? Initial comparisons between Landsat MSS and TM to AVHRR derived land cover products have demonstrated a poor one-to-one correspondence (Stoms and Davis, 1994). However, data accuracy cannot be determined by directly comparing land cover derived from higher resolution Landsat TM and MSS sensors to that from coarse resolution AVHRR data. The large differences in positional accuracy and spatial resolution render a direct comparison meaningless. Instead, the goal is to determine how to apply these two data sets in a complementary approach to optimize both spatial resolution and categorical accuracy.

What is the best approach for assessing the error associated with the fused or "higher order" products that integrate source data with disparate resolutions? One possible approach is to apply the following two-step model. First, apply the higher spatial resolution categorical data (i.e., forest polygons derived from Landsat MSS or TM) to establish the spatial distribution and effectively reduce the minimum mapping unit of the coarser resolution data set. By subsetting the coarser resolution data into the general land cover categories of interest (i.e., forested land cover, the errors of omission and commission are effectively established to coincide with the accuracy and precision of the higher resolution data. Secondly, subsequent to the labeling of forest polygons by forest community types and densities using the coarser spatial resolution data, polygons can then be selected on a random or stratified random basis for ground data collection to determine the predominate forest association and average densities for the construction of standard error matrices to perform an accuracy assessment of the "higher order" land cover products. The first model step has been illustrated in this chapter.

By first performing a general characterization of land cover types using remote sensor data at the appropriate spatial resolution required to delineate land cover types of interest (e.g., using Landsat MSS or TM data), the spatial distribution of forests can be determined at a known level of accuracy and precision. Subsequent to the assessment of categorical accuracy of the "higher order" data fusion products, their application can be evaluated for global change research applications in support of terrestrial C inventory and carbon flux studies. Future studies will address the application of these products for national and regional scale forest C inventories.

6.0 SUMMARY

With the launch of Landsat 1 on July 23, 1972 the United States initiated the capability for land resource monitoring from space. Over a 20-year period, the Landsat MSS sensor collected data documenting land cover conditions over the majority of the globe. The global change research community has prioritized reducing the uncertainty associated with land cover change as a major constituent of importance to balancing the global carbon cycle. The MSS archive currently represents the best available, continuous public source of relatively high resolution imagery for the monitoring of land cover over the 1972–1992 period. The North American Landscape Characterization project was designed to exploit this archive by providing standardized satellite data sets to support land cover change analysis. Land cover categorization products derived from MSS data alone provide only a general characterization of land cover condition and change. A data fusion approach using a post-classification technique is presented as a cost-effective method for delineating land cover at appropriate thematic and spatial resolutions to support a quantitative inventory of forest carbon stocks. Issues related to assessing the accuracy of carbon inventory data sets are presented and a two-step model proposed for accuracy assessment.

Landsat MSS data sets provide a one-of-a-kind record of land cover change over a two-decade period from 1972–92. Previous limitations to the use of Landsat MSS archive for large-area studies have included: inconsistencies in data quality; changes in Landsat satellite orbital configurations; differing data formats; and the high cost of data acquisitions. The NALC project Landsat MSS "triplicate images" represent a substantial contribution to a wide variety of users that will extend well into the twenty-first century. The low cost, consistent data processing, and multitemporal data coverage for major portions of North America will potentially benefit global change research, natural resource managers, and land use planners.

ACKNOWLEDGMENTS

The authors thank Dr. Roman Alvarez, Dr. Jose Luis Palacio, and Mr. Roberto Bonifaz of the Institute of Geography, National University of Mexico for their numerous contributions in support of research and development efforts in Mexico. Also, the authors would like to acknowledge Ms. Karen H. Lee for the development of numerous figures contained in this manuscript. The U.S. Environmental Protection Agency (EPA) partially funded this research.

REFERENCES

Anderson, J.R., E.E. Hardy, J.T. Roach, and R.E. Witman. A Land Use and Land Cover Classification System for Use with Remote Sensor Data. U.S. Geological Survey Professional Paper, Number 962, 1976.

Bara, T.J. and D.M. Shaw. *Multi-Resolution Land Characteristics Consortium: Documentation Notebook*. Internal Report, U.S. EPA, Research Triangle Park, NC, 1995.

Brown, D.E., C.H. Lowe, and C.P. Pase. A digitized classification system for the biotic communities of North America, with community (series) and association examples for the Southwest. *Journal of Arizona-Nevada Academy of Science*, 14(1), 1979.

Cairns, M.A., J.K. Winjum, D.L. Phillips, T.P. Kolchugina, and T.S. Vinson. Terrestrial carbon dynamics: a case study in the former Soviet Union, the conterminous United States, Mexico and Brazil. *Mitigation and Adaptation Strategies for Global Change*, 1, pp. 363–383, 1997.

Clinton, W.J. and A. Gore. *The Climate Change Action Plan*. U.S. Government Printing Office, October 1993

Committee on Earth and Environmental Sciences (CEES). *Our Changing Planet*. The fiscal year 1993 U.S. Global Change Research Program, 79 p., 1993.

Cowardin, L.M., V. Carter, F.C. Golet, and E.T. LaRoe. *A Classification of Wetlands and Deep Water Habitats of the United States*. Office of Biological Services, Fish and Wildlife Service, U.S. Department of the Interior, Washington, DC, 103 p., 1979.

Dixon, R.K., S. Brown, R.A. Houghton, A.M. Solomon, M.C. Trexler, and J. Wisniewski. Carbon pools and flux of global forest ecosystems. *Science*, 263, 185–90, 1994.

Eidenshink, J.C., R.E. Burgan, and R.H. Haas. Monitoring fire fuels conditions by using time-series composites of Advanced Very High Resolution Radiometer data. *Proceedings of Resource Technology International Symposium in Natural Resources Management*, Washington, DC, November 1990, pp. 68–82, 1991.

Guindon, B. Utilization of Landsat Pathfinder data for the creation of large area mosaics. *Proceedings of the ACSM/ASPRS Annual Convention*, Charlotte, NC, pp. 144–153, 1995.

Guindon, B. Design and development of a digital satellite image atlas. *Geo-Informations-Systeme*, 9(6), 2–8, 1996.

Haynes, R.W., R.J. Alig, and E. Moore. *Alternative Simulations of Forestry Scenarios Involving Carbon Dioxide Sequestration Options: Investigation of Impacts on Regional and National Timber Markets.* General Technical Report PNWGTR-335, U.S. Department of Agriculture, Forest Service, Pacific Northwest Research Station, Portland, Oregon, 66 p., 1994.

Hopkins, P.F., A.L. Maclean, and T.M. Lillesand. Assessment of Thematic Mapper imagery for forestry applications under lake states conditions. *Photogrammetric Engineering and Remote Sensing*, 54(1), 61–69, 1988.

Houghton, J.T., B.A. Callander, and S.K. Varney, Eds. *Climate Change 1992: The Supplementary Report to the IPCC Scientific Assessment.* Intergovernmental Panel on Climate Change, WMO/UNEP, Cambridge University Press, UK, 200 p., 1992.

Houghton, R.A. *A Blueprint for Monitoring the Emissions of Carbon Dioxide and Other Greenhouse Gases from Tropical Deforestation.* The Woods Hole Research Center, Woods Hole, Massachusetts, 29 p. and appendices, 1992.

Jensen, J.R., E.J. Christensen, and R. Sharitz. Nontidal wetland mapping in South Carolina using airborne multispectral scanner data. *Remote Sensing of Environ.*, 16(1), 1–12, 1984.

Keller, M., E. Veldkamp, A.M. Weitz, and W.A. Reiners. Effects of pasture age on trace gas emissions from a deforested area in Costs Rica. *Nature*, 36(2), 44–46, 1993.

Lunetta, R.S., R.G. Congalton, L.K. Fenstermaker, J.R. Jensen, K.C. McGwire, and L.R. Tinney. Remote sensing and geographic information system data integration: Error sources and research issues. *Photogrammetric Engineering and Remote Sensing*, 57(6), 677–687, 1991.

Lunetta, R.S., J.G. Lyon, J.A. Sturdevant, J.L. Dwyer, C.D. Elvidge, L.K. Fenstermaker, D. Yuan, S.R. Hoffer, and R. Werrackoon. *North American Landscape Characterization: Research Plan.* EPA/600/R-93/135, July 1993, 419 p., 1993.

Lyon, J.G., D. Yuan, R.L. Lunetta, and C.D. Elvidge. A change detection experiment using vegetation indices. *Photogrammetric Engineering and Remote Sensing*, 64(2), 143–150, 1998.

Nelson, R. and N. Horning. AVHRR-LAC estimates of forest area in Madagascar. *Int. J. Remote Sensing*, 14(8), 1463–75, 1990.

Nelson, R., N. Horning, and T.A. Stone. Determining the rate of forest conversion in Mato Grosso, Brazil using Landsat MSS and AVHRR data. *Int. J. Remote Sensing*, 8(12), 1167–84, 1987.

Nelson, R. and B. Holben. Identifying deforestation in Brazil using multiresolution satellite data. *Int. J. Remote Sensing*, 7(3), 429–48, 1986.

Ormsby, J.P. and R.S. Lunetta. Whitetailed deer food availability maps from thematic mapper data. *Photogrammetric Engineering and Remote Sensing*, 53(11), 1585–89, 1987.

Skole, D. and C. Tucker. Tropical deforestation and habitat fragmentation in the Amazon—satellite data from 1978 to 1988. *Science*, 260, 1905–1909, 1993.

Stoms, D.M. and F.W. Davis. *Synoptic National Assessment of Comparative Risks to Biological Diversity and Landscapes: Pilot Studies for Southern California.* Summary Progress Report Submitted to the U.S. Environmental Protection Agency, May 1994, 88 p., 1994.

United Nations Workshop. *Report of the UNEP-HEM/WCMC/GCTE Preparatory Workshop on Vegetation Classification.* Charlottesville, Virginia, January 24–26, 21 p., 1993.

Zhu, Z. and D.L. Evans. U.S. Forest Types and Predicted Percent Forest Cover from AVHRR Data. *Photogrammetric Engineering and Remote Sensing,* 60(5), 525–531, 1994.

CHAPTER 4

Classification-Based Change Detection: Theory and Applications to the NALC Data Set

Vittorio Castelli, Christopher D. Elvidge, Chung-Sheng Li, and John J. Turek

1.0 INTRODUCTION

The North American Landscape Characterization (NALC) data set is ideally suited for land cover/land use change studies due to its extent of spatial coverage, the temporal span represented (early–mid 1970s to early 1990s), the highly accurate image coregistration, and the ancillary data provided with each data "triplicate" set (Lunetta et al., 1993). Also, the relatively easy access and low cost of the NALC data make it an ideal data set for the study of land cover change across the continental United States and Mexico corresponding to a two-decade epoch beginning in 1972 through 1992. The purpose of this chapter is to analyze the application of change detection techniques based on statistical classification of a NALC data set.

The post-classification change detection technique is the simplest type of a classification-based change detection analysis. Applied to two coregistered images this approach consists of a labeling step that produces two classified maps, followed by a comparison step that identifies areas of change as differences in class labels. In contrast, the image-differencing approach utilizes image data to directly analyze spectral difference between two image dates to identify areas of change. In this chapter we discuss the main difficulties of these approaches, both in generic terms and specific to the NALC case, and propose improvements and new algorithms for land cover change analysis.

In this chapter we describe four basic approaches to the classification-based land cover change detection problem, schematically described in Figure 4.1. We identify three main causes of inaccuracy and inefficiency; namely, the intrinsic limitations of the data available for the study, the methodological shortcomings resulting from simplifying constraints and assumption, and the computational complexity of many of the algorithms. The chapter addresses methods for improving the accuracy and efficiency of change detection. A review of methods for reducing the error rate of classifiers by including ancillary information, such as digital elevation model (DEM) and normalized difference vegetation index (NDVI) data coverages is presented. We analyze in detail a preclassification method that relies on automatic scattergram-controlled regression, which improves the accuracy and increases processing efficiency by identifying potential areas of change before the classification step, and improves efficiency, by discarding large portions of images where in all likelihood no change occurred. We then propose a method for generating training sets using the information contained in the United Stated Geological Survey (USGS) land use/land cover (LULC) maps using an automated approach, and conclude

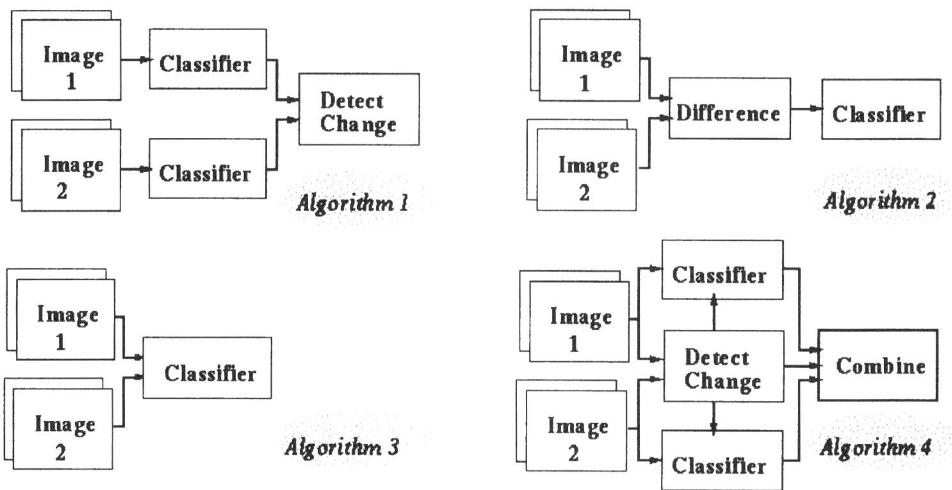

Figure 4.1. Basic approaches to classification-based change detection.

with the detailed description of a neural-network base architecture that implements the methodologies described in the chapter.

2.0 BACKGROUND

2.1 Classification-Based Change Detection

The most common approach to classification in the remote sensing community is based on pixel-wise labeling of spectrally unique and identifiable land-cover classes (Richards, 1993; Paola and Schowengerdt, 1995). Here, each pixel from an n-band sensor is represented as a point in an n-dimensional Euclidean space, with the ith coordinate value equal to the brightness of the pixel in the ith spectral band. Supervised image classification is customarily performed using training sets constructed either by collecting ground truth for each example (i.e., class) or by photointerpreting the image. In the latter case, an expert partitions the reflectances of the pixels into numerous groups, or clusters, with the aid of an unsupervised classification algorithm, and then labels each cluster individually (Dubes, 1993; Kaufman and Rousseuw, 1990; Duda and Hart, 1973). This method of constructing classifiers is widely supported by image processing packages developed for remote sensing applications.

Classification-based change detection suffers from several shortcomings, some general in nature, others specific to the NALC and similar data sets, that we will define and discuss while suggesting options that may be used to improve the final results of a land cover change analysis. In particular we can distinguish between data, computational, and methodological limitations.

Data-Related Limitations

The data available for a given application intrinsically limits the accuracy of classification-based change detection. All approaches are affected to a different extent by the following shortcomings.

- ***Limited spectral separation of classes***. Publicly available satellite data (including the NALC Landsat MSS triplicates) generally provide data in three to seven spectral bands. Consequently, different land-cover classes sometimes have similar spectral signatures.
- ***Limited spatial resolution***. In a satellite image individual pixels represent a specific area or instantaneous field of view (IFOV), rather than a point on the surface of the earth. Even at moderately high resolution many pixels correspond to areas with heterogeneous land cover. This condition is commonly referred to as the "mixed pixel" problem.
- ***Errors in coregistration***. If the images are not properly coregistered, the change detection algorithm produces incorrect results around the boundaries of homogeneous regions. Unless otherwise accounted for, errors resulting from the inaccurate co-registration of the imagery are attributed to thematic or categorical errors.

Methodological Limitations

Detecting change from satellite data is a complex task. To make the problem manageable, often we are forced to make simplifying assumptions or operational decisions which reduce the accuracy of the procedure. Moreover, semantic constraints are often imposed on the procedure that do not have an actual correspondence in the data: this is the case when we are interested in land use rather than land cover. Finally, the labeling step has intrinsic theoretical limitations that impose a lower bound on the attainable accuracy.

- ***The "statistical independence" assumption***. The pixel-wise approach to classification is based on the assumption that the DNs of neighboring pixels are independent. This assumption is frequently violated in the analysis of remotely sensed images, especially when data are obtained using medium-high resolution sensors, where connected regions of homogeneous land cover type can easily span several hundred pixels.
- ***The separation of the classification and change detection steps***. When classification is separate from the change detection stage, the change analysis is based on processed information from the two images and does not use the original data. There is thus a reduction of information content that can result in loss of accuracy.
- ***The construction of training sets***. The construction of reliable training sets for a particular image or set of images is a rather involved task. Obtaining the class labels for spectral clusters by collecting ground truth observations is expensive, time-consuming, and sometimes impossible. Additionally, care must be taken in recording the exact location of each observation. Constructing a training set from photointerpretation is a less expensive alternative, but the resulting classifier is at best as accurate as the human expert who constructed the training set. Finally, even in medium resolution images, numerous pixels represent regions of heterogeneous land cover type.
- ***The classification taxonomies***. Often classification taxonomies may be based on land use, which often cannot be identified using remote sensing, rather than land cover types. For example, in many instances it is impossible to spectrally discriminate between reservoirs and lakes. Also, taxonomies often aggregate different types of land cover under a single class; the typical example is the "urban" class, which in reality is an aggregation of barren terrain, vegetated land of various kinds, water, etc.
- ***The intrinsic limitations of classifiers***. Consider a classification taxonomy containing C classes. If the probability densities $\mathbf{f_i(x)}$ are continuous, and the prior probabilities $\boldsymbol{\pi_i}$, $\mathbf{i=1,...,C}$, of the C classes are known, the Bayes decision rule is the optimum classifier. The Bayes decision rule computes the conditional probability of each class j given the observation $\mathbf{X}$ (the posterior probability) and selects for $\mathbf{X}$ the label with maximum posterior probability. The probability of error (or risk) of the Bayes decision rule, called the Bayes risk, is denoted by $\mathbf{R^*}$ and is given by:

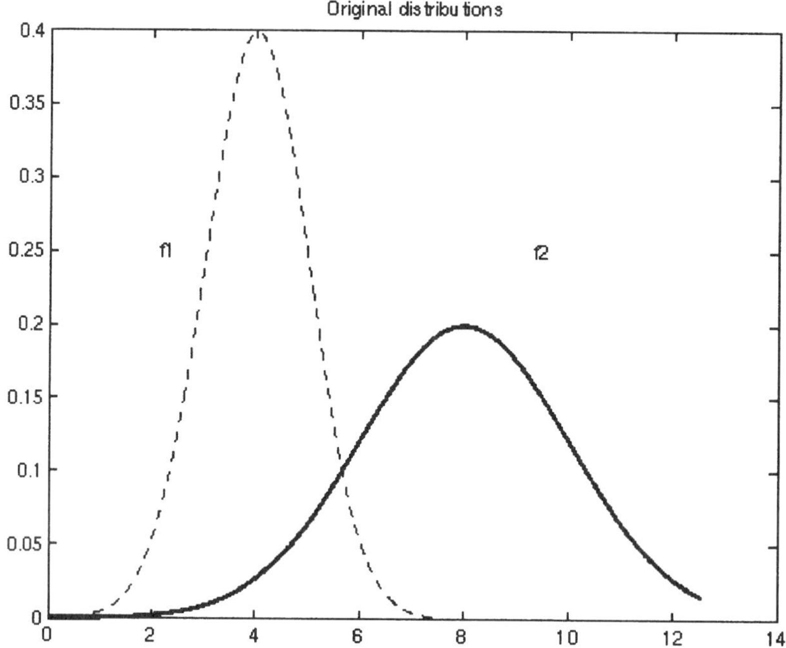

Figure 4.2. The distributions of the two classes of samples used in the example.

$$\mathbf{R}^* = 1 - \int \max_i(\boldsymbol{\pi}_i\, \mathbf{f}_i(\mathbf{x}))\, d\mu(x) \qquad (1)$$

The Bayes risk is the minimum probability of achievable error. In practice, for most cases $\mathbf{R^*>0}$, thus, even the optimum classifier is not error-free. Also, the prior probabilities $\boldsymbol{\pi}_i$ and the distributions $\mathbf{f}_i(\mathbf{x})$ are usually not known; then the classifiers must learn from examples (the training set), the resulting partition of the observation space is always suboptimal, and the risk is greater than $\mathbf{R}^*$. All the above limitations apply to the post-classification approach.

2.2 Detecting Change by Classifying the Difference Images

The image difference method offers a solution to the separation of classification and change detection, but introduces two major sources of imprecision compared to the post-classification approach, those being a reduction in the accuracy of the classification process and limited ability of providing information regarding the land cover classes involved in the change. The reduction in accuracy is intuitively illustrated by considering a Bayesian classifier that discriminates between two classes of samples (class 1 and class 2). Let the prior probabilities and the densities of the classes be $\boldsymbol{\pi}_1=\boldsymbol{\pi}_2=1/2$ and $\mathbf{f}_1(\mathbf{x})$ and $\mathbf{f}_2(\mathbf{x})$ (e.g., consider the one-dimensional Gaussians depicted in Figure 4.2). Assume that the probability that the class label for any given change pixel changes is a known constant; say 0.5 for sake of simplicity. Consider a pixel location, let i and j be the class labels in the first and second image, call $\mathbf{X}_1$ and $\mathbf{X}_2$ the corresponding pixel values, and let $\mathbf{D}$ denote their difference, $\mathbf{D}=\mathbf{X}_1-\mathbf{X}_2$. We have therefore four different combinations of the values of *i* and *j*, and, correspondingly, four distinct conditional distributions for $\mathbf{D}$ with densities denoted by $\mathbf{f}_{i,j}(\mathbf{x})$, i,j=1,2, (Figure 4.3). From the $\mathbf{f}_{i,j}(\mathbf{x})$ we can then derive the conditional densities of the differences $\mathbf{D}$ given change and no change in land

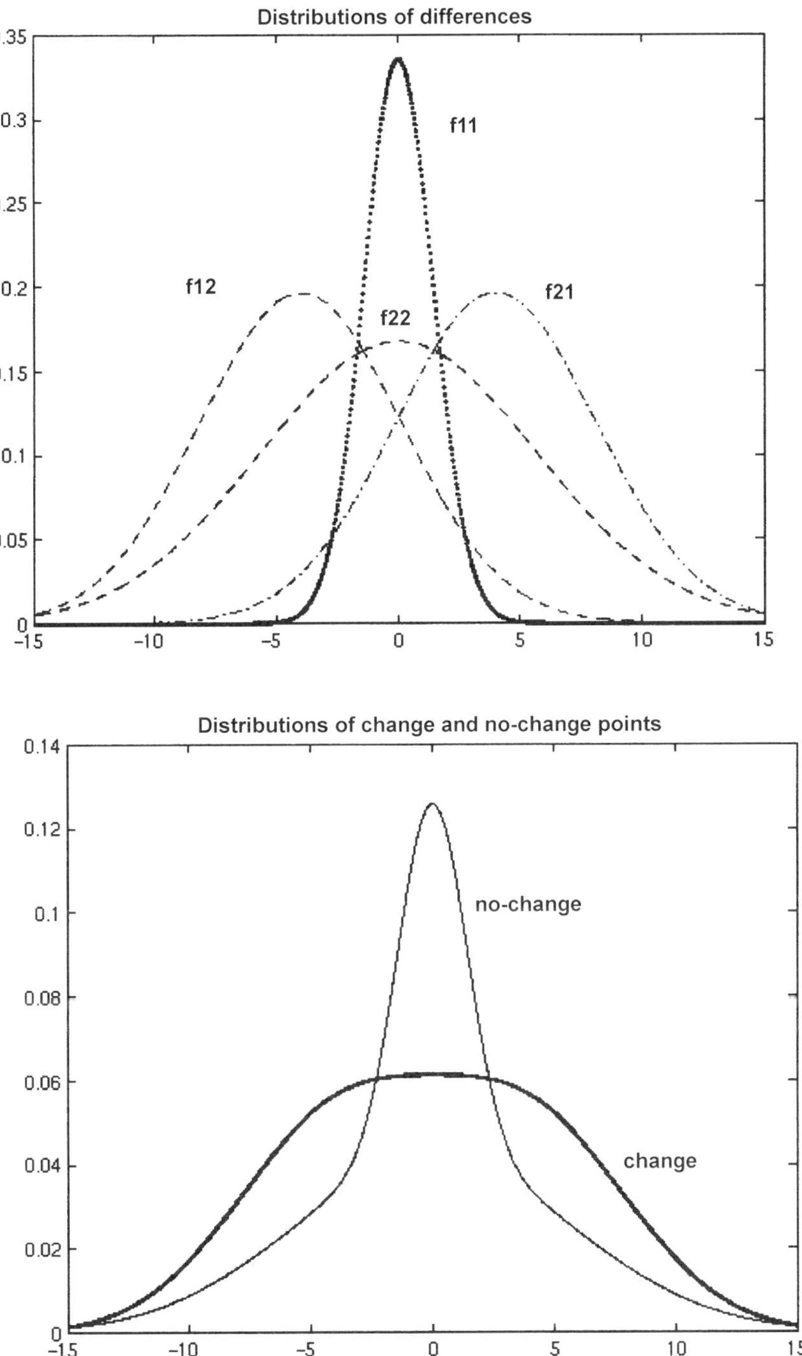

Figure 4.3. Top: the distributions of the four cases described in the example; bottom: the distributions of differences corresponding to regions of change and regions of no change.

cover (Figure 4.3). We can now compare the Bayes decision rules (the optimum classifiers) for the post-classification and the image-difference approaches. Consider first the post-classification method: here the change detection algorithm makes a mistake when either $\mathbf{X}_1$ is classified incorrectly and $\mathbf{X}_2$ correctly, or when $\mathbf{X}_1$ is classified correctly and $\mathbf{X}_2$ incorrectly. An individual classification error occurs with probability **R*** (Equation 1), and a correct classification occurs with probability **1-R***: thus the probability of change detection error is **2R*(1-R*)**. The quantity **2R*** is the area below the minimum of the two densities on the right-hand side of color Plate 7, labeled "method 1". The probability of change detection error of the image-difference method is just the Bayes risk **R*** of the corresponding classifier, which is given by the area under the minimum of the two densities on the left-hand side of Plate 7 (denoted by "method 2"). The areas in Plate 7 are at the same scale, and it is immediately apparent that the probability of error of the image-difference approach is significantly larger than the probability of error of the post-classification method.

The second problem is relevant when the classification taxonomy contains more than two classes. Change detection analysis should not only identify regions of potential change, but also return information on what land cover classes were involved in the change. It is easily seen that a classifier based on difference images does not produce reliable information regarding the original and new classes of change pixels.

2.3 Detecting Change with a Single Classifier

In the classification-alone method, a single classifier uses data from both images simultaneously to produce the land cover classes and to detect change. The main limitation here is the so-called "curse of dimensionality'' (Cherkassky et al., 1993). We shall here briefly explain this problem. If the images have ***b*** bands and the classifiers rely only on spectral information from each pixel location, then the classifier takes as input a vector of 2***b*** digital numbers, which is twice as long as the input of the individual classifiers in the other approaches. The effects of additional dimensions on the training set are dramatic: as new dimensions are added, the observation space becomes exponentially more sparsely populated; in other words, points are farther apart from each other. Thus, to describe appropriately a high dimensional space, an exponentially larger number of training set examples are required.

A second limitation of the approach is the proliferation of the class labels: if the taxonomy used contains ***k*** classes, then labeling jointly two images implies increasing the number of classes to k^2, one per each pair of original classes. Thus, the training set must now contain enough reliable examples from each of the k^2 classes.

2.4 Ancillary Information for the NALC Data Set

Using NALC triplicates for change detection offers numerous advantages; in particular, the selection of the highest quality original images from the MSS archive, the sophistication of the reprojection methodology, and the coregistration precision virtually eliminate many of the data-related limitations discussed above. Nevertheless, the limited number of bands can impair the accuracy of pixel-wise classification. Consequently, spectrally similar classes are difficult to discriminate using this data set. One can construct classifiers for simple taxonomies (i.e., the USGS Level-1), where pure classes can typically be separated in the four-dimensional reflectance space. Additionally, the NALC data set contains an NDVI band and a DEM, which can help in improving the accuracy of a change detection algorithm.

The NDVI can increase the discrimination ability of a classifier in regions of the sample space where all the bands have low DNs. Similarly, using the DEM the classifier can distinguish between spectrally similar types of land cover that occur at different altitudes. As an

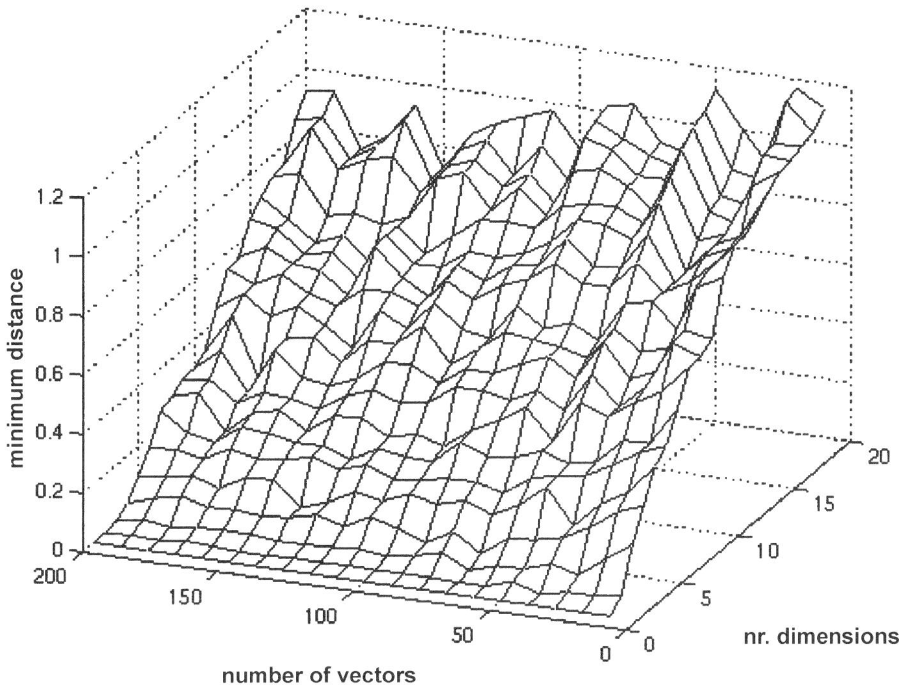

Figure 4.4. Sample image used in the experiment.

example, consider the image in Figure 4.4. Here, agricultural terrain and wet rangeland have overlapping spectral signatures. In this case both classification and regression tree classifiers (Breiman et al., 1984) are trained with the same training set. In the first case, the training examples contain only the DNs from the spectral bands, while in the second case the NDVI is added. As a result, the second classifier is more accurate, as can be seen by inspecting the agricultural fields area in the lower right portion of the classified maps (see Plate 8).

Also, the NDVI and DEM data can find uses in non-classification-based change detection techniques not described in this chapter. For example, the NDVI from Landsat data had been used as primary feature for vegetation change detection studies in the boreal forest-tundra vegetation transitional area by Awaya and Tanaka (1993).

3.0 PROGRESSIVE CLASSIFICATION AND CHANGE DETECTION

3.1 The Progressive Classification Framework

Classification-based change detection of a satellite image requires significant computational resources. The feature extraction process can be the most time-consuming step, if complex texture features (as described by Petitt et al., 1993; Philpot and Chavarria, 1994; Lohmann, 1994; Miller et al., 1995; Carr, 1996), or Markov random fields (as described by Bartlett, 1975; Guyon, 1995; Yamazaki and Gingras, 1995) are used by the classifier to capture the statistical dependence of neighboring pixels. The cost of the labeling stage depends on the type of classifier. Some algorithms, such as the linear discriminant classifier (Fung, 1995), CART (Breiman et al., 1994) and the learning vector quantizer (LVQ) (Kohonen, 1988; Kosmatopoulos and

Christodoulou, 1996) are appealing from the viewpoint of speed, and perform reasonably well in terms of accuracy. Other nonparametric methods, such as the nearest neighbor (Dasarathy, 1991; Cover and Hart, 1967) or more sophisticated algorithms, such as the neural networks (Haykin, 1994; Chettri and Cromp, 1992; Salu and Tilton, 1993; Paola and Schowengerdt, 1995) or probabilistic neural networks (Chettri and Cromp, 1993) and even parametric classifiers, such as the Bayesian parametric classifier (Duda and Hart, 1973) can be significantly slower, especially when the dimensionality of the sample space is large.

We propose here an approach to classification that results in faster and more accurate classifiers by relying on properties of appropriate representations of the images resulting, for instance, from compression algorithms. This framework termed "progressive classification" can be used in conjunction with essentially every standard classifier for change detection purposes, and offers an appealing solution to the concerns raised by both the high cost of the labeling step and the statistical independence assumption.

Image compression is often performed by first decorrelating the data using a transform, quantizing the coefficients, and lossless coding the result (Pennebaker and Mitchell, 1993). From some transforms, such as the wavelet transform (WT) described by Daubechies (1992), one can easily construct a multiresolution pyramid that is a sequence of approximations to the image that start from a very coarse representation, becoming progressively finer and more accurate, and end with the original data. In this dyadic multiresolution pyramid, each approximation is obtained from the immediately finer by filtering with a low pass filter and discarding every other row and every other column: consequently, at each step the amount of data is reduced by a factor of four. In simple terms, a multiresolution pyramid can be obtained by interactively averaging (using a weighted averaging filter) and reducing the original image.

Consider a region of homogeneous land cover and assume that the intensity values could be modeled as independent and identically distributed (i.i.d.) random variables. Then, elementary statistics show that the distribution of the lower resolution coefficients has a mean equal to the mean of the distribution of the original samples times the sum of the averaging coefficients $\mathbf{w_i}$ and variance equal to the original variance times the sum of the squares of the averaging coefficients. Assume WLOG (Without Loss Of Generality) that $\mathbf{\Sigma w_i}$=1, and that $\mathbf{\Sigma w_i^2}$<1. We then expect the distributions of the lower-resolution coefficients coming from homogeneous regions to be more separate than the distributions of the DNs in the original image. Figure 4.5 depicts scatterplots of three spectral bands of samples taken from the same region at full resolution and at level 2 of the wavelet transform containing samples from five different spectral classes. Note that the separations between vegetated wetland (labeled 2) and agricultural terrain (labeled 1), and between rangeland (labeled 3) and vegetated wetland are much wider in the wavelet transformed data.

3.2 The Progressive Classification Algorithm

The averaging process has the effect of increasing the separation between pure spectral classes. At the same time, many lower-resolution coefficients are obtained from portions of the image containing two or more different types of land cover. To address this issue, the progressive classification algorithm introduces one additional class, containing all these "mixed" coefficients. For each desired level of the multiresolution pyramid except the full resolution, a training set including both samples from "pure class" and from "mixed" coefficients is constructed, and a corresponding classifier is trained. The progressive classification algorithm (Plate 9) starts by analyzing the approximation of the image at the lowest suitable resolution. Each coefficient is classified and if the class label is one of the "pure classes," the entire block of the

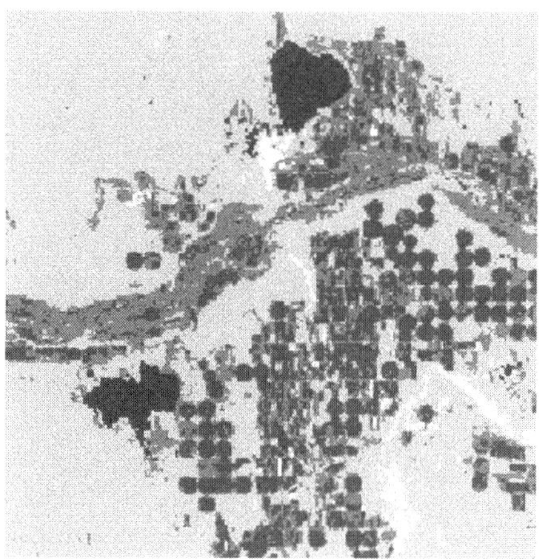

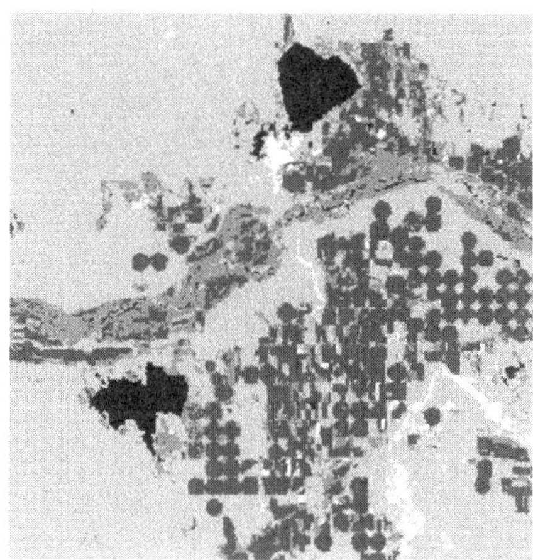

Figure 4.5. 3-D Scatterplot of random samples from NALC images. Left: samples at original resolution, Right, samples from the level-2 transform coefficients. The wavelet used in the transform is the Daubechies Biorthogonal Symmetric of order 2,2. Cluster no. 1 is agricultural land, cluster no. 2 is forested wetland, cluster no. 3 is rangeland, cluster no. 4 is barren terrain, and cluster no. 5 is water.

original image corresponding to the coefficient is classified in one step with this label. If, however, the class label is "mixed" class, then the multiresolution pyramid is inverted locally, four coefficients at the immediately higher resolution are produced, and the process is repeated until the entire image is classified.

The immediate advantage of the progressive classification scheme is its speed. Inverting the pyramid is in general much less computationally expensive than classifying an individual pixel, and the progressive classifier readily identifies portions of the images with homogeneous land cover. Since classifying a pixel at the first multiresolution level is equivalent to classifying four pixels at full resolution, the expected speedup is significant, especially when the labeling process is complex; for example, when using the nearest neighbor or neural networks algorithms. This speedup is documented by Castelli et al. (1996b) where the experiments are conducted on NALC triplicates. Surprisingly, the accuracy of the classifier also increases. The mathematical justifications for this behavior are discussed in a technical report by Castelli et al. (1996a). The statistical dependencies of both labels and DN of neighboring pixels are captured without resorting to additional features (such as texture features), and the accuracy increases. Note that the size of the neighborhood analyzed by the first stage of the classifier increases with the starting level of the pyramid. Classifying the image at Level-3 is equivalent to labeling blocks of 8×8 pixels in each elementary operation. Thus, care must be used in selecting an appropriate starting level for a given instrument. This algorithm is well suited for medium and high resolution satellite imagery including the NALC "triplicates."

The result of an experiment on the image of Figure 4.4 is shown in Figure 4.6. Here, the classified map produced by a progressive classifier is compared to the classified map obtained from the nonprogressive counterpart. The accuracy of the progressive classifier is better, as is easily seen in the agricultural areas in the lower-left portion of the image. Also, the progressive classifier produces smoother classification results, due to its lower sensitivity to sensor noise and to the overlap of the distribution of different classes, as is evident, for example, by inspecting the region of barren terrain in the lower left corner of the image.

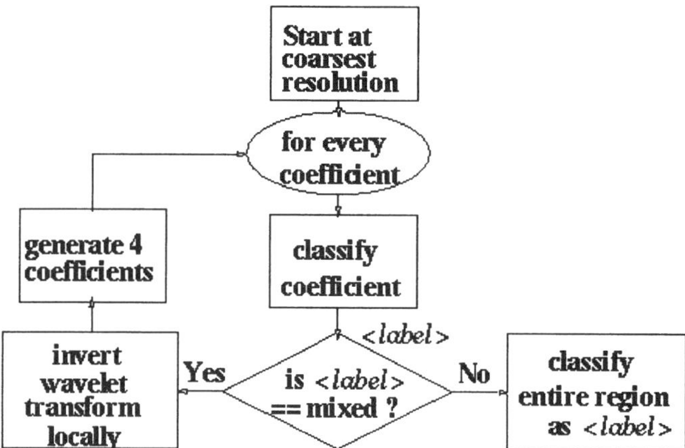

Figure 4.6. Classification results using a progressive classifier. From left to right: of standard classification without NDVI, result of progressive classification without NDVI. In the classified maps different gray levels represent different land use classes.

4.0 ASCR-BASED CHANGE DETECTION

4.1 Relative Radiometric Normalization

In this section, we show how to combine radiometric normalization with the identification of areas of potential change. In particular, we base our discussion on a radiometric normalization method developed specifically for the NALC data set, called Automatic Scattergram-Controlled Regression (ASCR) (Elvidge et al., 1995). This approach produces two benefits. Methodologically, the change detection algorithm provides both the original image data and the processed results from the classifiers. Computationally, the cost associated with the classifier is reduced by labeling only the candidate regions of change detected by the ASCR step.

Many relative radiometric normalization techniques are based on a linear model as described by Jenson (1983). If we let $\mathbf{x}_v(\mathbf{k})$ and $\mathbf{x}_v^n(\mathbf{k})$ denote the pixel values (DN) in image **X** at location $v = (i,j)$ and in band $\boldsymbol{k}$ before and after normalization, then the simplest approach to the problem consists in remapping $\mathbf{x}_v(k)$ into $\mathbf{x}_v^n(k)$ via a linear transformation:

$$\mathbf{x}_v^n(k) = a(k)\ \mathbf{x}_v(k) + \mathbf{b}(k) \tag{2}$$

To radiometrically correct **X** relative to **Y**, then the parameters $\mathbf{a}(k)$ and $\mathbf{b}(k)$ must be estimated from the two images **X** and **Y**. The simplest approach to deriving Equation 2 is to assume that **X** and **Y** are related by a linear model of the form $\mathbf{Y}_v(k) = \mathbf{a}_0(k)\mathbf{X}_v(k) + \mathbf{b}_0(k)+\mathbf{n}(k)$, where $\mathbf{n}(k)$ is uncorrelated zero-mean additive noise, and to estimate $\mathbf{a}(k)$ and $\mathbf{b}(k)$ with simple regression (Neter et al., 1985). However, this approach is not robust with respect to the presence of outliers (i.e., clouds, cloud shadows, and land cover change).

Elvidge et al. (1995) improved the simple regression by adding the automatic identification and elimination of the outliers and estimating the parameters via linear regression on the remaining data. Here, we discuss how to use this resulting method, ASCR, to identify regions of possible change.

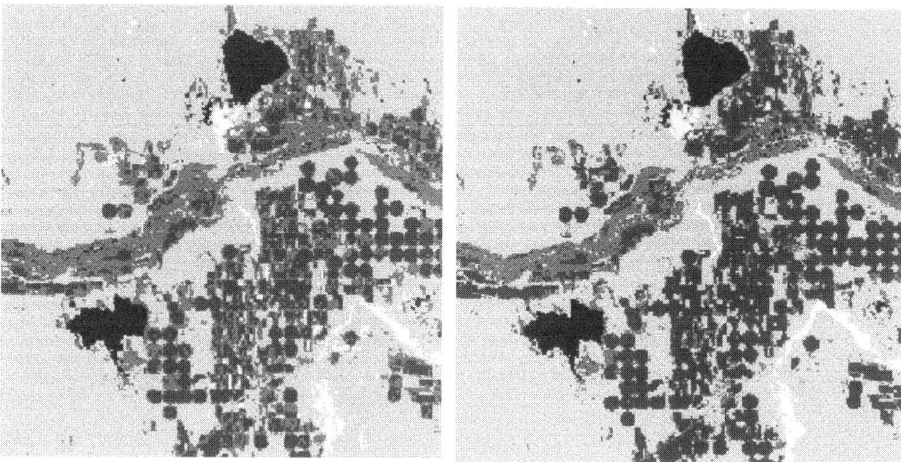

Figure 4.7. Histograms of band 3 (left) and band 4 (right), showing the maxima corresponding to water and land.

4.2 Automated Scattergram-Controlled Regression

The ASCR procedure relies on the observation that the histograms of MSS bands three and four display two prominent peaks (Figure 4.7). The first, located at very low DN values, is due to water pixels; the second, located slightly above the center of the full dynamic range of the sensors, is due to land. In the original ASCR algorithm the two local maxima of the scatterplot are used to identify a "no-change" pixel set (Figure 4.8) and we refer the interested reader to the cited article for the details. Consider the scatterplot of band three. The algorithm finds the two maxima points clearly identified by crossing lines in Figure 4.8 and constructs the line **l1** joining the two maxima points (the dashed line in Figure 4.8). The half width of the region of no-change HWNC is then automatically found, and the parallels **l2** and **l3** to the line **l1** at distance *HWNC* (the solid lines in the figure) are constructed. All the pixel locations corresponding to points outside the strip delimited by **l1** and **l2** are discarded. The previous steps are repeated for the scatterplot of band four (all the pixel locations outside the two delimiting lines were discarded). The algorithm then retains only the pixel locations that are not discarded in both scatterplots.

ASCR uses then this set of "no change" pixels to construct the best linear transformation in the mean squared error sense for each band. Let $\{x_i^k\}_{k=1,..,K}$ denote the set of DNs in band *i* extracted from the "no change" region of the reference image, and let $\{y_i^k\}_{k=1,..K}$ denote the corresponding set of DNs for the target image. Then the refined ASCR estimates of the linear regression coefficients are the values α_i and β_i by minimizing $SSE^i = \Sigma_{k=1}K\ (y_i^k - \alpha_i\ x_i^k - \beta_i)^2$. The coefficients α_i and β_i are then used in Equation 2 to normalize the target image with respect to the reference image.

4.3 Identifying Region of No-Change

The first advantage of using ASCR normalization in change detection studies is that a classifier developed for the reference image can be used on the target image after relative normalization with a negligible loss of accuracy. Here, we show how to use the algorithm to identify a

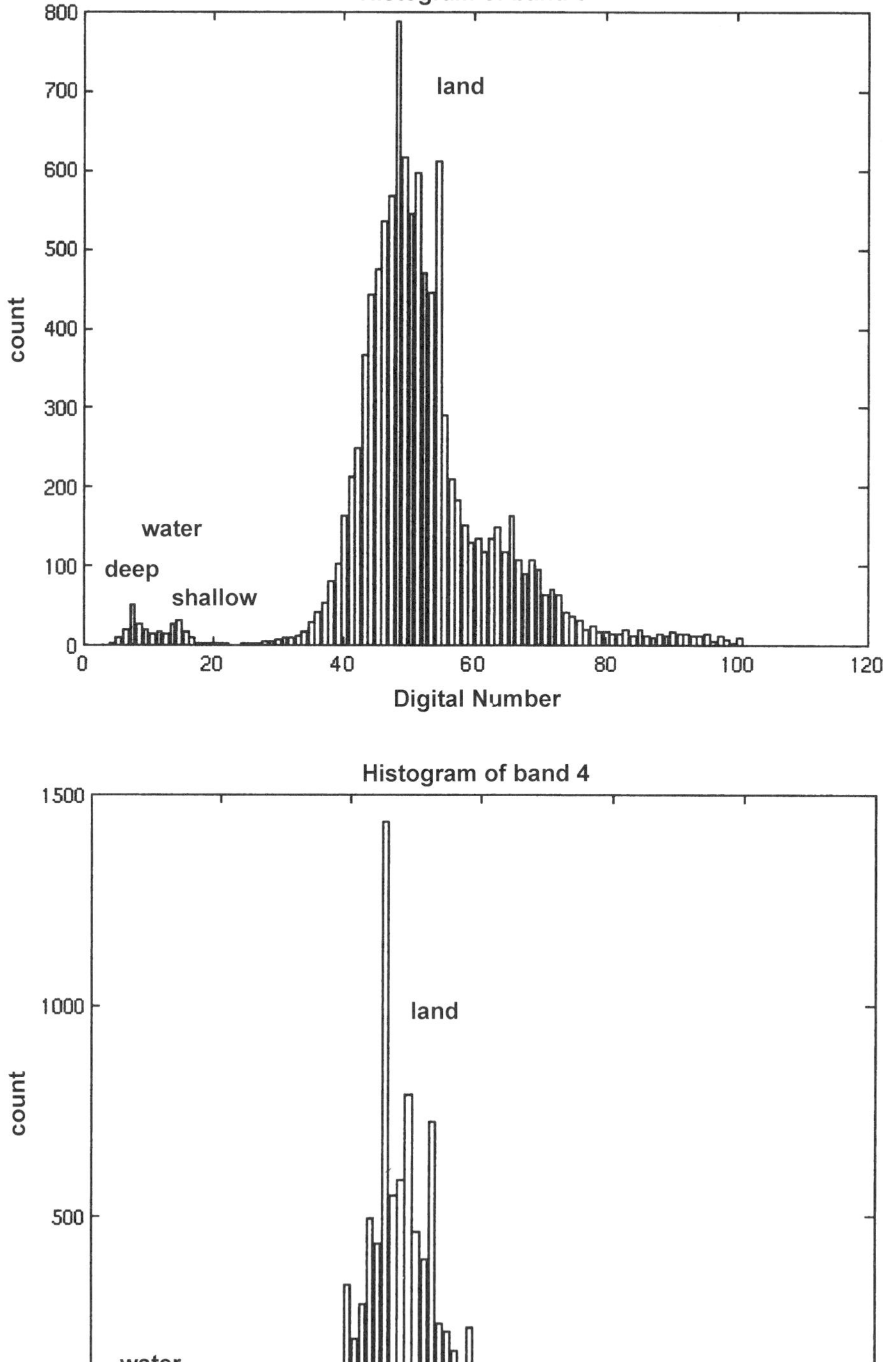

Figure 4.8. Constructing the first approximation of the region of no-change.

candidate region of change before the classification step. Only then pixels that fall within the identified region are analyzed, thus computational cost is significantly reduced and so is the probability of erroneous detection of change due to sensor noise.

Consider the regression coefficients $\{\alpha_i\}_{i=1,..,4}$ and $\{\beta_i\}_{i=1,..,4}$ minimizing the SSE^i and fix a value T_0 and define a thresholds $T_i = T_0/\cos(\arctan(\alpha_i))$ for each spectral band. For a pixel location let $x=[x_1,...,x_4]$ and $y=[y_1,...,y_4]$ denote, respectively, the vector of DN in the reference image and in the target image. Then, the pixel belongs to the candidate region of change if $y_i > \alpha_i x_i + \beta_i + T_i$ or $y_i < \alpha_i x_i + \beta_i - T_i$ in at least one band *i;* otherwise we assume that no change occurred at its location. Only the pixels in the candidate area of change are classified, and the resulting labels are compared to produce the final result, namely, the detected area of change. While the choice of T_0 could be left to the user, we can also determine it using statistical procedures.

4.4 Benefits of the Approach

An example of application of this simple algorithm is given in Figure 4.9 where the reference image is on the left-hand side, the target image is on the right-hand side, and the candidate region of change is in the middle. The older image obtained in 1986 shows more vegetation (darker color in the false color representation) and the reservoir contains more water (black in the image). This example is typical in that the candidate areas of change cover a small portion of the image. Only pixel belonging to the candidate areas of change are further analyzed by the classifier and compared by the final change detection step. The reduction in the computational costs of the feature extraction and the labeling steps is thus significant, especially when complex texture features and powerful classifiers (such as neural networks) are used.

5.0 LEARNING FROM UNRELIABLE TRAINING SETS

5.1 USGS Land Use and Land Cover Maps

Among the ancillary data sets that can be used for land-usage/land-cover change studies in conjunction with the NALC triplicates, the USGS LULC data files (Anderson et al., 1976; Mitchell et al., 1997; USGS, 1986 and 1991; Loelkes et al., 1983) provide a source of very valuable information. This data set is organized in map sheets that coincide with the standard 1:250000-scale USGS topographic maps, and have pixel resolution of 200 by 200 meters. These LULC data show man-made features covering at least 4.0 hectares and having a minimum width of 200 meters and natural occurring features covering at least 16 hectares and of minimum width of 400 meters. The information recorded in this data product was obtained from manual photointerpretation of aerial photographs collected during the '60s and early '70s and subsequently corrected, using data from field surveys and previous land use maps. The taxonomy used in the maps is based on the USGS Level-2 classification codes, and comprises 37 distinct classes (Anderson et al., 1976). This taxonomy is finer than the one we can extract from the MSS satellite imagery alone, since some of the land usage/land use classes have identical spectral signatures and texture appearance. The USGS LULC maps of the USA are available in digital format in a Universal Transverse Mercator (UTM) projection. Thus, they offer the same coverage of the NALC "triplicates" using the same projection at a similar resolution and represent LULC for dates which coincide closely with the 1970s images in the NALC data set.

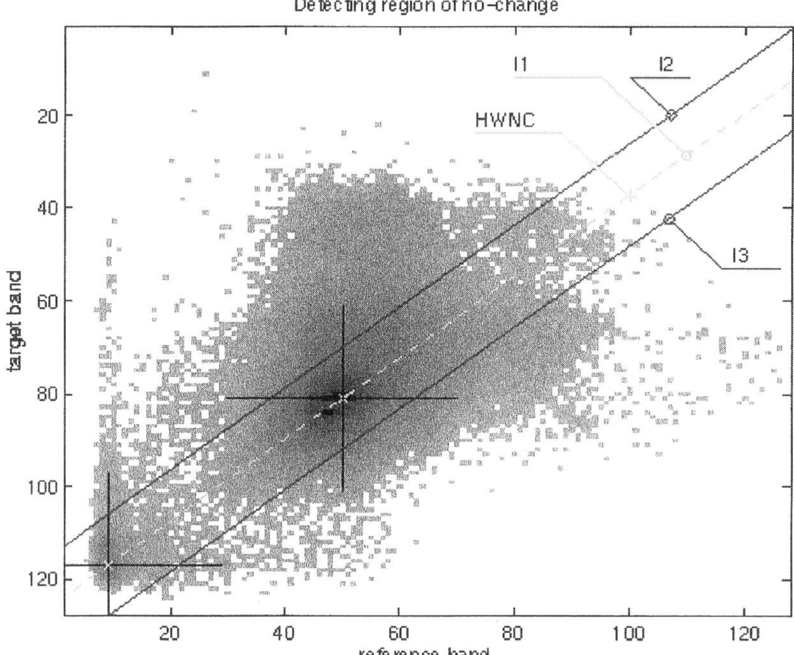

Figure 4.9. 1986 reference image (left), candidate region of change (center), and 1992 target image (right). Note that 1992 was a drier year than 1986; the reservoir is smaller and there are less vegetated areas.

The USGS LULC maps clearly provide important ancillary information for change detection studies. Here we discuss their use as ground truth in the automatic construction of classification algorithms for change detection. The potential benefits of this approach are twofold: (1) the ability to automatically construct a scene-dependent classifier essentially eliminates the cost associated with the preparation of reliable training sets; and (2) the combination of the resulting classifier and of the ancillary data allows the use of more detailed land cover taxonomies than those distinguishable using the image data alone. Regrettably, the LULC maps cannot be used directly as ground truth even for the '70s NALC images: experimental results show that classifiers trained in this fashion are extremely unreliable due to the different scale of the NALC and LULC data sets, the minimum size of the recorded regions, the different times of acquisition of the data, the fact that maps and images are not coregistered, and errors in the LULC maps (see Plate 10).

Here we propose and discuss a methodology for generating training sets using unreliable information, and we discuss the applications to the NALC data set. By using this method on the '70s images, then using the '70s images as reference in the ASCR normalization (see Section 4), we can automate a significant part of the preliminary work in change detection.

5.2 Systematic Errors

Systematic errors in geographic registration and photo interpretation occur along the boundaries of homogeneous regions due to coarse resolution and to changes over periods of time and make it impossible to use the LULC maps as sources of reliable ground truth. The effects

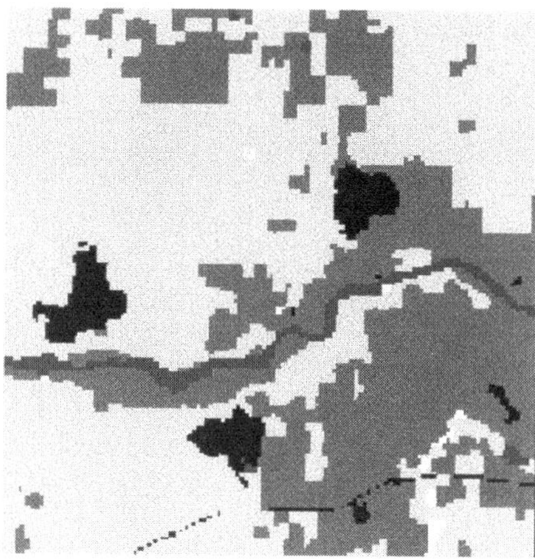
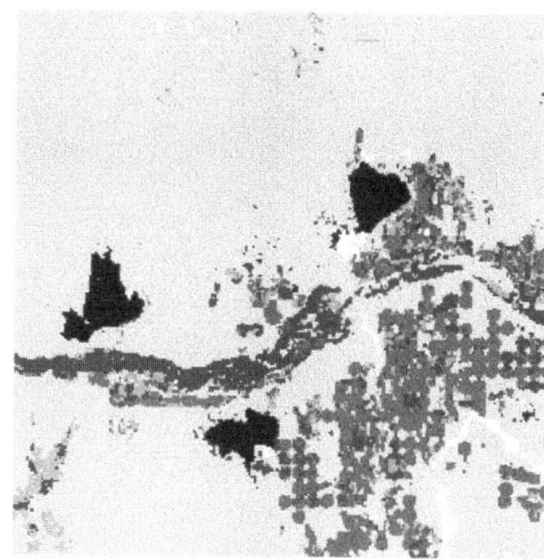

Figure 4.10. Scatterplots of bands 2 and 3 of a NALC image from 1972. The top plots depict the (support of the empirical) distribution of vegetated areas; the bottom plots depicts the distribution of nonvegetated areas. The plots on the right used the results of photointerpretation as ground truth; the plots on the left, the USGS maps. Note that the overlaps of the distributions obtained from photointerpretation are limited, while the distribution of the vegetated areas as described by the USGS maps significantly overlaps with the distribution of the nonvegetated areas.

resulting from relying on the USGS LULC maps in the construction of training sets can be modeled mathematically, but a detailed discussion is beyond the scope of this chapter. Practical tests showed that an instance of CART trained with the training set of Figure 4.10 and the USGS LULC labels is approximately five times less accurate than the corresponding instance of CART trained using the labels derived from photointerpretation.

We propose here a methodology that reduces the effects of systematic labeling errors on the accuracy of the resulting classifier. For illustration purposes we divide the range of bands two and band three (Figure 4.10) into 10 bins. For each bin, count the number of vegetated and nonvegetated samples and report all the nonempty bins as depicted in Table 4.1. In the table we can distinguish three kinds of fuzzy bins, or clusters, depending on the proportion of the labels. Illustrated are bins where one class is clearly preponderant, two classes are in similar proportion, and where the proportions are not similar but no class is clearly prevalent. If a cluster contains similar proportions of the two classes, it corresponds to a region of the observation space where the two underlying distributions are closely related. This implies that even the optimal classifier is bound to make a significant number of mistakes when labeling observations that fall within this cluster. In addition, even the optimal classifier would not produce significantly more accurate results than tossing a coin and labeling the samples accordingly. If, however, a cluster contains a preponderant number of observations from one class, then the probability of observing a sample of this class is much greater. This implies that even the optimal classifier will not produce significantly better results than a classifier that labels all the points as belonging to the more common class. For clusters where neither of the above behaviors is observed, the optimal classifier can perform significantly better than tossing coins or using a single label for the entire region.

This discussion suggests a family of strategies to improve the accuracy of the learning stage by effectively cleaning the training set (Figure 4.11). Bootstrap learning starts by partitioning a training set and a test set using clustering techniques. The training set is provided as input to an

Table 4.1. Distribution of Vegetated (V) and Nonvegetated (N) Pixels Across 63 Clusters, According to Actual and USGS LULC Labels.

Clus	Actual		USGS		Clus	Actual		USGS		Clus	Actual		USGS	
nr	V	N	V	N	nr	V	N	V	N	nr	V	N	V	N
1	3	0	3	0	2	35	0	35	0	3	0	1	0	1
4	1	0	1	0	5	23	0	22	1	6	0	2	1	1
7	0	1	0	1	8	0	8	2	6	9	0	48	3	45
10	0	51	2	49	11	0	33	2	31	12	0	14	0	14
13	0	4	0	4	14	0	6	0	6	15	0	7	0	7
16	0	1	0	1	17	2	0	2	0	18	4	0	4	0
19	4	2	5	1	20	1	5	1	5	21	0	13	3	10
22	0	37	6	31	23	0	61	8	53	24	0	71	6	65
25	0	33	0	33	26	0	8	0	8	27	0	11	0	11
28	0	6	0	6	29	3	5	4	4	30	21	4	14	11
31	34	32	28	38	32	6	60	21	45	33	0	18	2	16
34	1	8	0	9	35	0	2	0	2	36	0	2	0	2
37	0	1	0	1	38	33	0	21	12	39	243	0	176	67
40	75	31	70	36	41	8	19	11	16	42	2	2	1	3
43	0	1	1	0	44	1	0	1	0	45	1	0	1	0
46	362	0	306	56	47	633	3	555	81	48	48	1	37	12
49	6	0	2	4	50	2	0	2	0	51	219	0	173	46
52	152	0	103	49	53	23	0	10	13	54	3	0	2	1
55	58	0	31	27	56	38	0	19	19	57	1	0	0	1
58	17	0	6	11	59	4	5	1	8	60	10	0	8	2
61	2	0	1	1	62	1	1	1	1	63	3	0	3	0

unsupervised learning step and the resulting unsupervised classifier is applied to both the training and test set. The individual clusters are then analyzed, statistics of the class labels are collected, and appropriate decisions are taken. Such decisions depend on the application domain, on the assumptions on the distributions of the classes, and may be based on heuristics. Examples of rules include removing all the samples of nondominant classes from a cluster, relabeling samples of nondominant classes, or removing clusters where no dominant class exists.

Early results show that the method results in a significant improvement of classification accuracy. It is apparent that the choice of decision rules must be driven by considerations on the actual type of classifier used and on the assumption regarding the distribution of the data. The success of the procedure depends on the ability of the classifier to "generalize," that is, to make inference regarding the labels of new samples that are not equal to any of the observations in the training set. The knowledge of the generalization capabilities of the classifier together with the distributional assumptions then leads to the selection of appropriate rules.

6.0 AUTOMATED CHANGE DETECTION

To conclude the chapter, we describe in detail an automatic land cover tracking system for the monitoring of the rate of land use change based on a neural network engine. Discussing the example in detail will allow us to address several issues: (1) how to select appropriate features to account for the statistical dependence of the DNs of neighboring pixels; (2) how to combine multiple classifiers to increase the accuracy of the results; and (3) how to use postprocessing to account for the statistical dependence of the labels of neighboring pixels.

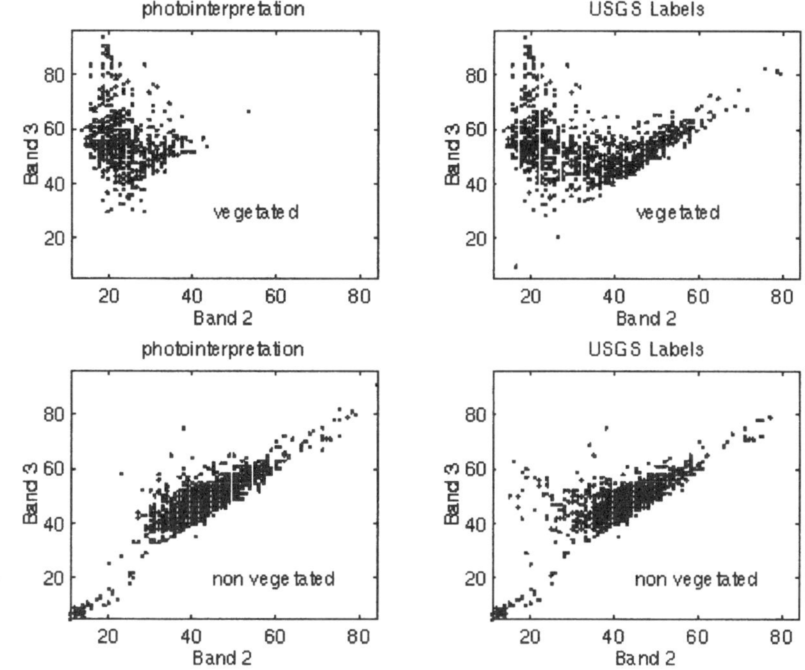

Figure 4.11. The bootstrap learning algorithm.

This case study relies on the USGS level 1 classification taxonomy (Anderson et al., 1976), that distinguishes between the following nine classes: urban, agricultural, rangeland, forest, water, wetland, tundra and perennial snow/ice.

6.1 System Architecture

The architecture of the automatic tracking system consists of control module, radiometric normalization engine, classification module, and change detection module. The operation of the system is controlled by the control module which interfaces with the data repository. Images are radiometrically co-normalized, with respect to a reference image. An example of normalization algorithm is the ASCR, previously discussed in detail in Section 4. The classification module is composed of preprocessing, image labeling, and postprocessing. During the preprocessing stage, cloud removal is performed and relevant features are extracted to be presented as input to the classifier. The classifier has a hybrid structure combining a three-layer feed-forward neural network with nine outputs, each corresponding to one land-cover class and a nearest neighbor classifier. The resulting classified map is then analyzed to reduce the effects of sensor noise and of classification errors, and finally combined by the change detection module. The change detection stage compares pixel-by-pixel the results of the classification on different images.

6.2 Preprocessing

The feature extraction stage processes the raw pixel DN values and produces quantities (features) that can be successfully used to discriminate between different land cover classes.

This stage is responsible for partially overcoming the independence assumption on which the traditional pixel-by-pixel approach to image classification relies. The preprocessor provides the classifier with information on the behavior of the image at and around each pixel location. There are two approaches to the problem. The first relies on reducing the description of each pixel neighborhood to few significant quantities. These quantities could correspond to visual properties of the image that a human expert would use in photointerpreting the data. Examples of this kind are the texture features. The advantage is the concentration of the information in few powerful descriptors which simplifies the learning stage; the disadvantage is that the process can potentially discard relevant information. The second approach operates a moderate reduction of the data and relies on the ability of the classifier to learn the complex dependencies between the descriptors and the class labels. Providing the classifier with numerous features obtained from the original data using limited transformations reduces the risk of discarding important information. At the same time, increasing the dimensionality of the sample space significantly complicates the training process and requires larger training sets. This problem, known as the "curse of dimensionality," can be to some extent alleviated by using parametric classifiers (i.e., linear discriminate classifier) or classifiers that internally operate a dimensionality reduction (i.e., neural networks).

Using a hybrid approach for each pixel we define a small neighborhood of size $\mathbf{M}\times\mathbf{M}$ and a large neighborhood of size $\mathbf{L}\times\mathbf{L}$. We extract information from each pixel location and code the DNs of the four spectral bands at the pixel location by remapping each digital number $\mathbf{DN}_{i,j}$ into $\mathbf{N}$ values $\{\mathbf{o}_n(\mathbf{DN}_{i,j})\}$, $\mathbf{n=1,...,N}$ through the activation functions $\mathbf{o}_n(x) = \exp\{-(x-\boldsymbol{\mu}_n)/2\boldsymbol{\sigma}_n^2\}$. The location parameters $\boldsymbol{\mu}_n$ where the activation functions attain their maximum can be equidistant or more concentrated in the areas where spectral classes are closer together and enhanced discrimination is desired. The dispersion parameters $\boldsymbol{\sigma}_n^2$ control the spread, and thus the selectivity of the activation functions. Coarse coding reduces the noise in the least significant bits. Also, for small values of the $\boldsymbol{\sigma}_n^2$, only few $\mathbf{o}_n(x)$ have values significantly greater than zero; the resulting set of input features are particularly well suited for use with a neural network.

For each pixel of the small neighborhood we compute the brightness defined for the NALC data set as $\mathbf{L}_{i,j} = (\boldsymbol{\Sigma}_{b=1,...,4}\,(\mathbf{DN}_{i,j}(\mathbf{b}))^2)^{1/2}$, where $\mathbf{DN}_{i,j}(\mathbf{b})$ is the DN in the **b**th spectral band, thus producing $\mathbf{M}^2$ coefficients. From the large neighborhood we extract **K** texture features; average brightness, coarseness, and entropy of the four MSS spectral bands. **If** $\mathbf{h(x)}$ denotes the histogram of the image pixels computed using the $\mathbf{L}\times\mathbf{L}$ window the average brightness is defined as $\mathbf{S}_L = \boldsymbol{\Sigma}_i\, \mathbf{i}\, \mathbf{h(i)/L}^2$, coarseness is defined as $\mathbf{C}_L = \mathbf{1-(1+D}_L)^{-1}$ (where the dispersion $\mathbf{D}_L$ is given by $\boldsymbol{\Sigma}_i\, (\mathbf{1} - \mathbf{S}_L)^2\, \mathbf{h(i)/L}^2$) and the entropy is $\mathbf{H}_L = \boldsymbol{\Sigma}_i\, \mathbf{h(i)/L}^2 \log_2 (\mathbf{h(i)/L}^2)$. Thus, for each pixel position, the preprocessing step takes $4\mathbf{L}^2$ inputs (DNs in the four spectral bands for each pixel within the $\mathbf{L}\times\mathbf{L}$ window) and produces $\mathbf{M}^2 + 4\mathbf{N} + \mathbf{K}$ features. The values of **M**, **N** and **L,** are determined experimentally by considering both accuracy and computational cost of the resulting classifier.

6.3 The Classifier

The automatic change detection system is a hybrid based system using a three-layer feed-forward neural network engine (as described by Haykin, 1994) and a nearest neighbor classifier. Neural network classifiers are able to "learn" rather complex decision surfaces without relying on parametric assumptions; however, they have high computational cost that can be overcome using dedicated hardware, and act essentially as a black box. When the classifier cannot produce a clear-cut answer, the problem is mitigated using a hybrid scheme based on a neural network and a different classifier that operate in parallel. If the neural network produces one and only one significant output label, then the label is used as output of the entire system, in all remaining cases the other classifier is invoked and its output used as final result.

The nearest neighbor classifier compares an input to all the examples in the training set and finds the "most similar" example (nearest neighbor). Alternatively, k nearest neighbors are found and their labels combined appropriately to produce the final result. This simple algorithm has rather surprising statistical properties (Cover and Hart, 1967) and can be efficiently implemented (Kim and Park, 1986).

Combining classifiers with totally different structure, as in the above hybrid approach, increases the accuracy. The decision regions of the two classifiers are substantially different in form and the nearest neighbor has the potential of being more accurate in those subsets of the sample space where the neural network displays poor discrimination capabilities.

The results of the classification are then smoothed by taking the majority vote of the labels over a moving $\mathbf{K}\times\mathbf{K}$ square window, where $\mathbf{K}$ is typically a small number such as 3. This has the effect of removing classification errors due to sensor noise or to overlapping distributions. The downside of the smoothing is that homogeneous regions that extend over very few pixels are discarded.

7.0 SUMMARY

The described algorithm is a generic automatic change detection structure that can be applied to data other than the NALC "triplicates." By applying a relative radiometric normalization the system requires the presence of only one training set per coregistered set of images. By extracting texture features, information on the neighborhood of the pixel under analysis is provided to the classifier. By coarse coding the DNs from the four spectral bands of the pixel under analysis, some of the effects of sensor noise are reduced and the sensitivity of the neural network is greatly enhanced. By smoothing the classification results the system removes local errors resulting from sensor noise or from the presence of spectrally overlapping land cover classes.

We estimate that the accuracy of the algorithm to be on the order of 80–85 percent for the USGS level-1 classification taxonomy. Most resultant errors are due to the difficulty in discriminating spectrally similar classes with only four spectral bands. Additional improvements are observed if the NDVI and the DEMs are used as input in conjunction with the original spectral bands.

ACKNOWLEDGMENTS

This work was funded in part by NASA/CAN Grant No. NCC5-101.

REFERENCES

Anderson, J.R., E.E. Hardy, J.T. Roach, and R.E. Witmer. A Land Use and Land Cover Classification System for Use with Remote Sensor Data. *Professional Paper 964*, U.S. Geological Survey, Reston, VA, 1976.

Awaya, Y. and N. Tanaka. Vegetation Change Detection with a Laplacian Filter Using LANDSAT Data, a Case Study in a Boreal Forest-Tundra Transition Area, in *Proc. of IGARSS '93 - IEEE International Geoscience and Remote Sensing Symposium*, 2, 743–746, 1993.

Bartlett, M.S. *The statistical analysis of spatial patterns.* John Wiley & Sons, 1975.

Breiman, L., J.H. Friedman, R.A. Olshen, and C.J. Stone. *Classification and Regression Trees.* Wadsworth & Brooks/Cole, 1984.

Carr, J.R. Spectral and Textural Classification of Single and Multiple Band Digital Images. *Comput. Geosci.*, 22(8), 849–65, 1996.

Castelli, V., I. Kontoyiannis, C.-S. Li, and J.J. Turek. Progressive Classification: A Multiresolution Approach. *IBM Research Report RC 20475*, 1996a.

Castelli, V., C.-S. Li, J.J. Turek, and I. Kontoyiannis. Progressive Classification in the Compressed Domain for Large EOS Satellite Databases. In *Proc. of 1996 IEEE Intern. Conf. Acoust. Speech Signal Process.*, 4, 2201–2204, 1996b.

Cherkassky, V.S., J.H. Friedman, and H. Wechsler. *From Statistics to Neural Networks: Theory and Pattern Recognition Applications*. Springer-Verlag, 1993.

Chettri, S.R. and R.F. Cromp. Probabilistic Neural Network Architecture for High-Speed Classification of Remotely Sensed Imagery. *Telematics and Informatics*, 10(3), 1993.

Chettri, S.R., R.F. Cromp, and M. Birmingham. Design of Neural Networks for Classification of Remotely Sensed Imagery. *Telematics and Informatics*, 9(3/4), 145–156, 1992.

Cover, T.M. and P. Hart. Nearest Neighbor Pattern Classification. *IEEE Trans. on Information Theory*, IT-13(1), 21–27, 1967.

Dasarathy, V. *Nearest Neighbor Pattern Classification Techniques*. IEEE Computer Society, edited by Belur, 1991.

Daubechies, I. *Ten Lectures on Wavelets*. Society for Industrial and Applied Mathematics, Philadelphia, PA, 1992.

Dubes, R.C. Cluster Analysis and Related Issues. In *Finding Groups in Data*. C.H. Chen, L.F. Pau, and P.S.P. Wang, Eds., World Scientific, New York, 1993.

Duda, R.O. and P.E. Hart. *Pattern Classification and Scene Analysis*. John Wiley & Sons, 1973.

Elvidge, C.D., D. Yuan, R.D. Weerackoon, and R.S. Lunetta. Relative Radiometric Normalization of Landsat Multispectral Scanner (MSS) Data Using an Automatic Scattergram-Controlled Regression. *Photogrammetric Engineering & Remote Sensing*, 61(10), 1225–1260, 1995.

Fung, W.K. Diagnostics in Linear Discriminant Analysis. *Journal of American Statistical Association*, 90(431), 952–956, 1995.

Guyon, X. *Random Fields on a Network. Modeling, Statistics, and Applications*. Springer Verlag, 1995.

Haykin, S.S. *Neural Networks: A Comprehensive Foundation*. Macmillan, 1994.

Iwahashi, M., H. Kiya, and K. Nishikawa. Subband Coding of Images with Circular Convolution. In *Proc. 1992 IEEE Intern. Symp. Circ. Sys.*, 3, 1356–1359, 1992.

Jenson, J.R. Urban/Suburban Land Use Analysis. In *Manual of Remote Sensing*. J.R. Jenson, Ed., American Society of Photogrammetry, pp. 1571–1666, 1983.

Kaufman, L. and P.J. Rousseuw. *Finding Groups in Data*. Wiley-Interscience Publication. John Wiley & Sons, 1990.

Kim, B.S. and S.B. Park. A Fast K Nearest Neighbor Finding Algorithm Based on the Ordered Partition. *IEEE Trans Pattern Anal Mach Intell*, PAMI-8(6), 761–766, Nov. 1986.

Kohonen, T. *Self-Organization and Associative Memory*. Second Edition, Springer Verlag, New York, 1988.

Kosmatopoulos, E.B. and M.A. Christodoulou. Convergence Properties of a Class of Learning Vector Quantization Algorithms. *IEEE Trans. Image Process.*, 5(2), 361–68, 1996.

Loelkes, G.L., G.E. Howard, E.L. Schwertz, P.D. Lampert, and S.W. Miller. *Land Use/Land Cover and Environmental Photointerpretation Keys*. Technical report, U.S. Geological Survey, Reston, VA, 1983.

Lohmann, G. Assessment of Textural Features for Remote Sensing Applications. In *Proc. SPIE—Int. Soc. Opt. Eng.*, 2357, pt. 2, 512–16, 1994.

Lunetta, R.S., J.G. Lyon, J.A. Sturdevant, J.L. Dwyer, C.D. Elvidge, L.K. Fenstermaker, D. Yuan, S.R. Hoffer, and R. Weerackoon. North American Landscape Characterization (NALC) Research Plan, USEPA, EPA/600/R-93/135, 1993.

Miller, D.M., E.J. Kaminsky, and S. Rana. Neural network classification of remote sensing data. *Comput. Geosci.*, 21(3), 377–86, 1995.

Mitchell, W.B., S.C. Guptill, K.E. Anderson, R.G. Fegeas, and C.A. Hallam. *GIRAS—A Geographic Information and Analysis System for Handling Land Use and Land Cover Data*. Professional Paper 1059, U.S. Geological Survey, Reston, VA, 1977.

Neter, J., W. Wasserman, and M.H. Kutner. *Applied linear Statistical Models*. Richard D. Irwin, Inc., 1985.

Paola, J.D. and R.A. Schowengerdt. A Detailed Comparison of Backpropagation Neural Network and Maximum-Likelihood Classifiers for Urban Land-Use Classification. *IEEE Transactions on Geoscience and Remote Sensing*, 33(4), 981–998, 1995.

Pennebaker, W. and J.L. Mitchell. *JPEG Still Image Data Compression Standard*. Van Nostrand Reinhold, New York, 1993.

Pettit, E.J., R.R. Bailey, R.L. Bowden, and R.C. Ashley. Performance Evaluation of Statistical and Neural Network Classifiers for Automatic Land Use/Cover Classification. In *Proc. SPIE—Int. Soc. Opt. Eng.*, 1838, 138–53, 1993.

Philpot, W. and V. Chavarria. Comparison of Two Spectral Texture Classification Algorithms. In *Proc. of IGARSS '94—1994 IEEE International Geoscience and Remote Sensing Symposium*, 2, 878–81, 1994.

Richards, J.A.. *Remote Sensing Digital Image Analysis, an Introduction*. Second Edition, Springer-Verlag, 1993.

Salu, Y. and J. Tilton. Classification of Multispectral Image Data by the Binary Diamond Neural Network and by Nonparametric, Pixel-by-Pixel Methods. *IEEE Transactions on Geoscience and Remote Sensing*, 31(3), 606–616, 1993.

United States Geological Survey (USGS). Land Use Land Cover Digital Data from 1:250,000 and 1:100,000-Scale Maps. Data User Guide 4, U.S. Geological Survey, Reston, VA, 1986.

United States Geological Survey (USGS). Land Use and Land Cover and Associated Maps Factsheet. Technical report, U.S. Geological Survey, Reston, VA, 1991.

Yamazaki, T. and D. Gingras. Image Classification Using Spectral and Spatial Information Based on MRF Models. *IEEE Trans. Image Process.*, 4(9), 1333–1339, 1995.

CHAPTER 5

An Evaluation of the CoastWatch Change Detection Protocol in South Carolina

John R. Jensen, David J. Cowen, John D. Althausen,
Sunil Narumalani, and Oliver Weatherbee

1.0 INTRODUCTION

Wetlands are recognized as a valuable natural resource (Podolsky and Conkling, 1991). They assimilate pollutants, provide flood control, and serve as breeding, nursery, and feeding grounds for fish and wildlife (Odum, 1989). Information on wetland distribution and condition is essential for their effective protection and management (Norton and Slonecker, 1990; Dobson and Bright, 1991). Unfortunately, wetlands present challenges to effective monitoring and quantification. For example, inland wetlands are found in diverse geographic areas ranging from small tributary streams, shrub/scrub and marsh communities, to open water lacustrine environments (Cowardin et al., 1979). In addition, the type and spatial distribution of wetlands can change dramatically between seasons, especially when non-persistent species are present (Mackey, 1990). For these reasons, remote sensing is often used to obtain important information on the spatial distribution and biophysical condition of wetlands (Jensen et al., 1991a; Roughgarden et al., 1991).

The conterminous United States lost 53 percent of its wetlands to agricultural, residential, and/or commercial land use from the 1780s to 1980s (Dahl, 1990). Oil spills occurring throughout the world continue to devastate coastal wetlands (Jensen et al., 1990). More abundant "greenhouse" gases in the atmosphere appear to be increasing the Earth's average temperature (Clarke and Primus, 1990). This may produce a significant rise in global sea level, eventually inundating much of today's coastal wetlands (Kana et al., 1984; Lee et al., 1992). The continued loss of coastal and inland wetlands may lead to the collapse of coastal ecosystems and associated fisheries (Haddad and Ekberg, 1989). Therefore, accurate and timely documentation of wet-

Reprinted with permission by the American Society for Photogrammetry and Remote Sensing, An Evaluation of the CoastWatch Change Detection Protocol in South Carolina. *Photogrammetric Engineering and Remote Sensing,* J.R. Jensen, D.J. Cowen, J.D. Althausen, S. Narumalani, and O. Weatherbee, Vol. 59(6): 1039-1046, 1993. Portions adapted from: Measurements of Seasonal and Yearly Cattail and Waterlily Changes Using Multidate SPOT Panchromatic Data. *Photogrammetric Engineering and Remote Sensing,* Jensen et al., Vol. 59(4): 519-525, 1993.

land gains and losses is critical to their conservation and management. To fulfill this need, the National Oceanic and Atmospheric Administration (NOAA) initiated the CoastWatch Change Analysis Project (C-CAP) which will utilize remote sensing technology to monitor changes in coastal wetland habitats and adjacent uplands on a cycle of 1 to 5 years (Kiraly et al., 1990).

There are four alternatives when collecting wetland information using remote sensing technology, including the use of (1) global positioning systems (GPS), (2) aerial photography, (3) aircraft multispectral scanner data, and (4) satellite derived remote sensor data. Each of these alternatives has advantages and disadvantages.

- In situ field investigation using global positioning systems (GPS). This method can provide detailed wetland information if the GPS data are differentially corrected and government "selective availability" is off (Shirer, 1991). Unfortunately, even when using GPS units, it is still difficult to traverse the exact perimeter of all the wetlands by foot, boat or helicopter to prepare a regional, planimetrically accurate inventory.
- Interpretation of color and color infrared aerial photography. Numerous organizations and individuals have demonstrated that aerial photography can be used to accurately map inland wetlands (Edwards and Brown, 1960; Welch et al., 1988; Wilen, 1990; Dahl and Johnson, 1991). Aerial photography can be acquired on demand when cloud cover conditions are ideal). However, wetlands may not be accurately inventoried using aerial photography when (a) certain film and film filter combinations are used (Dahl and Johnson, 1991), (b) relief displacement is present in the scene which can cause inaccurate area estimates, and (c) significant vignetting is present which can cause photointerpretation inconsistencies. Metric aerial photography may also be expensive to acquire when large regions must be inventoried (Nohara, 1991). Furthermore, the interpreted data must be transferred to a planimetric basemap and subsequently digitized into a geographic information system (GIS) to be of quantitative value (Jensen et al., 1991b).
- Analysis of high resolution aircraft multispectral scanner (MSS) data. Such data can provide accurate inland wetland information over small geographic areas on demand. However, the data are expensive to acquire, must undergo substantial radiometric and geometric preprocessing, and the areal coverage is limited (Jensen et al., 1984; 1986).
- Digital analysis of satellite remote sensor data. Satellite imagery such as that acquired by the Landsat Thematic Mapper (30- by 30-m spatial resolution), SPOT multispectral (20- by 20-m) and panchromatic (10- by 10-m) sensor systems can be analyzed to yield inland wetland information (Gao and Coleman, 1990; Jensen et al., 1990; 1991a; Podolsky and Conkling, 1991). While the spatial resolution of such data is not as good as the aforementioned data types, the radiometric and geometric attributes of the data sets are conducive to regional wetland inventories if a more coarse minimum mapping unit is acceptable. Satellite sensor systems which provide pointable, off-nadir viewing (e.g., SPOT) increase the probability of obtaining cloud-free imagery.

Practical procedures must be established before C-CAP can become an operational program producing a comprehensive, nationally standardized database on coastal habitat change. To achieve this, NOAA sponsored a series of workshops focused on developing regional operational protocols applicable to coastal uplands as well as wetlands (Cross, 1991). The results of these workshops and other interagency meetings are summarized in the revised C-CAP protocol dated 4 December 1992 (Dobson et al., 1992).

This research evaluated several elements of the protocol by applying them to coastal wetland (Fort Moultrie) and inland wetland (Kittredge) study areas in South Carolina (see color Plate 12). The chapter reports on (a) useful multiple date image classification logic, (b) appropriate change detection logic, and (c) preliminary findings concerning the effect of tidal stage when performing coastal change detection using satellite remote sensor data.

2.0 SOUTH CAROLINA COASTWATCH STUDY AREAS

2.1 Fort Moultrie, South Carolina

The Fort Moultrie, South Carolina, USGS 7.5-minute quadrangle encompasses several diverse land cover types including built-up beach front, undeveloped beach front, extensive salt and brackish marshes (dominated by smooth cordgrass, *Spartina alterniflora),* mature maritime forest, upland pine, and cultivated land (Plate 12c). Urban features include the towns of Mount Pleasant, Sullivan's Island, and Isle of Palms. The proximity of these communities to the growing metropolitan area of Charleston has led to an increase in the rate of urbanization and infrastructure development in recent years.

2.2 Kittredge, South Carolina

The Kittredge, South Carolina, USGS 7.5-minute quadrangle is located 40 miles inland from Charleston, South Carolina along the Cooper River (Plate 12a). It consists of extensive stands of upland pine, bottomland hardwoods, numerous oxbow lakes, emergent and submergent riverine aquatic beds, and cultivated land. The Cooper River is tidally influenced in this study area. Although several small towns exist within the area, the largest urban feature in the quadrangle is a U.S. Naval Reserve.

3.0 DATA SOURCES AND PREPROCESSING

Six Landsat Thematic Mapper (TM) images were analyzed in this study (Table 5.1). The November 9, 1982 and December 19, 1988 TM images were used in the evaluation of the change detection methodologies. All six TM scenes were used to investigate the effect of tidal stage on image classification and change detection.

1:58,000-scale National High Altitude Photography (NHAP) was acquired on March 10, 1983 and more recent 1:40,000-scale National Aerial Photography Program (NAPP) data were obtained on February 10, 1989 (Table 5.1). The leaf-off, color-infrared photography corresponds closely with the November 9, 1982 and December 19, 1988 TM images and was used to evaluate classification and change detection error.

3.1 Image Rectification

CoastWatch deliverables are, first and foremost, change detection products, the accuracy of which is largely dependent on the precise geometric registration of multitemporal remote sensor data sets. For this reason, image-to-map and image-to-image rectification error must be minimized.

A subset of the November 9, 1982 TM data containing both the Fort Moultrie and Kittredge study areas was rectified to a Universal Transverse Mercator (UTM) projection using 81 ground control points, nearest-neighbor resampling logic, and a root-mean-square error (RMSE) of less than ±1.0 pixel (±30 m). The remaining TM scenes were registered to this geometrically corrected image with an average RMSE of < 0.75 pixel. This "image-to-map" then "image-to-image" procedure allowed many more usable ground control points to be identified in the multiple dates of TM data and greatly improved the multiple date image-to-image registration which is so important when performing change detection.

Table 5.1. Study Area and Data Sources.

Study Areas	Satellite Remote Sensor Digital Data	Aerial Photography	Digital National Wetlands Inventory
Fort Moultrie, SC	TM Nov 9, 1982	NHAP Mar 10, 1983	
• coastal wetland	Mar 4, 1987	(1:58,000)	
	Oct 14, 1987		
	Dec 19, 1988		Yes
Kittredge, SC	Oct 6, 1990	NAPP Feb 10, 1989	
• inland wetland	Dec 9, 1990	(1:40,000)	

4.0 CLASSIFICATION OF MULTIPLE DATES OF IMAGERY

A portion of the tentative CoastWatch Classification Scheme is summarized in Table 5.2 (Klemas et al., 1992). The C-CAP protocol requires maps of the coastal zone which identify "from-to" changes in land cover based on this scheme. It is not sufficient to simply identify "change" versus "no-change" pixels in the map. Rather, specific elements of a change detection matrix such as the one shown in Figure 5.1 must be able to be selected and portrayed in map format, e.g., a pixel changed from water to estuarine emergent wetland. "Post-classification comparison" change detection logic is one of the most appropriate methods for providing such specific cartographic and statistical information (Jensen, 1986). Therefore, this research evaluated methods of minimizing the errors associated with "post-classification comparison" change detection techniques. The methods included (1) the classification of individual dates of remote sensor data, (2) the application of various post-classification spatial filters to improve classification accuracies, and (3) the use of image differencing techniques to create a "change/no-change" binary mask to exclude areas identified as "unchanged" from further analysis.

The independent classification of the November 9, 1982 and December 19, 1988 TM datasets were produced using iterative, "cluster busting" unsupervised classification logic (Jensen et al., 1987). Five spectral bands were used to classify the 1982 image (bands 1 to 5) and six (bands 1 to 5 and band 7) were used for the 1988 scene.

4.1 Classification of the Fort Moultrie Study Area

The rectified TM data were classified using a maximum-likelihood sequential clustering algorithm to derive 150 spectral clusters which were plotted in two-dimensional (red versus near-infrared) feature space (Jensen, 1996; Hodgson and Plews, 1989). Pixels represented by these clusters were labeled based on (a) their position in feature space, and (b) their spatial location when overlaid onto a color composite of the rectified imagery. Those clusters that could not be readily classified (usually mixed pixels) were used to create a mask to extract the corresponding areas of confusion in the original, rectified remote sensor data. The clustering algorithm was then applied to only the confused pixels to obtain additional clusters. A total of 186 clusters was used to produce final classification of the November 9, 1982 image. This "cluster-busting" procedure was iterated three times to classify the 1988 image into 176 clusters. The final clusters for each date were recoded using the Tentative CoastWatch Classification Scheme (Table 5.2) to produce the Fort Moultrie classification maps shown in Plate 13c and 13d. The three "woody" forest classes listed in Table 5.2 were mapped, but were recoded

Table 5.2. Tentative CoastWatch Classification Scheme (Klemas et al., 1992).

Class Number	Class Name
1.0/5.0	Developed/Exposed Land
2.0	Cultivated Land (agriculture)
3.0	Herbaceous Grassland
4.1	Woody Deciduous[a]
4.2	Woody Evergreen[a]
4.3	Woody Mixed[a]
7.26	Estuarine Emergent Wetland
7.34	Riverine Aquatic Beds[b]
7.98	Palustrine Forested Wetland
8.0	Water
8.22	Estuarine Unconsolidated Bottom

[a] Merged to "Upland Forest" for display purposes.
[b] Found only in the Kittredge study area.

into a single "Upland Forest" class for presentation purposes (and because the specific types of intraforest change are not that important to C-CAP objectives). All error evaluations presented were based on the disaggregate forest classes.

There were some problems associated with the classification of the Fort Moultrie study area on both dates. First, it was necessary to combine developed and exposed land (bare soil) into a single class. This was primarily due to the high sand content of soils in this area, which results in bare soil having approximately the same reflectance characteristics as urban concrete. Second, cultivated land was not always separable from other classes. The draft protocol suggests that the dates of imagery be selected to optimize the separability of wetland vegetation which, for the southeastern United States, corresponds to the winter and early spring months (Jensen et al., 1987). Therefore, the dates selected were not ideal for classifying cultivated land because the fields may be fallow or overgrown with short grass during this time of the year. Ideally, a third date of imagery acquired in the summer months could be used to identify all cultivated land. When this is not possible, on-screen digitizing of "consistent" cultivated land might be appropriate such as described in Jensen et al. (1992).

A method based on the work of Martin (1989) was selected to evaluate the accuracy of the individual thematic maps. Samples from the high resolution NHAP and NAPP color-infrared aerial photographs were selected through the use of a grid overlay and the generation of random *i,j* coordinates. A class was assigned to each sample based on photointerpretation and a limited amount of fieldwork. These random samples were then located in the digital classification map only if there was a homogeneous 3 by 3 block of pixels at that location. This methodology resulted in 86 "ground reference" samples for the 1982 classification and 84 for the 1988 classification. The application of randomly selected samples for accuracy assessment is a useful methodology as long as the samples are no larger than 10 pixels in size (Congalton, 1988). The 1982 classification had an overall accuracy of 82.91 percent (Kappa of 0.803) while the 1988 classification had an overall accuracy of 84.78 percent (Kappa of 0.822). When the three "woody" forest classes were merged into a single "upland forest" class, the classification accuracy of the individual dates increased to 86.29 percent (Kappa of 0.834) in 1982 and 87.94 percent (Kappa of 0.85) in 1988.

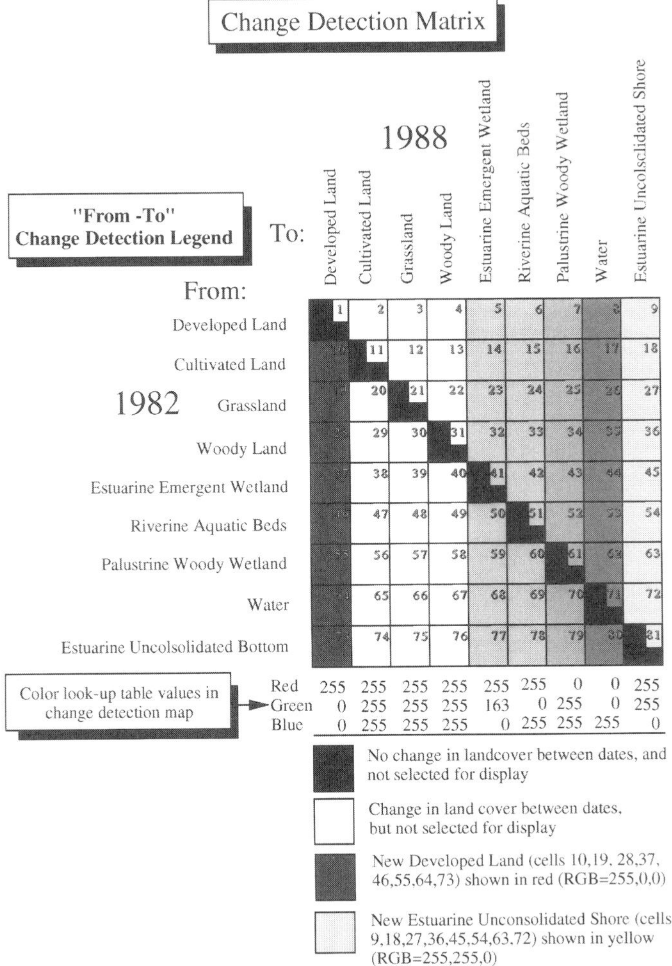

Figure 5.1. This diagram summarizes the basic elements of a change detection matrix which can be used to select specific "from-to" classes for display in a change detect/on map. There are $(n^2 - n)$ off-diagonal possible change classes which may be displayed in the change detection map (72 in this example) although some may be highly unlikely. The off-diagonal cells shaded in this diagram were used to produce the Kittredge and Fort Moultrie, South Carolina change maps in Plates 11a and 11b. For example, any pixel in the 1982 map that changed to Developed/Exposed Land in 1988 is red (RGB = 255, 0, 0). Any pixel that changed into Estuarine Emergent Wetland in 1988 is orange (RGB = 245, 163, 0). Individual cells can be color coded in the change map to identify very specific "from to" changes. A color version of the matrix can be used as a change detect/on legend.

The application of post-classification "majority" filtering techniques has increased the accuracy of land-cover mapping in certain instances (Kenk et al., 1988). To test the ability of majority spatial filtering to improve the TM classifications, four 3 by 3 majority filters with different threshold values (ranging from 3 to 6) were applied to both classifications. Accuracy statistics (Kappa coefficient of agreement and overall) for all treatments were calculated and summarized in Table 5.3. In every case, the accuracy of the 1982 and 1988 classification improved through the application of majority filtering techniques. The greatest improvements were achieved using a 3 by 3 matrix filter with a majority threshold of 3 (Table 5.3).

Table 5.3. Classification Accuracy of Individual Dates of Landsat TM Data Using Unsupervised Classification Logic and Various Post-Classification Majority Filters (Fort Moultrie Study Area).

	Nov 9, 1982 Classification		Dec 19, 1988 Classification	
Treatment	Kappa	Overall (%)	Kappa	Overall (%)
Original	0.803	82.91	0.822	84.78
Filter - Threshold 6	0.821	84.45	0.834	85.80
Filter - Threshold 5	0.839	86.03	0.857	87.85
Filter - Threshold 4	0.843	86.45	0.881	89.39
Filter - Threshold 3	0.846	86.67	0.881	89.90
"Merged Upland Forest"	0.834	86.29	0.850	87.94

4.2 Classification of the Kittredge Study Area

The original, rectified TM data of the Kittredge study area were classified using the same methodology. The "cluster busting" procedure was iterated three times for the classification of both dates and resulted in 284 clusters for the 1982 scene and 241 for the 1988 scene. The final classification maps are shown in Plates 13a and 13b.

Problems similar to those encountered during the classification of the Fort Moultrie TM imagery occurred when attempting to classify the two Kittredge TM images. There was difficulty distinguishing between developed and exposed land (e.g., bare soil). Cultivated land was confused with palustrine forest in certain instances. Different dates of imagery or "on-screen" digitizing of consistently cultivated land may be required to successfully distinguish between these phenomena.

Ninety-three and 84 ground reference samples, respectively, were used to determine the accuracy of the 1982 and 1988 Kittredge classification maps. Accuracy statistics (Kappa and overall) for the original classification and a 3 by 3 majority filtered map (threshold of 3) were calculated and summarized in Table 5.4. As expected, the application of a majority filter yielded superior results on both dates. In fact, the filtered map data on both dates were > 90 percent accurate.

5.0 CHANGE DETECTION

5.1 Treatments

Four variations of "post-classification comparison" change detection were evaluated for the Fort Moultrie study area (Table 5.5): (1) a traditional method which compared the original classifications directly, (2) the comparison of two classifications which had undergone majority filtering with a 3 by 3 pixel filter and a threshold value of 3, (3) the use of a "change/no-change" mask applied to the original classifications prior to any comparison, and (4) the application of the same mask to the majority filtered classifications. Traditional classification and majority filtering have already been discussed. It is instructive to describe the nature of the "change/no-change" mask treatment.

The accuracy of any post-classification change detection is strongly influenced by the accuracies of the independent classifications (Jensen, 1986). Classification error in either of the dates will result in an erroneous indication of change. The draft protocol suggests that one

Table 5.4. Classification Accuracy of Individual Dates of Landsat TM Data Using Unsupervised Classification Logic and a 3 by 3 Post-Classification Majority Filter (Kittredge Study Area).

	Nov 9, 1982 Classification		Dec 19, 1988 Classification	
Treatment	**Kappa**	**Overall (%)**	**Kappa**	**Overall (%)**
Original	0.858	88.30	0.840	86.80
Filter - Threshold 3	0.901	92.30	0.884	91.40

Table 5.5. Results of the Fort Moultrie Post-Classification Change Detection Comparisons Using 1982 and 1988 Landsat Thematic Mapper Imagery: Kappa Coefficients and Overall Classification Accuracy.

Treatments	Original Unsupervised "Cluster Busting" Classification	Majority Filter Applied to Individual Classification Maps
Original Unsupervised "cluster busting" classification	Kappa = 0.65 Overall % = 69.51	Kappa = 0.82 Overall % = 85.17
1982 vs. 1988 "Change/no-change" mask applied	Kappa = 0.55 Overall % = 61.43	Kappa = 0.72 Overall % = 75.75

possible method of minimizing such errors is to use the spectral information of the raw images to differentiate between areas of change and no change. Therefore, a "change/no-change" mask was created and applied using the following logic.

Band 5, November 9, 1982 data for the entire subset (encompassing both study areas) were algebraically differenced with band 5 data obtained on December 19, 1988 to identify those pixels in the scene that changed. The same procedure was applied using TM band 3 data. Using classical GIS overlay logic, the TM band 5 "change/no-change" pixels were allowed to "dominate" all upland pixels found within the 1982 classification map while the TM band 3 derived "change/no-change," pixels were allowed to dominate for the tidally influenced areas. The union of these two operations was a single change/no-change binary mask which was applied to the December 19, 1988 classification map prior to the post-classification change detection procedure. This resulted in the theoretical removal of all pixels in the December 19, 1988 classified map which had not changed since November 9, 1982 according to the "change/no-change" mask. It was hypothesized that the removal of such information would reduce errors of commission in the change detection as suggested by Pilon et al. (1988). An alternative is to apply the mask to the raw 1988 data prior to classifying it. Theoretically, the results of the final change detection would be the same.

5.2 Fort Moultrie Study Area Change Detection

A change detection map of the Fort Moultrie study area derived from 1982 and 1988 Landsat TM data is shown in Plate 11b. The 1982 and 1988 classification maps were compared on a pixel by pixel basis using an n by n GIS "matrix" algorithm whose logic is shown in Figure 5.1. This resulted in the creation of a "change image (map)" consisting of brightness values from 1 to 81. The analyst then selected specific "from-to" classes for emphasis. Only a select number of the 72 ($n^2 - n$) possible off-diagonal "from-to" land-cover change classes summarized in Figure 5.1 were selected to produce the change detection map. For example, all pixels which changed from any land cover in 1982 to "Developed/Exposed Land" in 1988 were color coded red (RGB = 255, 0, 0) by selecting the appropriate "from-to" cells in the change detection matrix (10, 19, 28, 37, 46, 55, 64, and 73). If desired, the analyst could highlight very specific changes such as all pixels which changed from "Developed/ Exposed Land" to "Estuarine Emergent Wetland" (cell "5" in the matrix) by assigning a unique color look-up table value (not shown). The color-coded change detection map revealed significant growth in developed/exposed land, palustrine forested wetland, and estuarine unconsolidated bottom.

Assessing the accuracy of a change detection map is no simple task. In fact, there is relatively little literature on the topic (Jensen and Narumalani, 1992). This study used the error evaluation methodology previously described except that the random samples selected in the NHAP and NAPP data had to coincide with "from-to" categories which could be located in *both* the 1982 and the 1988 digital classification maps. Despite the generation of hundreds of random *i,j* coordinates, only 25 homogeneous, 3 by 3 pixel samples were obtained for the Fort Moultrie quadrangle. These reference data were used to calculate the overall and Kappa coefficient change detection map accuracy (Table 5.5). The highest change detection accuracy for the Fort Moultrie study area was obtained by the direct comparison of the filtered classifications (Kappa statistic of 0.82; overall accuracy of 85.17 percent). The lower accuracy of the "change/no-change" mask treatment was due to the underestimation of classes of change versus unchanged. In other words, the "change/no-change" mask previously discussed was too conservative. Despite this shortcoming, the application of this method to the filtered classifications had a higher accuracy than the traditional comparison of the original classification (Kappa of 0.72; overall accuracy of 75.55 percent). An additional analysis was performed with the three "woody" classes merged into a single "upland forest" class. This resulted in a change detection Kappa coefficient of 0.852 and an overall accuracy of 90.59 percent (not shown in table).

5.3 Kittredge Study Area Change Detection

A map of selected classes of change in the Kittredge study area is shown in Plate 11a and is based on the change detection matrix logic presented in Figure 5.1. Of particular interest are the extensive areas of new estuarine emergent wetlands and riverine aquatic beds. Change detection maps should be interpreted cautiously, however, because the maps are absolutely a function of which classes in the change detection matrix are selected for display. For example, in order to identify both the losses and gains of riverine aquatic beds from 1982 to 1988, additional classes in the matrix would have to be selected (compare Plate 13a and 13b to see where the beds were gained and lost). Therefore, great care must be exercised when selecting which change classes to display from the change detection matrix.

The selection of hundreds of random *i,j* coordinates within the NHAP and NAPP aerial photography resulted in 23 homogeneous 3 by 3 pixel samples that could be used to evaluate the change detection error in the Kittredge study area. The overall accuracy was 76.6 percent with a Kappa of 0.721.

Table 5.6. Tidal Stage and Classified Using Landsat TM Data of Fort Moultrie, South Carolina.

Date	Tidal Stage (cm)[a]	Estuarine Emergent Wetland
Dec 19, 1988	1	3,514.59
Nov 9, 1982	44	3,359.70
Dec 9, 1990	72	3,061.08
Oct 14, 1987	83	3,026.61
Mar 4, 1987	133	2,728.44
Oct 6, 1990	211	2.075.40

[a] Centimeters above Mean Low Tide (MLT) derived from NOAA tide tables.

6.0 THE EFFECT OF TIDAL STAGE ON WETLAND CLASSIFICATION IN THE FORT MOULTRIE STUDY AREA

The tentative C-CAP tidal protocol for selecting satellite remote sensor data is (a) "mean low tide" preferred, (b) 30 to 60 cm (1 to 2 feet) acceptable, and (c) 90 cm (3 feet) or more unacceptable. Unfortunately, only tangential empirical research has been conducted to determine the significance of tidal stage variation (or flooding) when detecting change in coastal wetlands (Madec, 1991; Williams and Lyon, 1991). Ideally, tidal stage would be held constant between dates, but this would greatly undermine the utility of satellite imagery by severely limiting the amount of available data.

A preliminary study sponsored by NOAA was initiated to determine the potential significance of tidal stage on wetland classification using multiple dates of Landsat TM data acquired along a continuum of tides (Table 5.6). These data were classified using the "cluster busting" technique described earlier into just four classes of information (maps not shown): estuarine emergent marsh, estuarine unconsolidated bottom, water, and upland. A National Wetlands Inventory (NWI) map of the region (not confined just to the Fort Moultrie quadrangle) was recoded to create a mask containing only these tidally influenced classes. This mask was applied to each classification map, guaranteeing a consistent geographic area for comparison. The total hectares of estuarine emergent wetland in each image were computed. The tide at the time of Landsat TM data acquisition was obtained from NOAA tide tables (Table 5.6).

The data were subjected to a regression model to determine the strength of the relationship between tidal stage and estuarine emergent wetland present. The resulting equation:

$$y = -6.953x + 3591.41$$

yielded an r^2 of 0.983 significant at the 0.001 level. We know that very little estuarine emergent marsh has changed in this region because of extensive biological research by Bradley et al. (1990) and the change detection research presented in this chapter. Therefore, as the tide comes in, it produces more mixed pixels which may be misclassified as water. The result may be a decrease in the amount of emergent wetlands reported and an inaccurate map. Spurious change would be depicted in change detection maps if these data were used. Additional research is being conducted to verify the consistency of the relationship and to suggest more rigorous C-CAP tidal protocol.

7.0 RECOMMENDATIONS

Below are recommendations based on an evaluation of C-CAP protocols in two South Carolina coastal environments.

7.1 CoastWatch Image Data Selection

- Landsat TM data were used to map the land cover of wetlands and adjacent upland areas. One image per change detection year may not be adequate to discriminate between certain classes, especially those confused with cultivated land.
- The middle infrared TM bands were particularly useful and accounted for much of the separability between wetland types.
- C-CAP tidal stage protocol appears to be too lenient and must be made more rigorous through additional investigation of the tidal influence on wetland classification.

7.2 Image Classification

- The tentative CoastWatch Classification Scheme was generally useful. However, attempts to completely distinguish between some classes (e.g., developed/bare soil, cultivated land, herbaceous) were not feasible using Landsat TM data. The land-cover classes should be prioritized according to their importance to C-CAP program objectives. This would eliminate the need to obtain information on certain classes and standardize the scheme used between regions and states.
- The spectral remote sensing data cannot differentiate between estuarine, riverine, or marina wetlands as identified in the complete CoastWatch Classification Scheme. "Saline" stratification must be made using ancillary GIS data sources.
- Unsupervised "cluster busting" techniques coupled with "post-classification majority filtering" yielded the most accurate individual date classification maps.
- The C-CAP classification accuracy protocol is 90 percent for all categories. This accuracy was not obtainable using Landsat TM data. The protocol should probably be reduced to 85 percent accuracy, similar to the USGS land cover mapping protocol.

7.3 Change Detection

- "Post-classification comparison" change detection logic is suitable for the C-CAP program if the individual dates of imagery are classified as accurately as possible. Such logic is essential if diverse "from-to" classes of interest are to be displayed.
- The use of a "change/no-change" mask based on multiple date image differencing of specific bands may be a useful preprocessing function if an appropriate "threshold" is selected. The mask can then be used to classify only those pixels which have changed on the latest date of imagery.

7.4 CoastWatch Products

- The change detection "from-to" classes of interest must be standardized so that those classes identified in South Carolina are equivalent to change detection classes for other states. The change detection legend must be standardized.
- Compressed data files provided to the user must be standardized and include (a) rectified remote sensor data for each date, (b) the raw final clusters obtained during the classification of each date,

(c) the recoded final classes for each date as specified in the CoastWatch Classification Scheme, and (d) the post classification comparison change detection file with integers representing the "from-to" classes.

- A history of the procedures used to create the C-CAP change detection map should be carefully documented in a "lineage" file as described by Jensen and Narumalani (1992).

8.0 SUMMARY

The NOAA sponsored CoastWatch Change Analysis Project (C-CAP) will utilize remote sensing technology to monitor changes in coastal wetland habitats and adjacent uplands on a cycle of 1 to 5 years. Two study areas in South Carolina were selected to test various C-CAP change detection protocols using near-anniversary Landsat Thematic Mapper data obtained in 1982 and 1988. Fort Moultrie (dominated by salt and brackish marsh) and Kittredge (40 river miles inland and dominated by bottomland hardwoods and riverine aquatic beds) study areas were used to evaluate a modified C-CAP classification scheme, image classification procedures, change detection algorithm alternatives, and the impact of tidal stage on coastal change detection. The modified CoastWatch Classification Scheme worked well and can be adapted for South Carolina with minor adjustments. Unsupervised "cluster-busting" techniques coupled with "threshold 3 majority filtering" yielded the most accurate individual date classification maps (86.7 to 92.3 percent overall accuracy; Kappa coefficients of 0.85 to 0.90). The best change detection accuracy was obtained when individual classification maps were majority filtered and subjected to "post-classification comparison" change detection (85.2 percent overall accuracy; Kappa coefficient of 0.821). Suggestions are made concerning appropriate change detection matrix logic and the format of change detection legends. The multiple date images selected for coastal change detection should meet stringent tidal stage guidelines which have yet to be fully documented

ACKNOWLEDGMENTS

This project was funded by the National Oceanic and Atmospheric Administration through the S.C. Sea Grant Consortium. Richard Lacy (S.C. Land Resource Conservation Commission), Bjorn Kjerfve, and Pradeep Talwani (Dept. of Geology, USC) made valuable contributions to this research. All image processing was performed using ERDAS software.

REFERENCES

Bradley, P.M., B. Kjerfve, and J.T. Morris. Rediversion Salinity Change in the Cooper River, South Carolina: Ecological Implications, *Estuaries.* 13(4):373–379, 1990.

Clark, J.A. and J.A. Primus. Sea-Level Changes Resulting from Future Retreat of Ice Sheets: An effect of CO_2 Warming of the Climate. *Sea-Level Changes* (M.J. Tooley and I. Shennan, Eds.) Basil Blackwell, Inc., Oxford. 1990, pp. 356–370.

Congalton, R.G. A Comparison of Sampling Schemes Used in Generating Error Matrices for Assessing the Accuracy of Maps Generated from Remotely Sensed Data. *Remote Sensing of Env*ironment, 37:45–46, 1988.

Cowardin, L.M., V. Carter, F.C. Golet, and T. La Roe. *Classif*ication *of Wetlands and Deepwater Habitats of the United States.* FWS/OBS-79/31, U.S. Fish & Wildlife Service, Washington, DC, 1979.

Cross, F.A. *Draft Protocol: Change Detection in Coastal Wetlands, Adjacent Uplands and Submerged Aquatic Vegetation.* NOAA National Marine Fisheries Services. Beaufort, SC, 1991.

Dahl, T.E. *Wetlands Losses in the United States 1780s to 1980s,* U.S. Fish & Wildlife Service. Washington, DC, 1990.

Dahl, T.E. and C. E. Johnson. *Status and Trends of Wetlands in the Conterminous United States. Mid-1970s to Mid-1980s.* U.S. Fish & Wildlife Service, Washington, DC, 1991.

Dobson, J.E. and E.A. Bright. CoastWatch—Detecting Change in Coastal Wetlands, *Geo Info Systems.* 1:36–40, 1991.

Dobson, J.E., R.L. Ferguson, D.W. Field, L.L. Wood, K.D. Haddad, H. Iredale, V.V. Klemas, R.J. Orth, and J.P. Thomas. *NOAA CoastWatch Change Analysis Project: Guidance for Regional Implementation,* NOAA Coastal Ocean Program, Washington, DC, 1992.

Edwards, R.W. and M.W. Brown. An Aerial Photographic Method for Studying the Distribution of Aquatic Macrophytes in Shallow Waters, *Journal of Ecology,* 48:161–163, 1960.

Gao, T. and T.L. Coleman. Use of Satellite Spectral Data for Mapping Aquatic Macrophytes and Nutrient Levels in Lakes. *Proceedings, International Geoscience & Remote Sensing Symposium.* (1):109–116, 1990.

Haddad, K.D. and D.R. Ekberg. Potential of Landsat TM Imagery for Assessing the National Status and Trends of Coastal Wetlands, *Proceedings, 5th Symposium on Coastal & Ocean Management,* American Society of Civil Engineers, New York, 5192–5201, 1989.

Hodgson, M.E. and R. Plews. N-Dimensional Display of Cluster Means in Feature Space, *Photogrammetric Engineering & Remote Sensing,* 55(5):613–619, 1989.

Jensen, J.R., E.J. Christensen, and R.R. Sharitz. Nontidal Wetland Mapping in South Carolina Using Airborne Multispectral Scanner Data, *Remote Sensing of Environment,* 18:1–12, 1984.

Jensen, J.R. *Introductory Digital Image Processing: A Remote Sensing Perspective,* Prentice-Hall, Englewood Cliffs, NJ, 1996.

Jensen, J.R., M.E. Hodgson, E.J. Christensen, H.E. Mackey, L.R. Tinney, and R.R. Sharitz. *Remote Sensing Inland Wetlands: A Multispectral Approach, Photogrammetric Engineering & Remote Sensing* 52(1):87–100, 1986.

Jensen, J.R., H. Lin, X. Yang, E. Ramsey, B. Davis, and C. Thoemke. The Measurement of Mangrove Characteristics in Southwest Florida Using SPOT Multispectral Data. *GEOCARTO International—A Multidisciplinary Journal of Remote Sensing,* 2:13–21, 1991a.

Jensen, J.R., S. Narumalani, O. Weatherbee, and H.E. Mackey, Jr. Remote Sensing Offers An Alternative for Mapping Wetlands, *Geo Info Systems,* (October):46–53, 1991b.

Jensen. J.R., D. Cowen, O. Weatherbee, J. Althausen. S. Narumalani, B. Kjerfve, P. Talwani, and R. Lacy. Change Detection Algorithm Evaluation: CoastWatch Protocol Development in South Carolina. *Technical Papers,* American Society for Photogrammetry & Remote Sensing, 1:120–129, 1992.

Jensen, J.R. and S. Narumalani. Improved Remote Sensing and GIS Reliability Diagrams, Image Genealogy Diagrams & Thematic Map Legends to Enhance Communication, *Archives,* International Society for Photogrammetry and Remote Sensing, Commission VI, 1992, pp. 125–132.

Jensen, J.R., E.W. Ramsey, J.M. Holmes, J. Michel, B. Savitsky, and B.A. Davis. Environmental Sensitivity Index (ESI) Mapping for Oil Spills Using Remote Sensing and Geographic Information System Technology, *International Journal of Geographic Information Systems,* 4(2):181–201, 1990.

Jensen, J.R., E.W. Ramsey, H.E. Mackey, E. Christensen, and R. Sharitz. Inland Wetland Change Detection Using Aircraft MSS Data. *Photogrammetric Engineering & Remote Sensing,* 53(5):521–529, 1987.

Kana, T.W., J. Michel, M.O. Hayes, and J.R. Jensen. The Physical Impact of Sea Level Rise in the Area of Charleston, South Carolina, *Greenhouse Effect and Sea Level Rise: A Challenge for this Generation* (M.C. Barth and J.G. Titus, Eds.), Van Nostrand Reinhold Co., New York, 1984, pp. 105–150.

Kenk, E., M. Sondheim, and B. Yee. Methods for Improving Accuracy of Thematic Mapper Ground Cover Classifications. *Canadian Journal of Remote Sensing,* 14(1):17–31, 1988.

Kiraly, S.J., F.A. Cross, and J.D. Buffington. Overview and Recommendation, *Federal Coastal Wetland Mapping Programs.* U.S. Dept. Interior, Washington, DC, Biological Report 90(18):1–7, 1990.

Klemas, V.V., S.R. Hoffer, R. Kleckner, D. Norton, and B.O. Wilen. A Wetland/Upland Land Cover Classification System for Use with Remote Sensors, *Forum on Land Use & Land Cover Summary Report,* U.S. Geological Survey, Reston, VA, 1992, pp. 65–69.

Lee, J.K., R.A. Park, and P.W. Mausel. Application of Geoprocessing and Simulation Modeling to Estimate Impacts of Sea Level Rise on the Northeast Coast of Florida, *Photogrammetric Engineering & Remote Sensing,* 58(11):1579–1586, 1992.

Mackey, H.E. Monitoring Seasonal and Annual Wetland Changes in a Freshwater Marsh with SPOT HRV Data, *Technical Papers,* American Society for Photogrammetry and Remote Sensing, 4:283–292, 1990.

Madec, V.A. Spatial Model of Emersion Rates Applied to the Intertidal Zone of the Bay of Mont Saint-Michel, *Oceanologica Acta,* 14(5):525–529, 1991.

Martin, L.R. Accuracy Assessment of Landsat-Based Visual Change Detection Methods Applied to the Rural-Urban Fringe, *Photogrammetric Engineering & Remote Sensing,* 55(2):209–215, 1989.

Nehara, S. A Study on Annual Changes in Surface Cover of Floating-Leaved Plants in a Lake Using Aerial Photography, *Vegetatio,* 97:125–136, 1991.

Norton, D.J. and E.T. Slonecker. The Ecological Geography of EMAP, *Geo Info Systems,* 1(1):33–43, 1990.

Odum, E.P. Wetland Values in Retrospect, *Freshwater Wetlands and Wildlife* (R.R. Sharitz and J.W. Gibbons, Eds.), Report #8603101, U.S. Department of Energy, Washington, DC, 1989, pp. 1–8.

Pilon, P.G., P.J. Howarth, R.A. Bullock, and P.O. Adeniyi. An Enhanced Classification Approach to Change Detection in Semi-Arid Environments, *Photogrammetric Engineering & Remote Sensing,* 54(12):1709–1716, 1988.

Podolsky, R. and P. Conkling. Satellite Search Aids Wetlands Visualization, *GIS World,* 4(9):80–85, 1991.

Roughgarden, J., S.W. Running, and P.A. Matson. What Does Remote Sensing Do for Ecology? *Ecology* 72(8):1918–1922, 1991.

Shirer, H.O. GPS and the U.S. Federal Radionavigation Plan, *GPS World* 2(2):32–37, 1991.

Welch, R., M.M. Remillard, and R.B. Slack. Remote Sensing and Geographic Information System Techniques for Aquatic Resource Evaluation, *Photogrammetric Engineering & Remote Sensing,* 54(2):177–185, 1988.

Wilen, B.O. U.S. Fish & Wildlife Service's National Wetlands Inventory, *Federal Coastal Wetland Mapping Programs,* Biology Report #90, U.S. Fish & Wildlife, Washington, DC, (18):9–20, 1990.

Williams, D.C. and J.G. Lyon. Use of a Geographic Information System to Measure and Evaluate Wetland Changes in the St. Mary's River, Michigan, *Hydrobiologia,* 291:83–95, 1991.

CHAPTER 6

Comparison of Methods for Detecting Conifer Forest Change with Thematic Mapper Imagery

Warren B. Cohen and Maria Fiorella

1.0 INTRODUCTION

Numerous methods have been applied to the problem of detecting forest cover changes with the aid of digital imagery. In their review of change detection methods, Coppin and Bauer (1996) recognize 11 distinct methods groups. Among these, some of the more commonly applied methods include image differencing, multitemporal linear data transformation, and composite analysis. In this chapter, these three methods are compared for detecting change in a conifer forest environment using Landsat Thematic Mapper imagery.

Image differencing involves simple image subtraction (Vogelmann, 1988; Price et al., 1992). For one image spectral band acquired at two separate dates over the same ground scene, image differencing explicitly captures the univariate magnitude of radiometric change. Similarly, the direction of change is captured as the sign (+ or –) of radiometric change. For multiple input spectral bands, difference images have a multivariate structure, with a calculated magnitude and direction for each temporal-pair of input bands. Malila (1980) describes a procedure whereby magnitude and direction are directly calculated in multivariate space. This procedure, known as change vector analysis (CVA), is an example of a multitemporal linear data transformation. Other examples are given by Fung and LeDrew (1987) and Collins and Woodcock (1994). In CVA, magnitude is computed as Euclidean distance and direction is computed as angle of change, with the former representing amount of land cover change, and the latter representing type. Composite analysis is little more than a classification procedure (e.g., unsupervised) applied to a layered, multitemporal image data set (Schowengerdt, 1983; Muchoney and Haack, 1994). Transformation of the original data into vegetation indices is common prior to composite analysis, but this is done for data reduction or to highlight certain vegetation cover features within the individual dates of imagery, not to highlight changes between the dates.

A perusal of the change detection literature supports Singh's (1989) claim that image differencing is likely the most widely used method, and in studies where image differencing has been compared with other methods, results generally indicate that image differencing exhibits superior performance (e.g., Singh, 1986; Muchoney and Haack, 1994; Coppin and Bauer, 1996). The ease with which images are subtracted from one another (or "differenced") and the inherent meaning such radiometric differences have in terms of cover change, both strongly contribute to the method's appeal and successful performance. Although CVA was introduced

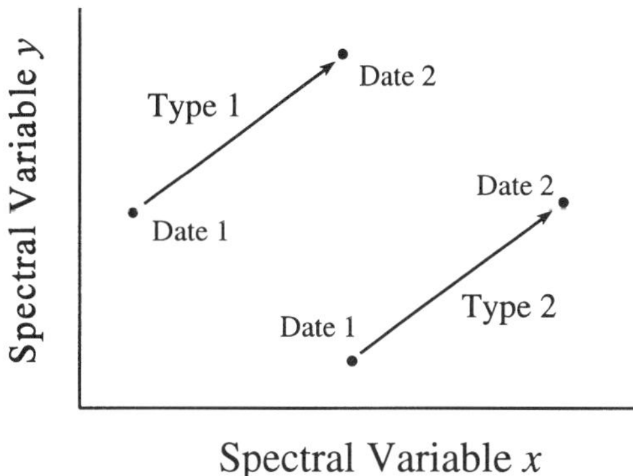

Figure 6.1. Two different types of land cover change with exactly the same amounts and directions of spectral change.

nearly two decades ago and has received little attention in the literature, it is likely to become the algorithm of choice for the land cover change product being developed for the MODIS sensor (Lambin and Strahler, 1994a). Because of this, CVA may be poised to become one of the most popular change detection algorithms over the next several years. Unlike image differencing and CVA, composite analysis might be considered undesirable because there is no distillation of original input bands into bands that explicitly characterize cover change.

One consideration in digital change detection that may be of major importance, but that has not received much attention, is the concept of a "reference image;" that is, a reference against which derived change information can be compared. This could be in the form of a land cover map at T_1 or T_2 in a two-date analysis, or a T_1 or T_2 spectral image that contains natural variations in reflectance of land cover categories. Composite analysis inherently involves reference imagery. In contrast, image differencing and CVA do not; that is, although they result in images that contain change information, all reference to original data are lost. This is a potential problem because two very different types of land cover change can have similar amounts and directions of spectral change, as illustrated in Figure 6.1. As such, there is likely under certain circumstances to be excessive confusion among cover change categories.

Recognizing the need for a reference image, Virag and Colwell (1987) used a land cover classification derived from the T_2 image as a reference for calculated change vectors. Collins and Woodcock (1996) extend the tasseled-cap transformation by defining a multitemporal tasseled-cap transformation (MKT) that has one stable, or reference output dimension, and one change output dimension for each input dimension. They then compared the MKT against Gramm-Schmidt orthogonalization in a study of forest mortality, and found that the MKT transformation gave superior results. Franklin et al. (1995) used discriminant function analysis to evaluate changes in forest cover due to insect defoliation. Discriminant function models based on a multidate, layered image data set (i.e., composite analysis) were 21 percent more accurate in predicting among three defoliation classes than were models based on image differencing. In the dense conifer forest condition of the region where this study was done, tasseled-cap wetness is very highly correlated with forest structure (Cohen and Spies, 1992; Cohen et al., 1995). For the first discriminant function of models based on composite analysis, 1988 wetness had a weighting of –1.50, with 1993 wetness having a relatively low weighting of –0.25. Tasseled-cap brightness and greenness were important contributors to this function only as temporal contrasts (–0.79 and 0.96 for 1988 and 1993 brightness, respectively; 0.46 and –0.44 for 1988 and 1993 greenness, respectively). This result suggests rather strongly that differences in multidate brightness and greenness were of great importance in distinguishing among defoliation classes, but only after natural variations in initial forest structure (prior to defoliation) were accounted for by a 1988 wetness reference image.

The purpose of this chapter is to initiate an exploration of the relative values of composite analysis, image differencing, and CVA for detecting changes in the dense forest environment of the Pacific Northwest region of the United States. This analysis involves a simple test of the three methods and includes some comparisons with and without a reference image, to determine its value in change detection. Descriptions of CVA in the literature do not appear adequate for fully understanding this potentially valuable change detection algorithm. Thus, prior to describing the test of methods, some theoretical considerations for change detection, with specific emphasis on CVA, are given. The description here is still inadequate, but we hope that it provides seed for additional insights by others.

2.0 THEORETICAL CONSIDERATIONS FOR CVA

Change vector analysis was developed for use with Landsat MSS data, in particular, for the two spectral dimensions of MSS data known as brightness and greenness (Kauth and Thomas, 1976). As described by Malila (1980), CVA is conceptually very simple (Figure 6.2). For a given image pixel, magnitude is calculated as the Euclidean distance between its location in brightness-greenness space on T_1 and its location on T_2:

$$d = \sqrt{(x_2 - x_1)^2 + (y_2 - y_1)^2}$$

where d=Euclidean distance, x_2 and y_2 are pixel brightness and greenness values for T_2, respectively, and x_1 and y_1 are pixel brightness and greenness values for T_1. The angle calculation requires establishment of a "baseline." Malila arbitrarily defined the baseline as parallel to the brightness axis (x), with its origin at the T_1 pixel vector and its terminus at the brightness value of T_2. The angle of spectral change is then measured relative to this baseline using standard trigonometric functions (e.g., sin, cos, or tan). In the resulting two-dimensional image, change is observed if the magnitude dimension exceeds a defined threshold.

The geometric concepts of CVA are applicable to any number of spectral bands, whether original scaled radiance, calibrated radiance, or transformed variables (e.g., reflectance, vegetation indices). Virag and Colwell (1987) present an analysis using three spectral dimensions, and they conceptually extend the procedure into n-dimensional space. For more than two spectral dimensions, they calculate magnitude of change as n-dimensional Euclidean distance:

$$d = \sqrt{\sum_{i=1}^{p} (DN_{2i} - DN_{1i})^2}$$

where i=spectral band number from 1.0 to p, DN_{2i}=pixel value at T_2 in band i, and DN_{1i}=pixel value at T_1 in band i. Rather than explicitly calculate the angular component of the change vector, however, Virag and Colwell approximate it using a multispectral positive and negative sector code accounting system. With this accounting system, 2^n sectors (directions) are possible: positive or negative spectral change in each of the n dimensions. This same sector coding

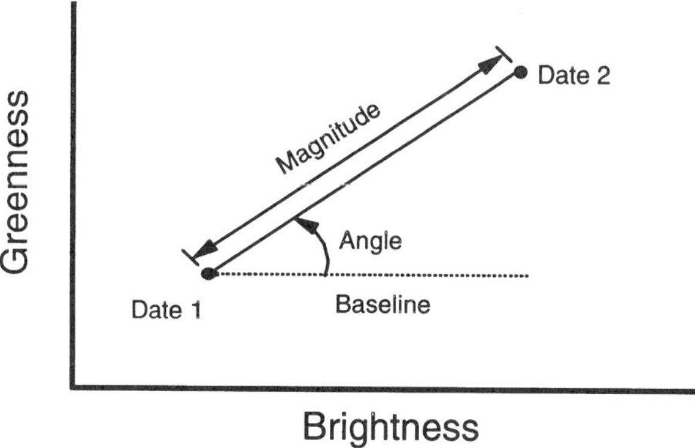

Figure 6.2. The concept of Change Vector Analysis in two spectral dimensions (adapted from Malila, 1980).

logic is used by Michalek et al. (1993) in their analysis of a coastal marine environment. Lambin and Strahler (1994b) use CVA with monthly composited AVHRR data. In their unique application of CVA, the 12 monthly NDVI observations of a given pixel for a given year constitute a 12-dimensional temporal vector for that pixel. Mathematically, the Lambin and Strahler (1994b) approach is exactly the same as that of Malila (1980) and Virag and Colwell (1987) for calculating change magnitude. Using two temporal NDVI vectors, one for T_1 and the other for T_2, magnitude of change between the dates is calculated as the Euclidean distance between the two 12-dimensional NDVI vectors. Lambin and Strahler (1994b) calculated neither the angle of change in any of the 12 dimensions, nor sector codes. Instead they chose to infer angularity from a principal components analysis on the 12 difference images derived from two-date NDVI data.

The approximations of directionality by Virag and Colwell (1987) and of Lambin and Strahler (1994b) are useful, but within the context of CVA there are other possibilities for more precisely describing the relative locations of two pixel vectors in n-dimensional space. One is to calculate more than a single angle. In the two-dimensional MSS tasseled-cap example of Malila (1980), magnitude and only one angle are required to precisely locate a T_2 pixel vector relative to a T_1 vector. This angle is measured in what Crist and Cicone (1984) refer to as the Plane of Vegetation (defined by the brightness and greenness axes). For the three primary dimensions of the Landsat TM tasseled-cap transformation, one also can measure an angle in the Plane of Soils (defined by the brightness and wetness axes), and in the Transition Zone (defined by the greenness and wetness axes). As each of these "views" of TM tasseled-cap data space contain different information about a ground scene, change vector angles measured within them should likewise contain different types of change information.

Another possibility is to explicitly define an angle of change relative to a new axis, or baseline k, as shown in Figure 6.3. This enables one to precisely define a T_2 pixel vector vis-a-vis a T_1 vector with two angles, ϕ and θ, and Euclidean distance magnitude d. The selection of x as the primary baseline is arbitrary, as use of y or z in combination with k would also precisely define the relative relationships of the two vectors. Furthermore, using ϕ and θ, d could be replaced by a distance measured along x, y, z, or k with no consequence.

Definition of a new baseline through original data space can be accomplished in more than one way. Figure 6.3 illustrates that one can define this baseline k within one of the existing planes, such as the Plane of Vegetation (x, y). An additional possibility is to define a baseline

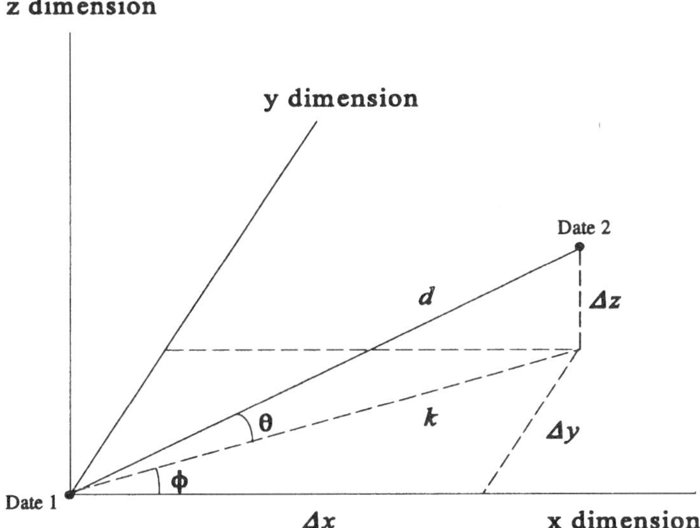

Figure 6.3. An example showing precise characterization of a change vector in three spectral dimensions. Required are establishment of a secondary baseline, *k*, in addition to the original baseline, *x*, and calculation of two angles, ϕ and θ, and distance *d*.

directly between a set of pixels that have changed, which (in Figure 6.3) is the same as defining the baseline along the distance *d*. A variant of this method is to use the Gramm-Schmidt orthogonalization procedure, as done by Collins and Woodcock (1994). In theory, a separate axis could be defined for each type of cover change of interest using this procedure. In such a case, one is less concerned about angles, as they are implicit in the baselines. Thus, the distances along these new baselines become the main source of change information.

3.0 PRELIMINARY TEST OF METHODS

To assist us in further understanding the relative values of composite analysis, image differencing, and CVA for detecting change in the dense conifer forest environment of the Pacific Northwest region of the United States, a simple test was conducted. This test had two specific objectives, the intent of which was to provide preliminary results that could be used to help design a more thorough and statistically rigorous study:

- Explore the relative values of composite analysis, image differencing, and CVA for detecting different degrees and types of forest disturbance and succession. These are compared in the context of both including and excluding a T_1 reference image.
- Determine the value of tasseled-cap wetness for forest cover change detection. Wetness has been valuable for forest cover mapping in the Pacific Northwest region (Cohen and Spies, 1992; Cohen et al., 1995), and recently, Collins and Woodcock (1996) found it to be the single most valuable indicator of forest change among several indicators tested in a different forest environment.

For this test, a 625 km^2 area in western Oregon near the H.J. Andrews Experimental Forest was selected. Many of the major conifer forest types of the central and northern Cascade Range are represented, including the western hemlock/Douglas-fir, Pacific silver fir, and mountain hemlock forest zones (Franklin and Dyrness, 1988). Historically, forests of this area have been largely managed for wood production using a forest clear-cutting harvest strategy, but some forest reserves are present. In general, closed canopy conifer forests present in the region have a high leaf area index which causes them to absorb much of the incident solar radiation (Cohen et al., 1995). When disturbed, leaf area indices are diminished, resulting in higher albedo. After a severe disturbance such as clear-cutting, a forest stand commonly goes through several intermediate seral stages (e.g., grass/forb/herbaceous, and hardwood and conifer brush) on its return to a closed conifer canopy state. In riparian zones, hardwood forest are common.

Two Landsat TM images (Path 46/Row 29) of the selected area were used: one acquired on August 30, 1988 and the other on August 29, 1993. The 1988 image was georeferenced and terrain corrected prior to our receiving it. For this study, the 1993 image was coregistered to the 1988 image using 30 ground control points, a second order polynomial, and nearest neighbor resampling (RMSE = 0.768). The subset area of 625 km^2 in the central Oregon Cascades Range was extracted from both image data sets, and these were transformed into tasseled-cap brightness (B), greenness (G), and wetness (W) images. From these two-date B, G, and W images (1-6) several new images were created, including B, G, and W differences (7-9), change vector magnitude in BG (10) and in BGW (11) data space, and change vector angles in BG, BW, and GW data space (12-14). To satisfy the two stated objectives, 10 combinations of the complete 14-image set were selected (Table 6.1). These were: (a) B and G differences and (b) B, G, and W differences; (c) BG change vector magnitude and angle; BGW magnitude with BG and BW angles (d), with BG and GW angles (e), with BW and GW angles (f), and with BG, BW, and GW angles (g); (h) d with 1988 reference B, G, and W images, (i) 1988 and 1993 B, G, and W images, and (j) 1988 and 1993 B and G images. The 1988 reference B, G, and W images were not combined with the B, G, and W difference images because mathematically, these comprise exactly the same data as the six original B, G, and W images (i), and would thus yield the exact same result as composite analysis.

The 10 selected band combinations were used to develop maps of forest cover change and then the results among the maps compared. For this comparison, two maps of forest cover developed from the two original TM images were used. The 1988 map was previously published (Cohen et al., 1995), and included three early-successional, mixed cover classes (open, semiclosed, and closed) and three closed-canopy conifer cover classes (young, mature, and old-growth). Subsequent to development of this map, a similar map was created from the 1993 data. The two maps were each formally assessed for errors and found to have accuracies in excess of 80 percent for the six forest cover classes. As an alternative to these two forest cover maps, the possibility of obtaining independent ground and airphoto reference data was considered, but given the preliminary nature of this study, the cost of obtaining forest change data for the specific time interval of interest was considered too costly.

To develop the forest cover change map for each band combination, a delta classification (Coppin and Bauer, 1996) developed from the 1988 and 1993 forest cover maps was used. This classification had 36 classes (the product of the number of classes in the two original maps), as shown in Table 6.2. Using the delta classification as a training set, each band combination to be tested was processed by a maximum likelihood supervised classification algorithm to develop a cover change map with 36 change classes. Then, the label of each cell of the delta classification was referenced to the label for each cell of the change map, resulting in a standard classification error matrix for each of the 10 change maps. We recognize that this had the effect of facilitating a positive accuracy bias, in that each map was nothing more than a simple reclassification of the training data. For this reason, these error matrices are referred to as "agreement" matrices. A potential source of "disagreement" in our test arises from the use of image data that

Table 6.1. Overall Percent Agreement for Different Numbers of Land Cover Change Classes and 10 Combinations (a-j) of Image Bands, Including Original (Reference) 1988 and 1993 Tasseled-Cap Brightness (B), Greenness (G), and Wetness (W) Images, and Those Derived from Change Vector Analysis (CVA) and Image Differencing. (Combinations below the dashed line include a reference image.)

Band Combination	3 Classes	7 Classes	36 Classes
Difference images			
(a) B,G	76.0	40.2	14.7
(b) B,G,W	76.4	48.7	19.1
Change vector analysis (CVA)			
(c) 2-dimensional (B,G)			
magnitude and angle BG	73.1	38.9	12.8
3-dimensional (B,G,W) magnitude and			
(d) angles BG and BW	71.6	45.8	15.2
(e) angles BG and GW	74.9	41.5	13.9
(f) angles BW and GW	72.9	45.6	14.2
(g) angles BG, BW, and GW	73.1	45.7	14.6
(h) d with 1988 reference B, G, and W	85.7	66.2	57.5
Composite Analysis			
(i) 1988 & 1993 B, G, and W	89.4	75.7	71.2
(j) 1988 & 1993 B and G	87.2	58.5	45.4

were not radiometrically normalized. But we do not consider this a great problem, as the ground scene was imaged through a clear atmosphere on both dates and the solar illumination angles were virtually identical between the two dates.

There was no formal characterization of errors in the delta classification, but it was assumed to be at least 60 percent accurate for the 36 classes, as each independent six-class map was at least 80 percent accurate and errors are expected to be multiplicative (Howarth and Wickware, 1981). Concerned that a delta classification with 36 classes was only minimally acceptable as a reference data source, the classification and agreement matrix process for each of the 10 band combinations was repeated with aggregated delta classification classes. The process was repeated once with a seven-class delta classification and once with a three-class delta classification (Table 6.2). Classes used for the seven-class strategy were: one no-change class (NC), three forest succession classes (i-iii), and three forest disturbance classes (iv-vi). For the three-class strategy, the classes were: no-change (NC), a single forest succession class (S), and a single forest disturbance class (D).

4.0 RESULTS AND DISCUSSION

The 10 band combinations used in this study can be compared using a summary of the agreement matrices for the 10 sets of change images created (Table 6.1). Not surprisingly, high levels of agreement were observed when only three classes of cover change were sought (no-change, succession, and disturbance), regardless of the band combination used (72–89 percent). However, those combinations excluding the 1988 reference brightness, greenness, and wetness image exhibited the poorest (72–76 percent), with all that included the 1988 reference image exhibiting the highest (86–89 percent), levels of agreement. All CVA combinations for which a reference image was excluded agreed slightly less with the delta classification than the combinations based solely on difference images (72–75 versus 76 percent). For all combina-

Table 6.2. The 1988 to 1993 Delta Classification Matrix. [Cell values are: original class number (1-36)—class label under grossest class aggregation (NC=no-change, S=succession, D=disturbance)—and class label for a more ecological significant class aggregation (NC, i-vi)].

1988 Vegetation	1993 Vegetation Cover Class					
Cover Class	**Open**	**Semi-Closed**	**Closed-Mix**	**Young Conifer**	**Mature Conifer**	**Old Conifer**
Open	1—NC—NC	2—S—i	3—S—i	4—S—ii	5—S—ii	6—S—ii
Semi-Closed	7—D—iv	8—NC—NC	9—S—i	10—S—ii	11—S—ii	12—S-ii
Closed-Mix	13—D—iv	14—D—iv	15—NC—NC	16—S—ii	17—S—ii	18—S—ii
Young Conifer	19—D—v	20—D—v	21—D—v	22—NC—NC	23—S—iii	24—S-iii
Mature Conifer	25—D—v	26—D—v	27—D—v	28—D—vi	29—NC—NC	30—S—iii
Old Conifer	31—D—v	32—D—v	33—D—v	34—D—vi	35—D—vi	36—NC—NC

tions, when three change classes were sought, the use of wetness in addition to brightness and greenness did not significantly increase agreement.

For seven change classes, overall agreements were below 50 percent for all combinations which excluded the use of a 1988 reference image. Of these seven combinations, those based only on brightness and greenness agreed least (39 and 40 percent). For the remaining five of these seven combinations in which wetness was included, image differencing agreed most (49 percent), with CVA combination agreeing between 42 and 46 percent. Use of the 1988 reference image to characterize seven forest change classes had a large impact on the level of agreement for CVA. The three-dimensional CVA based on the brightness-greenness and brightness-wetness angles resulted in 20 percent higher agreement when the 1988 reference image was used than when it was not (66 versus 46 percent). Using the six original bands (composite analysis), equivalent to using the difference images in combination with the 1988 reference image, resulted in a 27 percent higher agreement (49 versus 76 percent) relative to the three-dimensional image differencing combination. Unlike three change classes, when seven classes were sought, the composite analysis combination excluding wetness resulted in a significant decrease in agreement compared to the combination that included wetness (76 to 59 percent).

The greatest contrast among the 10 combinations occurred when evaluating the full suite of 36 change classes. For this number of cover change classes, both the relative and complementary values of a reference image, difference images over CVA images, and of wetness in addition to brightness and greenness is most apparent. All combinations that excluded the use of a reference image, exhibited less than 20 percent agreement. Use of the reference image increased agreement of CVA from 15 to 58 percent. For composite analysis, equivalent to image differencing with the use of a 1988 reference image, agreement increased from 19 to 71 percent. For the composite analysis combination excluding wetness, agreement was only 45 percent, compared to 71 percent when wetness was included. When no reference image was used, CVA had a slightly lower level of agreement than did image differencing.

Given that the use of a reference image in combination with CVA (h) resulted in a significantly greater level of agreement than did CVA without a reference image (d-g), it now becomes important to more closely compare the combination of CVA and a reference image (h) with composite analysis (i). As part of this further comparison it is important to better evaluate the value of wetness in composite analysis (i versus j) for change detection in a forest system. As such, these three band combinations are summarized in Table 6.3 by change class, for the seven-class example. In this comparison, CVA had higher agreement for all three succession classes than for the no-change and disturbance classes, with the highest agreement for Class ii (the progression from early-successional nonconifer forest to conifer forest). For this class, CVA had a slightly higher agreement (89 percent) than did composite analysis (86 percent). For composite analysis, the highest agreement, 89 percent, was for Class v (conifer forest changed to early-successional forest; i.e., forest clear-cut). This is in contrast to the relatively poor agreement of CVA for clear-cut mapping (68 percent). For all disturbance classes (iv-vi), composite analysis had about 20 percent greater agreement than did CVA, whereas for succession classes the differences in levels of agreement were significantly less pronounced and inconsistent. For the no-change class, composite analysis had about 10 percent higher agreement than CVA when a reference image was included, but for both, it was one of the more difficult classes.

The value of wetness was highly variable among cover change classes (Table 6.3). Although there was some value to the use of wetness for detecting changes associated with the nonconifer forests classes (i, ii, iv, and v), the primary importance of wetness was for detecting changes

Table 6.3. Percent Agreement for Each of Seven Vegetation Cover Change Classes for Three of the Band Combination Evaluated (h-j, Table 6.1). [Class i=succession within the nonconifer forest; Class ii=succession from nonconifer to conifer; Class iii=succession within the conifer forest; Class iv=disturbance within the nonconifer forest; Class v=disturbance from conifer to nonconifer forest; and Class vi=disturbance within the conifer forest (see Table 6.2)].

Class Label	CVA (3-Dimensional Magnitude, BG & BW Angles) with 1988 Reference Images (B,G,W)	Composite Analysis (B,G,W)	Composite Analysis (B,G)
Succession			
i	74.6	79.5	79.5
ii	88.9	86.1	82.1
iii	78.6	85.7	49.8
No-Change	60.5	69.9	54.5
Disturbance			
iv	42.3	63.1	52.5
v	68.4	89.3	80.6
vi	60.1	79.5	35.5

within the closed canopy conifer forest (Classes iii and vi). For composite analysis, there was a 36 percent difference in levels of agreement for succession within the conifer class (iii), depending on whether wetness was included or excluded. For disturbance within the conifer class (vi), the difference in agreement was 44 percent. This observation is consistent with Collins and Woodcock (1996), who found that in their study, wetness was the single most important indicator of conifer forest change.

5.0 CONCLUSIONS

Numerous methods exist for change detection using digital image data. In this study, three methods for detecting changes in a conifer forest environment with Landsat TM data were evaluated: image differencing, change vector analysis (CVA), and composite analysis. As commonly used, all three methods have several procedures in common. Needed are decisions concerning which original input bands to use (e.g., DN, radiance reflectance, vegetation indices), what type of classification algorithm to apply (e.g., supervised, neural-net), and a strategy for error assessment. Where they differ is in how the input bands are used prior to classification. Image differencing and CVA involve transformation of input bands into temporal change vectors, with the former being a band-by-band temporal subtraction, and the latter requiring derivation of spectral change magnitude and angle. Composite analysis uses the input bands directly in classification.

Although difference images and CVA magnitude and angle images represent direct characterizations of spectral change over time, they contain no reference to location within the original input data space. In contrast, because composite analysis uses input bands directly, they do contain this reference information. As such, natural variability in original and final (i.e., T_1 and T_2, respectively) land cover classes is directly incorporated into the change classification procedure. In this study, a T_1 reference image was combined with CVA magnitude and angle images, for comparison with CVA magnitude and angle images used alone in a classification of conifer forest change. Because image differencing used in combination with input T_1 images is mathematically equivalent to composite analysis, there was no reason to combine these images for further analysis.

The most important finding of this study was that use of a reference image was extremely important for accurate characterization of forest change. When no reference image was used, image differencing performed better than CVA, but only minimally so. Even in combination with a reference image, CVA did not compare well to composite analysis, with the disparity between them increasing as the number of cover change classes increased. The importance of using a reference image is illustrated in Figure 6.4. If only a brightness difference image is used in change detection, forest clear-cuts (Class v) are readily separable from all other change classes; however, all other classes are nearly impossible to separate accurately along a brightness difference axis alone. Including a 1988 reference brightness image with the brightness difference image enables accurate separation of the change class associated with early-successional forest progressing to conifer forest (Class ii), and to a somewhat lesser extent, the classes associated with succession within nonconifer forest (Class i). Other classes remain confused with brightness and brightness difference data alone, but become more separable with the addition of greenness and greenness difference, and wetness and wetness difference. Recall that the use of a reference image in conjunction with difference images provides precisely the same result as composite analysis when used in a change classification algorithm.

The tasseled-cap wetness feature has been important for separating different classes of conifer forest in the Pacific Northwest region of the United States, and has been shown to be important for conifer forest change detection in conifer forests elsewhere. In this study, wetness significantly improved the results of change detection analyses, especially for changes occurring within a closed conifer forest condition.

The existing literature on CVA falls short of discussing the complex nature of angle measurements in three or more spectral dimensions. As CVA magnitude is measured in terms of spectral Euclidean distance, its calculation is straightforward. Faced with calculating CVA angle, however, one has several choices. For just three spectral dimensions, x, y, and z, one can define an angle in x-y, x-z, and y-z space. Also, an angle can be calculated relative to one of the planes formed by these three axis pairs. For each additional spectral band, additional angle calculations are possible. In theory, only two angles in combination with magnitude are required to precisely locate a T_2 pixel vector vis-a-vis a T_1 vector in three dimensions; but as each angle provides a somewhat different view of the change data space, each angle likely contains different types of change information.

6.0 SUMMARY

Image differencing, change vector analysis (CVA), and composite analysis were compared for detecting changes in conifer forest cover using Landsat TM data. The concept of using a T_1 reference image in a two-date analysis was discussed, and results from use of a reference image were compared to results without the use of such an image. The importance of the tasseled-cap wetness feature in conifer forest change detection is evaluated. In a test of methods, composite analysis performed significantly better overall than either of the other two methods, with image differencing yielding slightly better results than CVA. Including a T_1 reference image with a two-date image difference data set is mathematically equivalent to composite analysis. Inclusion of a reference image with CVA, however, greatly improved the performance of CVA. Tasseled-cap wetness improved overall results when included with tasseled-cap brightness and greenness images, with the most marked improvement evident for

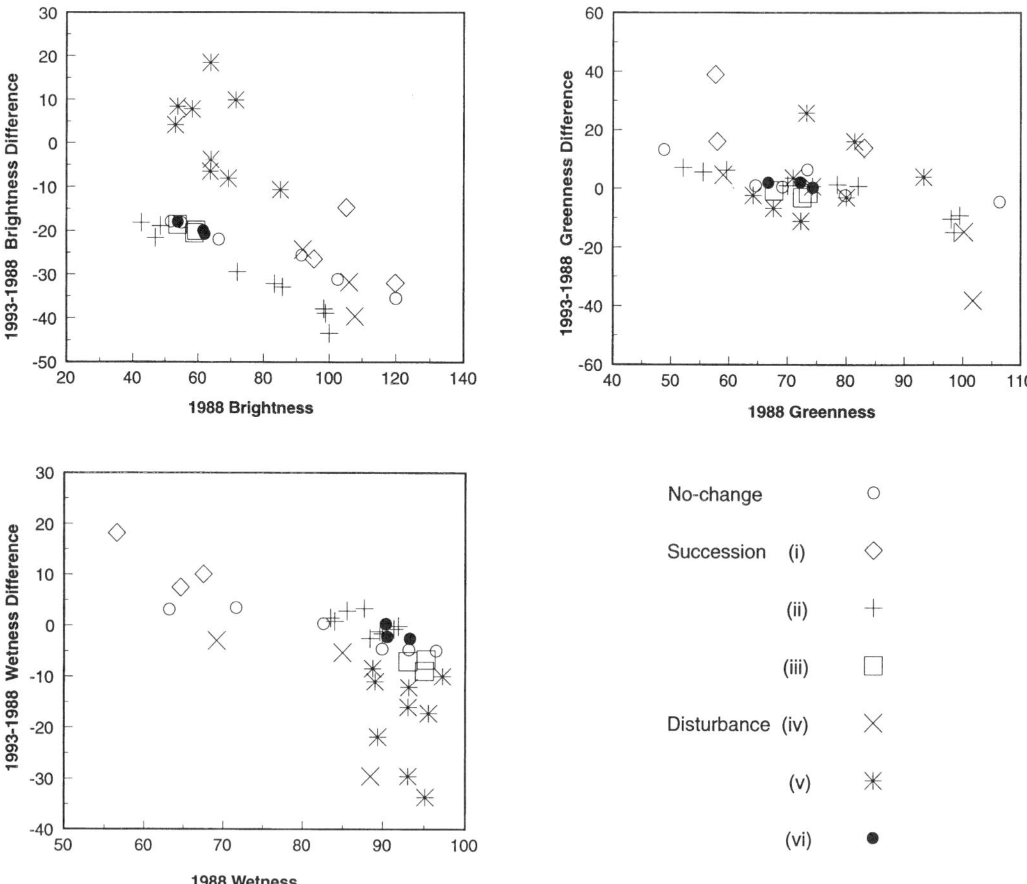

Figure 6.4. Mean temporal spectral differences for 36 forest cover change classes, as a function of original (1988 reference) mean spectral values for the Tasseled-Cap vegetation indices brightness (top, left), greenness (top, right), and wetness (bottom, left). The 36 cover change classes are grouped ecologically into seven classes: no-change, 3 succession classes, and 3 disturbance classes, as given in Table 6.2.

changes within the closed canopy conifer condition. CVA is an important change detection method that is inadequately described in the literature; in particular, the angular component of CVA appears not to be thoroughly appreciated. A discussion of CVA angle characterization with the intent of inspiring a closer evaluation of this potentially important change detection algorithm was presented.

ACKNOWLEDGMENTS

This research was funded in part by the Terrestrial Ecology Program, Office of Mission to Planet Earth, NASA (W-18,020), and by the National Science Foundation-sponsored H.J. Andrews Forest LTER Program (BSR 90-11663). We gratefully acknowledge stimulating discussions with Don Sachs, Rick Lawrence, Phil Sollins, and Bill Langford. A special thanks to Eileen Helmer and Tom Maiersperger for valuable feedback on a draft manuscript.

REFERENCES

Cohen, W.B. and T.A. Spies. Estimating structural attributes of Douglas-fir/western hemlock forest stands from Landsat and SPOT imagery. *Remote Sens. Environ.,* 41, 1–17, 1992.

Cohen, W.B., T.A. Spies, and M. Fiorella. Estimating the age and structure of forests in a multi-ownership landscape of western Oregon, U.S.A. *Int. J. Remote Sens.,* 16, 721–746, 1995.

Collins, J.B. and C.E. Woodcock. Change detection using the Gramm-Schmidt transformation applied to mapping forest mortality. *Remote Sens. Environ.,* 50, 267–279, 1994.

Collins, J.B. and C.E. Woodcock. An assessment of several linear change detection techniques for mapping forest mortality using multitemporal Landsat TM data. *Remote Sens. Environ.,* 56, 66–77, 1996.

Coppin, P. and M. Bauer. Digital change detection in forested ecosystems with remote sensing imagery. *Remote Sensing Reviews,* 13, 207–234, 1996.

Crist, E.P. and R.C. Cicone. A physically-based transformation of thematic mapper data—the TM tasseled cap. *IEEE Transactions on Geoscience and Remote Sensing,* GE-22, 256–263, 1984.

Franklin, J.F. and C.T. Dyrness. *Natural Vegetation of Oregon and Washington.* 2nd edition, Oregon State University Press, Corvallis, OR, 1988.

Franklin, S.E., R.H. Waring, R.W. McCreight, W.B. Cohen, and M. Fiorella. Aerial and satellite sensor detection and classification of western spruce budworm defoliation in a subalpine forest. *Canadian Journal of Remote Sensing,* 21, 299–308, 1995.

Fung, T. and E. LeDrew. Application of principal components analysis to change detection. *Photogrammetric Engineering and Remote Sensing,* 53, 1649–1658, 1987.

Howarth, P.J. and G.M. Wickware. Procedures for change detection using Landsat. *Int. J. Remote Sens.,* 2, 277–291, 1981.

Kauth, R.J. and G.S. Thomas. The tasseled cap—a graphic description of the spectral-temporal development of agricultural crops as seen by Landsat. In: *Proceedings, Second Annual Symposium on Machine Processing of Remotely Sensed Data*, Purdue University Laboratory of Applied Remote Sensing, West Lafayette, IN, June 6–July 2, 1976.

Lambin, E.F. and A.H. Strahler. Indicators of land-cover change for change-vector analysis in multitemporal space at coarse spatial scales. *Int. J. Remote Sensing,* 15, 2099–2119, 1994a.

Lambin, E.F. and A.H. Strahler. Change-vector analysis in multitemporal space: A tool to detect and categorize land-cover change processes using high temporal-resolution satellite data. *Remote Sens. Environ.,* 48, 231–244, 1994b.

Malila, W.A. Change vector analysis: an approach for detecting forest changes with Landsat. Proceedings Sixth Annual Symposium Machine Processing of Remotely Sensed Data, in: *Soil Information Systems and Remote Sensing and Soil Survey*, P.G. Burroff and D.B. Morrison, Eds., Purdue University Laboratory of Applied Remote Sensing, West Lafayette, IN, June 3–6, pp. 326–335, 1980.

Michalek, J.L., T.W. Wagner, J.J. Luczkovich, and R.W. Stoffle. Multispectral change vector analysis for monitoring coastal marine environments. *Photogrammetric Engineering and Remote Sensing,* 59, 381–384, 1993.

Muchoney, D.M. and B.N. Haack. Change detection for monitoring forest defoliation. *Photogrammetric Engineering and Remote Sensing,* 60, 1243–1251, 1994.

Price, K.P., D.A. Pyke, and L. Mendes. Shrub dieback in a semiarid ecosystem: The integration of remote sensing and geographic information systems for detecting vegetation change. *Photogrammetric Engineering & Remote Sensing,* 58, 455–463, 1992.

Schowengerdt, R.A. *Techniques for Image Processing and Classification in Remote Sensing.* Academic Press, New York, 1983.

Singh, A. Change detection in the tropical forest environment of northeastern India using Landsat. In: *Remote Sensing and Tropical Land Management*, M.J. Eden and J.T. Parry, Eds., John Wiley & Sons, New York, 1986, pp. 237–254.

Singh, A. Digital change detection techniques using remotely-sensed data. *Int. J. Remote Sensing,* 10, 989–1003, 1989.

Virag, L.A. and J.E. Colwell. An improved procedure for analysis of change in Thematic Mapper image-pairs. In: *Proceedings, Twenty-First International Symposium on Remote Sensing of Environment*, ERIM, Ann Arbor, MI, October 26–30, 1987, pp. 1101–1110.

Vogelmann, J.E. Detection of forest change in the Green Mountains of Vermont using Multispectral Scanner data. *Int. J. Remote Sensing*, 9, 1187–1200, 1988.

CHAPTER 7

Wildfire Detection with Meteorological Satellite Data: Results from New Mexico During June of 1996 Using GOES, AVHRR, and DMSP-OLS

Christopher D. Elvidge, Dee W. Pack, Elaine Prins, Eric A. Kihn,
Jackie Kendall, and Kimberly E. Baugh

1.0 INTRODUCTION

Wildfires are a primary force shaping terrestrial landscapes in terms of ecology (Whelan, 1995), hydrology (Scott and van Wyk, 1992), and nutrient cycling (Overby and Perry, 1996). In addition, fires play a critical role in the global carbon cycle by transferring carbon from the land surface to the atmosphere (Crutzen and Andreae, 1990). Even human health can be impacted by smoke released from wildfires (Lipsett et al., 1994).

While fires are a natural part of many ecosystems, they are actively suppressed or managed in the United States at a cost of approximately $600 million (the exact dollar amount varies substantially from year to year). Total national losses related to wildfires due to loss of lives, dwellings, facilities, and natural resources are much higher. Satellite remote sensing has been used successfully for tracking ecological impacts from fires and for the estimation of greenhouse gas emissions (e.g., Setzer and Pereira, 1991). These are applications which can be performed adequately by aggregating coarse resolution satellite imagery from sensors like the NOAA Advanced Very High Resolution Radiometer (AVHRR) through a burn season or through the analysis of high spatial resolution imagery (e.g., Landsat) acquired some time after the burn season. Timeliness in reporting, spatial resolution, and fire detection accuracy become limiting factors determining the benefits of satellite remote sensing for fire control and management (Rauste, 1996).

It has been known for more than 20 years that meteorological satellite sensors have an ability to detect wildfires (see Croft, 1973). These instruments include the NOAA Advanced Very High Resolution Radiometer (AVHRR), the NOAA Geostationary Operational Environmental Satellite (GOES), and the Defense Meteorological Satellite Program Operational Linescan System (DMSP-OLS). All were designed with the primary function of observing clouds for use in weather prediction. None were designed with any expectation that the data would be used for fire detection. An ardent student of satellite fire detection could certainly design sensors far more capable of detecting and characterizing fires than the currently available meteorological systems (e.g., Levine et al., 1996). However, several features of the meteorological satellites sensors make them a practical choice for near real-time satellite detection of active fires: (1) multiple daily coverages, (2) low data cost, (3) reasonably low data volume, (4) potential for real-time products, and (5) decentralized data access via direct broadcast (only the DMSP direct broadcast is encrypted). The logic behind use of meteorological satellite data for fire detection becomes apparent if you consider the prospects for using Landsat data for the same task: (1) coverage repeats once per 16 days at the equator, (2) high data cost, (3) high data volume, (4) little potential for real-time products, and (5) high potential for downtime due to lack of backup satellite.

The major source of economy in the meteorological satellite systems comes from the relatively coarse spatial resolution, measured in kilometers rather than meters. Fires are so unusual in terms of their spectral emissions that it is possible to detect fires that are much smaller than the pixel size.

Over the years there have been several attempts made to use NOAA-AVHRR satellite data for the early detection of fires in the western U.S. (e.g., Prevendel, 1995); however, each of these attempts has failed to satisfy the end users within the U.S. Forest Service and Bureau of Land Management.

Perhaps the expectations placed on the fire detection capabilities of the AVHRR have been too high? In fact, the fire detection capabilities of AVHRR, GOES and DMSP-OLS have not been well defined in terms of minimum fire detection capabilities and probability of detection. There are currently several teams of investigators developing and testing algorithms for detecting fires with the meteorological sensors. However, these groups tend to work independently rather than focusing their studies on a common experiment.

We report on the detection of active fires using meteorological satellite data during June of 1996 for a study site in southwest New Mexico. This test was a cooperative effort between multiple federal and nonfederal participants. The purpose of the experiment has been to define the relative fire detection capabilities of the systems and to determine what synergy might be obtained through combined use of the meteorological assets.

2.0 EXPERIMENT DESCRIPTION

A fire detection test area was established within New Mexico in a box spanning the region from 32–34.5°N and 106–109°W (see Figure 7.1). This area included the Gila National Forest and Wilderness area which encompass 3.3 million acres of juniper and pinyon pine forest, combined with mixed rangeland areas of forest and grass savanna and grasslands. Elevations within the National Forest range from 4,500 ft in the desert lowlands, to almost 11,000 ft at the crest of the mountain Whitewater Baldy. Other Forest Service lands which overlap the test area are portions of the Apache and Cibola National Forests. Also included within the defined test area are large private land holdings, many of which are grassland, as well as the vast federally owned areas of White Sands Missile Range (WSMR) and White Sands National Monument. A time period from June 3, 1996 to June 18, 1996 was chosen for the two-week test. These dates often are a period of peak lightning activity which historically has given rise to many fires.

To acquire data suitable for intersystem comparison and algorithm testing, data from all satellites capable of fire detection were recorded during the two-week test period. Data were gathered from the following meteorological satellite sensors: (1) NOAA Geostationary Operational Environmental Satellites (GOES), sensor - imager, (2) NOAA Polar Orbiting Environmental Satellites (POES), sensor - Advanced Very High Resolution Radiometer (AVHRR), (3) Defense Meteorological Satellite Program (DMSP) spacecraft, sensor - Operational Line Scan-

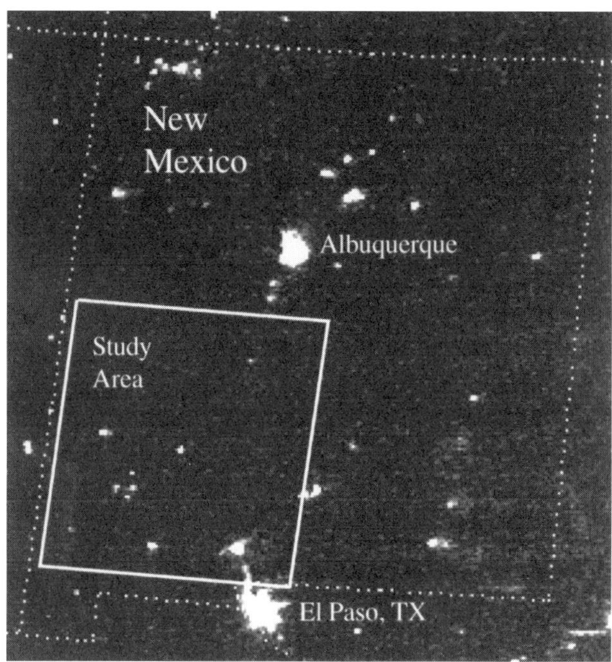

Figure 7.1. Location of the study area in southwest New Mexico outlined on a DMSP-OLS nighttime image acquired June 11, 1996.

ner (OLS). In addition to the satellite data, Forest Service fire reports were archived by the staff at the Gila National Forest for later comparison to space-based fire detection results.

The beginning and end of the two-week period was extremely cloudy. To reduce the analysis effort to the period least impacted by clouds, a subset of the test period was specified for a detailed intersensor fire detection comparison. The days targeted for comparison between the satellites for the New Mexico experiment were June 9–13, which corresponded to the peak period of fires and lowest level of cloud cover during the study period. Only fires with final burn sizes of at least 10 acres were actively tracked in the intercomparison, though fires of smaller size were occasionally detected. A total of 10 fires met this size criteria. The names, locations, initial report times, control or out times, and final burn sizes of the 10 fires are listed in Table 7.1.

Given that the GOES, AVHRR, and DMSP satellite data were processed in multiple locations using different software systems, it was not possible to generate products using a consistent map projection. Instead, each of the groups processing satellite data were requested to provide images showing fire detects, cloud cover, USFS fire locations, and an outline of the study area for each satellite pass of interest. By standardizing the output features contained in the output products, it was possible to compare results from the three systems.

3.0 SENSOR CHARACTERISTICS AND APPLICATION OF FIRE DETECTION ALGORITHMS

Table 7.2 compares the instrument characteristics and orbital parameters from the different satellite sensors.

Table 7.1. U.S. Forest Service Fire Reports.

Fire Name	Start Date & Time	Location	Control Date & Time	Final Size (acres)
Four-Five	06/07/96 08:40	32.95, –108.19	06/08/96 01:00	5
Ring/Hells	06/07/96 20:57	33.28, –108.40	06/15/96 00:32	365
Tadpole	06/07/96 00:55	33.02, –108.36	06/09/96 18:05	60
Lookout	06/08/96 17:59	33.22, –108.60	06/14/96 20:00	100
Pelona	06/08/96 20:04	33.67, –108.12	06/09/96 18:00	60
Montana	06/08/96 23:31	33.70, –107.86	06/11/96 00:00	250
San Pedro	06/08/96 00:00	33.91, –106.85	06/16/96 06:00	5900
Black	06/09/96 20:47	33.28, –108.52	06/11/96 00:00	30
Saddle	06/09/96 21:05	33.26, –108.71	06/13/96 22:00	5
Percha	06/10/96 20:15	32.96, –107.75	06/13/96 18:00	12
Brother	06/10/96 20:20	32.87, –107.73	06/15/96 17:47	50
Cabin Springs	06/10/96 20:50	32.93, –108.12	06/12/96 19:40	30

3.1 GOES

In April 1994, the first in the series of NOAA's next generation of geostationary satellites (GOES-8) was launched, providing a new capability for diurnal monitoring of subpixel fire activity associated with wildfires, prescribed burns, and agricultural applications throughout the Western Hemisphere (Menzel and Prins, 1996). For the first time, the GOES-8 imager makes it possible to monitor diurnal variability in fire intensity by providing multispectral imagery every 15 minutes over the continental United States and half-hourly elsewhere at a much higher spatial resolution. The visible band on GOES-8 is available at 1 km resolution, the 3.9, 10.7, and 12.0 μm bands have a spatial resolution of 4 km, and the 6.7 μm band has a spatial resolution of 8 km. In addition, the saturation temperature (Tmax) in the 3.9 mm band is set at 338 K for GOES-8, which is 18 K higher than on previous GOES satellites and results in fewer saturated fire pixels.

Detection of active fires with GOES data relies on the enhanced emission of thermal radiation in the 3.9 μm region relative to the 11.2 μm region (Weaver and Purdom, 1995). The GOES data of New Mexico were processed by Elaine Prins of NOAA/NESDIS at the UW-Madison Cooperative Institute for Meteorological Satellite Studies (CIMSS). The GOES Automated Biomass Burning Algorithm (ABBA) was originally developed at CIMSS to locate subpixel size fires and to estimate temperature and areal extent in an effort to determine trends in biomass burning throughout the Amazon Basin during the 1980s utilizing the GOES Visible Infrared Spin Scan Radiometer (VISSR) Atmospheric Sounder (VAS) archive. The GOES ABBA is a dynamic multispectral thresholding algorithm which utilizes regional thresholds derived from the satellite data (visible, 4, and 11 μm bands) to screen out cloud contamination and to locate fire pixels. The algorithm is based on the sensitivity of the 4 μm band to high temperature subpixel anomalies and is an expanded version of a technique originally developed by Matson and Dozier (1981) for NOAA AVHRR. The shortwave region is especially useful in detecting subpixel fire activity, since the peak of the Planck function shifts toward shorter wavelengths as temperature increases, causing the radiance to increase more rapidly at 4 mm than 11 mm. With temperatures in the 500 to 900 degree centigrade range, fires can be located by identifying anomalously high temperature sources in the shortwave IR window. Once the GOES ABBA locates a fire pixel it incorporates ancillary data to correct for water vapor at-

Table 7.2. Sensor Characteristics.

GOES - Imager					
Spectral Bands (μm):	0.55–0.75	3.80–4.00	6.50–7.00	10.20–11.20	11.50–12.50
Nadir Footprint (km):	1 x 1 km	4 x 4 km	8 x 8 km	4 x 4 km	4 x 4 km
Measurement Range:	1.6-100%A*	4-338 K	4-320 K	4-320 K	4-320 K
Signal Quantitization:	10 bit (all channels)				
Satellite Orbits:	Geosynchronous, 135° W and 75° W				
AVHRR - LAC and HRPT					
Spectral Bands (μm):	0.56–0.68	0.73–1.1	3.55–3.93	10.3–11.3	11.5–12.5
Nadir Footprint (km):	1 x 1 km	1 x 1 km	1 x 1 km	1 x 1 km	1 x 1 km
Measurement Range:	1.6-100%A*	4-325 K	4-325 K	4-325 K	4-325
Signal Quantitization:	10 bit (all channels)				
Satellite Orbits:	Polar, 98.86° inclination, 7:30 am and 1:40 pm equatorial crossings				
OLS - Normal Daytime Operation					
Spectral Bands (μm):	0.40–1.10	10.0–13.4			
Nadir Footprint (km):					
smoothed data	2.75 x 2.75 km	2.75 x 2.75 km			
real-time data	0.55 x 0.0.55 km	0.55 x 2.75 km			
Measurement Range:	1.6-100%A*	4-310 K			
Signal Quantitization:	6 bit	8 bit			
OLS - Normal Nighttime Operation					
Spectral Bands (μm):	0.47–0.95	10.0–13.4			
Nadir Footprint (km):					
smoothed data	2.75 x 2.75 km	2.75 x 2.75 km			
real-time data	0.55 x 2.75 km	0.55 x 0.0.55 km			
Measurement Range:	0-64 counts	4-310 K			
Signal Quantitization:	8 bit	6 bit			
Satellite Orbit:	Polar, 98.8° inclination, 05:30 am and 09:30 am equatorial crossings				

* A = Albedo

tenuation, surface emissivity, and solar reflectivity and then uses numerical methods to solve for subpixel size and mean fire temperature. The GOES ABBA is described in more detail in Prins and Menzel (1992; 1994).

Multiyear (1983, 1988, 1989, 1991, 1994, and 1995) applications of the GOES ABBA in South America have indicated the unique ability of the geostationary platform to monitor diurnal signatures in fire activity as well as regional variability in fire activity from year to year. Over the past three years CIMSS scientists have used the GOES-8 imager to detect and monitor diurnal variations in fire activity throughout North, Central, and South America including numerous wildfires in the forests of the United States and Canada; the Long Island fire in New York; wildfires in Texas, Oklahoma, and Kansas; agricultural burning in the Yucatan Peninsula in Mexico, Guatemala and Belize; and fires associated with deforestation and agricultural applications in South America (Menzel and Prins, 1996; Prins and Menzel, 1996; Bywaters and Prins, 1996).

The GOES ABBA was recently redesigned to take advantage of the improved capabilities of the GOES-8 instrument. The new GOES-8 ABBA enables monitoring throughout South America and includes an expanded surface vegetation classification scheme and associated 4 and 11 mm emissivity factors. It incorporates GOES-8 visible data in an effort to screen subpixel cumulus cloud contamination. In addition, NMC model estimates have been incorporated into the algorithm to account for water vapor attenuation in the 4 and 11 mm bands (Prins et al., 1997). The GOES ABBA is currently being adapted at CIMSS for operational use in North America.

Although limited in scope, validation of the GOES-8 ABBA was done in association with the NASA-sponsored Smoke Clouds and Radiation Experiments in California (SCAR-C, 1994) and Brazil (SCAR-B, 1995). In both experiments, half-hourly GOES-8 ABBA results were compared with ground truth measurements made by the USFS (Menzel and Prins, 1996; Prins et al., 1997). GOES-8 ABBA results were within 20 percent of ground truth observations made by the USFS for a 48-acre prescribed burn in the Pacific Northwest. The minimum size detected by GOES-8 for this fire was approximately 10 acres. During SCAR-B the GOES-8 ABBA detected and monitored a USFS prescribed burn of 2–3 acres in size. More validation studies are anticipated.

3.2 NOAA-AVHRR

The Advanced Very High Resolution Radiometer (AVHRR) is a scanning radiometer on board NOAA's Polar Orbiting Environmental Satellites (POES) which measures reflected and emitted radiation at approximately 1-kilometer spatial resolution in five channels ranging from 0.5 to 12.5 mm. Data from POES-AVHRR have been used for biomass burning analyses for more than a decade. The potential for identifying actively burning fires with AVHRR was first documented in a theoretical paper by Jeff Dozier in 1981. Dozier showed that subpixel high temperature targets can be identified in the shortwave infrared data (channel 3 in AVHRR) because these features have greater effect in the 3.5–4.0 μm wavelength band than in the longer wave infrared bands (Dozier, 1981). Using Dozier's principles, a number of studies have demonstrated the capability of AVHRR to identify industrial high temperature sources and map fire activity (Matson and Dozier, 1981; Muirhead and Cracknell, 1984; Matson et al., 1984, 1987). Successful automatic procedures for hot spot detection have been developed to enable operational monitoring of active fires with AVHRR data (Flannigan and Vonder Haar, 1986; Lee and Tag, 1990). Examples of operational AVHRR fire products which are being used to characterize seasonal and interannual fire activity exists for several regions including South America and Africa (Setzer and Pereira, 1991; Kennedy et al., 1993; Arino et al., 1993; Justice and Malingreau, 1996).

While the presence of fires with spatial characteristics well below the full resolution of the POES system have been identified in data from its infrared sensors, detection of subpixel fires is nevertheless problematic. In general, the active fire detection algorithms proposed for operational use with the POES satellites have been designed and fine-tuned for use at the regional scale. A summary of the contributions and limitations of POES-AVHRR active fire detection may be found in several articles including Langaas (1995), and Dowty (1996). Since 1992, the IGBP-DIS Fire Algorithm Working Group (FWG) has been reviewing current satellite fire detection and laying the groundwork for global fire processing with AVHRR data from the IGBP-DIS Global 1 km Program (Justice and Dowty, 1994). A community consensus on AVHRR active fire detection appropriate for large-scale application has been reached through the FWG (Justice and Malingreau, 1996). Slightly different versions of the proposed algorithm have been successfully applied to AVHRR data for Southern Africa (Justice et al., 1996) and Central Africa (Flasse and Ceccato, 1995) and GOES data for South America (Prins and Menzel, 1992). Following recommendations of the FWG, recent work at Goddard Space Flight Center was undertaken to establish the limits of detectability of active fires in AVHRR under different environmental conditions and to evaluate the performance of several proposed algorithms proposed for global application (Giglio et al., 1997).

The contextual fire algorithm for 1-kilometer AVHRR data implemented at the Goddard Space Flight Center was first proposed by Justice et al. during the initial IGBP-DIS Fire Algorithm Workshop in February, 1991. This algorithm, which has been documented in several different forms (Justice and Dowty, 1994; Justice et al., 1996) was originally developed for use in tropical biomes. During the last two years, however, the algorithm has been modified for global application (Giglio et al., 1997). Described below is the most recent version of the algorithm which was applied to the AVHRR data for New Mexico provided by the University of Colorado at Boulder for use in the NOAA intersensor fire comparison study. A detailed evaluation of this and other AVHRR active fire detection algorithms, including implications for global application, using both simulated and actual AVHRR data may be found in a recent paper by Giglio et al. (1997).

The algorithm begins with a no-fire prescreen, used primarily to reduce processing time by eliminating obvious nonfire pixels from additional processing. T3 = brightness temperature in Band 3 (3.9 micrometers), T3–4 = difference in Band 3 brightness temp and Band 4 (11.0 micrometers) brightness temp (i.e., T3 – T4), r2 = Rho (reflectance, 0.25 = 25 percent) in Band 2 (near-IR, 0.86 micrometers). A pixel is flagged as a potential fire pixel if $T_3 > 310$ K, $T_{3\text{-}4} > 6$ K, r2 < 0.25 ; otherwise it is classified as *nonfire*.

If the pixel has been identified as a potential fire, an attempt is made to use the neighboring pixels to estimate the brightness temperatures of the potential fire pixel in the absence of fire. Valid background pixels in a window centered on the potential fire pixel are flagged, where a valid background pixel is defined as those pixels that are (1) land (i.e., neither cloud nor water identified with external cloud and water masks), and (2) are not potential background fire pixels. Potential background fire pixels are those background pixels having $T_3 > 318$ K and $T_{3\text{-}4} >$ 12 K. This test is intended to prevent fire pixels in the background from biasing the background brightness temperature estimates.

The window starts as a 5 by 5 pixel square ring about the potential fire pixel (the 8 pixels surrounding the potential fire pixel are excluded) and increases up to 21 by 21 as necessary, until at least 25 percent of the background pixels in the window are valid, and the number of valid background pixels is at least 6. If an insufficient number of valid background pixels are identified, the pixel is classified as *unknown*, otherwise the following statistics are computed: $T_{3\text{-}4B}$ (mean $T_3 - T_4$), $d_{3\text{-}4B}$ (mean absolute deviation of $T_3 - T_4$), T_{4B} (mean T_4), and d_{4B} (mean absolute deviation of T_4). Let DT represent the larger of 2.5 $d_{3\text{-}4B}$ and 4 K. If $T_{3\text{-}4} > T_{3\text{-}4B} + DT$ and $T_4 > T_{4B} + d_{4B} - 3$ K, the potential fire pixel is classified as a *fire*, otherwise it is classified as *nonfire*.

3.3 DMSP-OLS

The DMSP maintains a constellation of two satellites in sun-synchronous, near-polar orbit at altitudes of approximately 833 km, an inclination of 98.8 degrees, and an orbital period of 102 minutes. One satellite is in a dawn-dusk orbit, the second in a day-night orbit. The DMSP platforms are three axis stabilized, with roll, pitch, and yaw variations kept to within ±0.01 degrees. It has been known since the early 1970s that fires can be detected in the nighttime visible band data generated by the Operational Linescan System (OLS) (Croft, 1973 and 1979). In 1992 a joint U.S. Air Force, NOAA, and Aerospace Corporation effort led to the inception of a digital DMSP archive at the NOAA National Geophysical Data Center.

The OLS is an oscillating scan radiometer designed for cloud imaging with two spectral bands (VIS and TIR) at 2.7 km pixel resolution and a swath of 3000 km. The "VIS" bandpass straddles the visible and near-infrared portion of the spectrum. The TIR band spans the 10.3–12.9 μm region. The TIR band is calibrated using an onboard blackbody source and views of deep space to provide 8 bit data with a temperature range of 190 to 310 degrees Kelvin, ideal for detecting and characterizing clouds. The wide swath widths provide for global coverage four times a day: dawn, day, dusk, night.

The VIS band signal is intensified at night using a photomultiplier tube (PMT), making it possible to detect faint VNIR emission sources. The PMT system was implemented to facilitate the detection of clouds at night using the visible band. With sunlight eliminated, the light intensification results in a unique data set in which city lights, gas flares, and fires can be observed. The first systematic inventory of fires with OLS data was accomplished by Cahoon et al. (1992), who manually digitized fire points from film produced from nighttime OLS orbits over Africa. More recently NGDC has developed a digital algorithm for the detection of fires with nighttime OLS data (Elvidge et al., 1996). This algorithm screens VNIR emissions observed on a single night against a known set of spatially stable light sources. The stable lights data set, indicating locations of cities, towns, gas flares, and industrial sites is derived by identifying persistent light sources in an OLS time series. Kihn (1996) developed a method for estimating subpixel burn size with OLS data.

There are three primary spatial resolution modes for OLS data (Figure 7.2). The instrument acquires data with 0.55 km resolution. Due to constraints in the onboard data storage capacity, OLS data are "smoothed" by averaging five by five blocks of pixels prior to recording for playback to a centralized receiving station. It is possible to acquire limited quantities of the 0.55 km "fine" resolution data. The DMSP data are also broadcast in real time and can be obtained using a direct readout station. Telemetry constraints on the direct readout systems do not permit both bands of the OLS to be broadcast at full resolution. Instead, one band is broadcast at full resolution, the other band is broadcast following the alongscan averaging of five pixel blocks, creating pixels which are 0.55 km wide and 2.7 km long. Typically the VIS band is broadcast at the fine spatial resolution during the day and the thermal band is broadcast with fine spatial resolution at night.

For the New Mexico experiment, two types of OLS data were analyzed. NOAA-NGDC analyzed 2.7 km resolution data received from the Air Force Global Weather Central. Analysis of the direct readout data, received at Los Angeles Air Force Base, was performed by the Aerospace Corporation. The direct readout data had fine resolution thermal band data and the 0.55 by 2.7 km resolution data in the visible band. NGDC processed the smoothed OLS data using the fire detection algorithm described by Elvidge et al. (1996). The direct readout data was analyzed largely using visual methods.

4.0 RESULTS

Each of the groups who analyzed meteorological satellite data were asked to produce images for the June 9–13 time period, indicating the location of cloud cover, the location of the USFS reported fires, and the location of fire detects. From these products, charts were prepared indicating the result of the satellite data observation for each of the analyzed overpasses as either: (1) fire detected, (2) cloud covered and no fire detected, or (3) clear and no fire detected. The charts indicate time on the horizontal axis. The known duration of each fire is shown with a horizontal line indicating the initial report time and the control or out time. The total estimated burn area is indicated in parentheses following the name of the fire (see interpretation key in

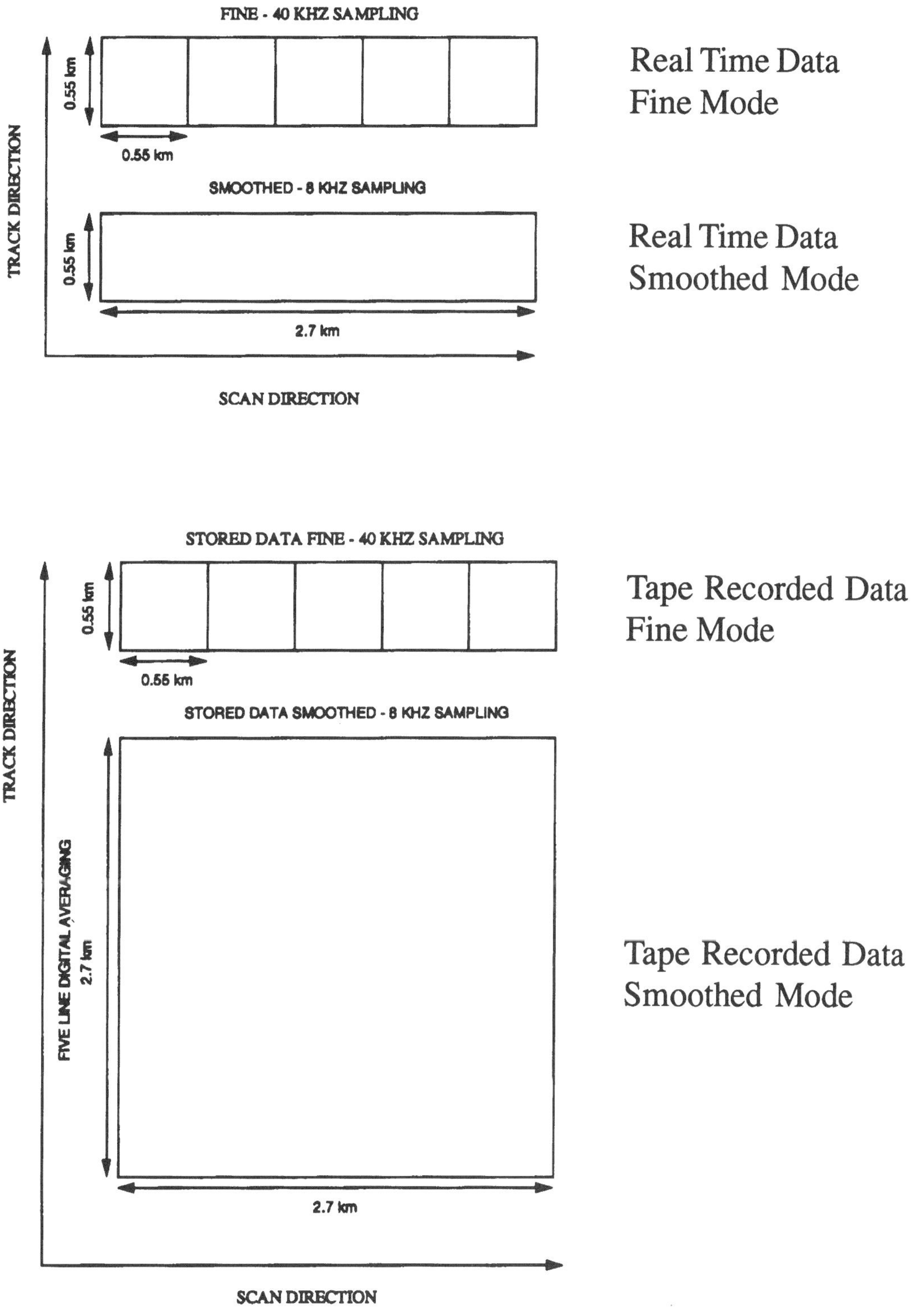

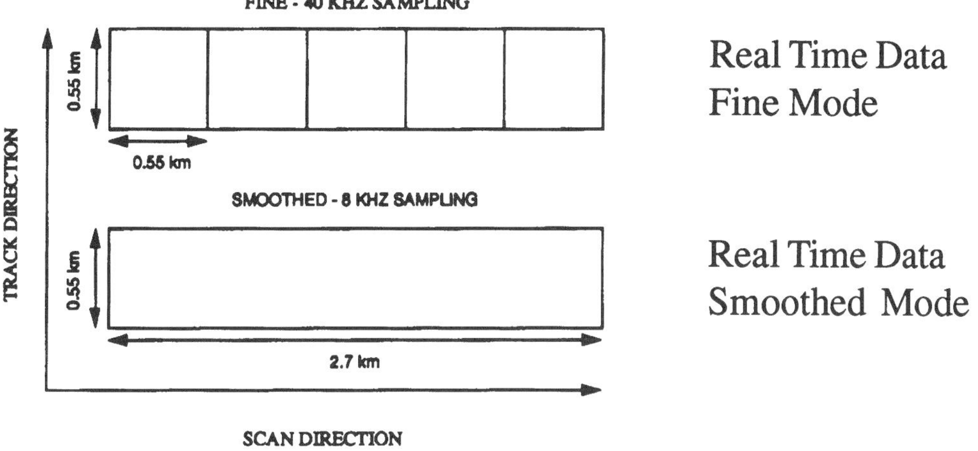

STORED DATA FINE - 40 KHZ SAMPLING

0.55 km

0.55 km

Tape Recorded Data
Fine Mode

STORED DATA SMOOTHED - 8 KHZ SAMPLING

TRACK DIRECTION

FIVE LINE DIGITAL AVERAGING
2.7 km

Tape Recorded Data
Smoothed Mode

2.7 km

SCAN DIRECTION

Figure 7.2. Comparison of the three spatial resolution modes available from the DMSP-OLS. Note that the real-time fine data and tape recorded fine data are equivalent.

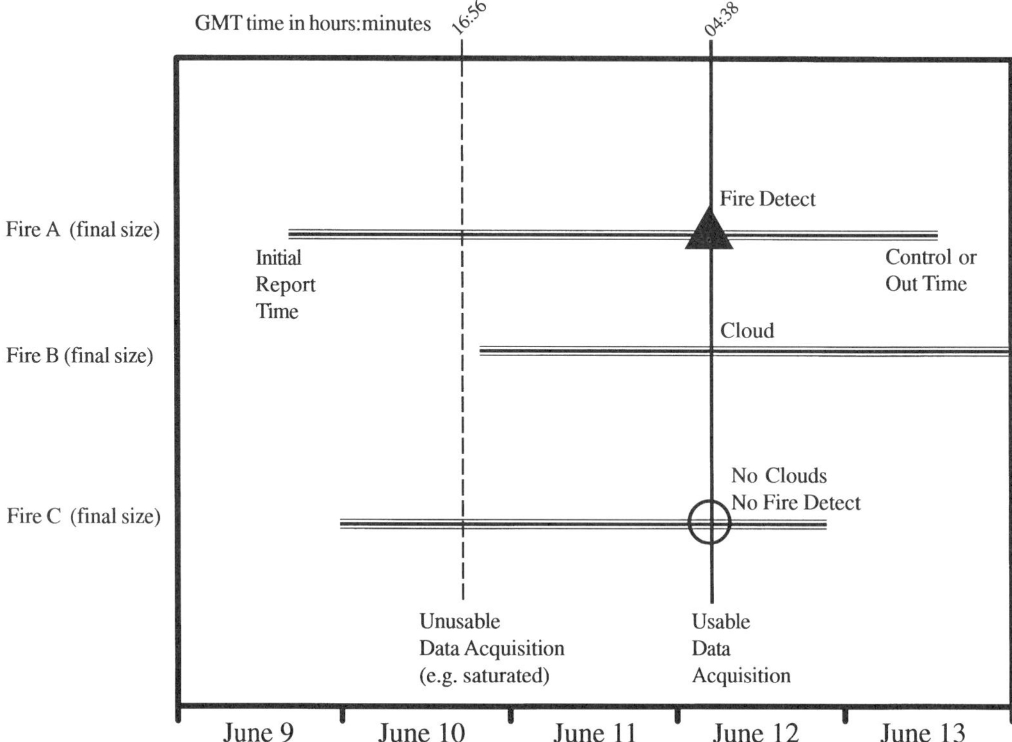

Figure 7.3. Key to the sensor intercomparison figures.

Figure 7.3). The relative fire detection capabilities for each system have been estimated using the following formula:

$$\text{Fire detection success rating} = \frac{\text{Number of fire detects}}{\text{Fire detects plus misses}} \times 100$$

The results obtained with the three sensors are summarized below.

4.1 GOES

Approximately 250 half-hourly multispectral GOES imagery (June 9–13, 1996) were processed at UW-Madison/CIMSS with an initial version of the North American GOES-8 ABBA. Throughout the study period the GOES-8 ABBA identified the San Pedro fire 27 times, while manual inspection of the GOES-8 data revealed 5 additional times with a noticeable fire signal in the 4 μm imagery. The Lookout fire was reported once by the GOES-8 ABBA on June 12, at 06:15 UTC with an additional 2 manual detections. The Black fire was identified by the GOES-8 ABBA on 15 occasions on June 11–13, which was after the USFS- reported end time of the fire. There were an additional 6 manual detections of the Black fire in the GOES-8 imagery. A smoldering/low intensity fire was identified by the GOES-8 ABBA at the location of the USFS-reported Saddle fire at 06:15 UTC on June 12, 1996. The GOES-8 ABBA also identified several other fires in the study regime during this time period. From 1700 to 2300 UTC surface

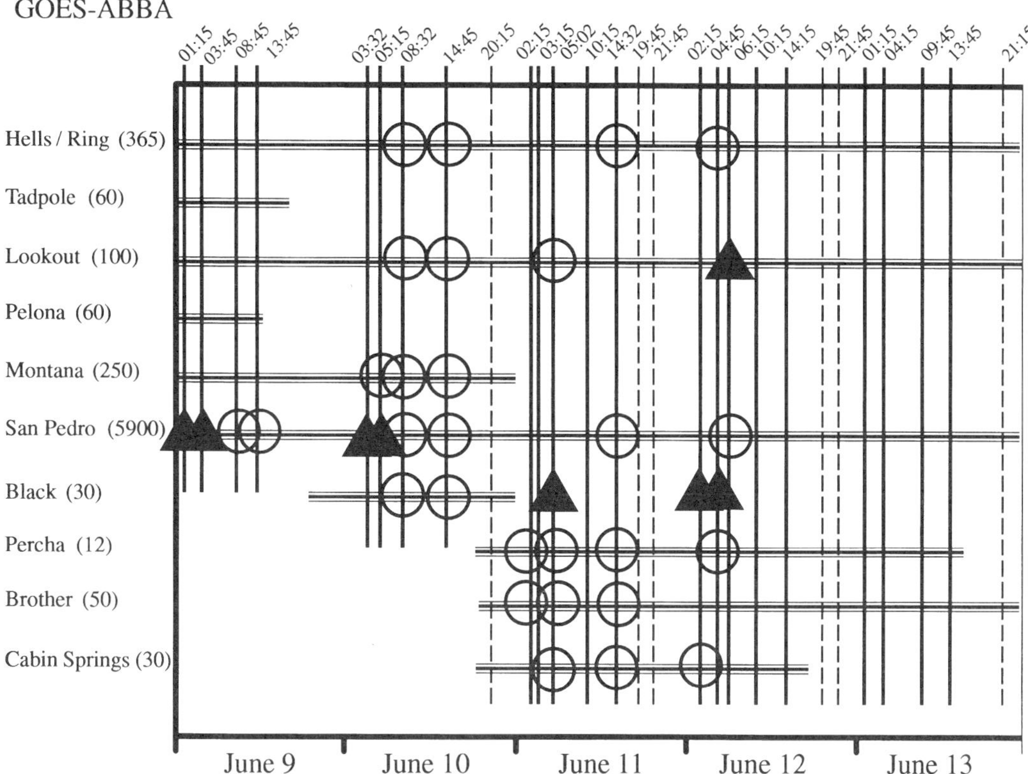

Figure 7.4. Fire detection results from GOES during the June 9 to 13, 1996 period.

heating and solar reflectance in the 4 um band made it difficult for the GOES-8 ABBA to separate fire activity from the hot background. Figure 7.4 shows the GOES-8 ABBA fire reports for those times that bracket the times of the DMSP and AVHRR data acquisitions. Only three of the 10 fires were detected with GOES for these selected times. The San Pedro fire was detected four times, but was also missed on 6 cloud-free data acquisitions. The Lookout fire was detected once and the Black fire was identified 3 times. With 8 fire detects and 28 misses, GOES ABBA had a fire detection success rating of 22 percent. Although it was only detected once, the smallest fire identified by the GOES ABBA was the Saddle fire, which had a final burn size of 5 acres.

4.2 AVHRR

A total of 20 AVHRR overpasses acquired of the study site between June 9 and June 13 were included in the intercomparison. However, thermal band data for six of these passes were unusable due to large areas of saturated data. AVHRR detected the Hells/Ring fire complex six times, Tadpole once, Lookout three times, San Pedro once, and Black once (Figure 7.5). With a total of 12 fire detects and 20 misses, AVHRR had a fire detection success rating of 37.5 percent. The smallest fire detected by AVHRR was the Black fire, which had a final burn size of 30 acres.

A notable result from AVHRR was the detection of an unreported fire on the White Sands Missile Range during four AVHRR passes acquired on June 5. A second unreported fire was

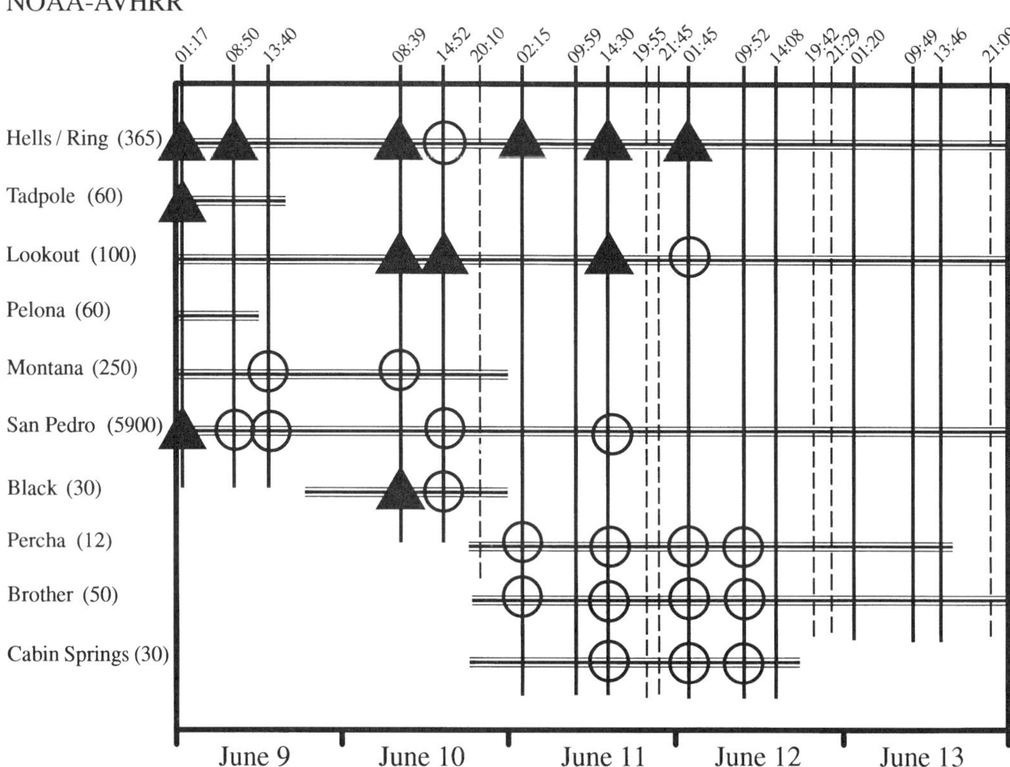

Figure 7.5. Fire detection results from the NOAA-AVHRR during the June 9 to 13, 1996 period.

detected twice with AVHRR data in a grassland area near the center of the study area (June 10 and 13). The Black fire was detected in the first available AVHRR overpass.

4.3 DMSP Smoothed

Twelve DMSP-OLS smoothed data sets from satellite F-12 were acquired and examined. Of these, five were daytime passes which were unsuitable for fire detection. Lunar illumination for the nighttime passes was 45 percent of the maximum on June 9 and gradually declined to 19 percent on June 13. From the seven nighttime passes the Hells/Ring fire complex was detected four times, Lookout fire was detected twice, San Pedro and Black fires were detected three times each, and the Percha and Cabin Springs fires were each detected once (Figure 7.6). With a total of 14 fire detects and seven misses, DMSP had a fire detection success rating of 66 percent.

The smallest fires detected with the DMSP smoothed data were the Saddle and Four Five fires, both with 5 acres burned. Regarding the early detection of fires, the Percha and Cabin Springs fires were detected on the second overpass after the fires started. The Saddle fire was detected in OLS data acquired on June 12 at 04:38, four hours prior to the USFS's initial report of the fire. The unreported fire detected with AVHRR data at WSMR on June 5 was detected the same night with DMSP-OLS data. In addition, the unreported grassland fire detected by AVHRR near the center of the study area on June 10 and 13 was detected with OLS data on June 11. Interestingly, the Black fire continued to be detected in OLS data for three nights in a row after it had been reported to be under control or out.

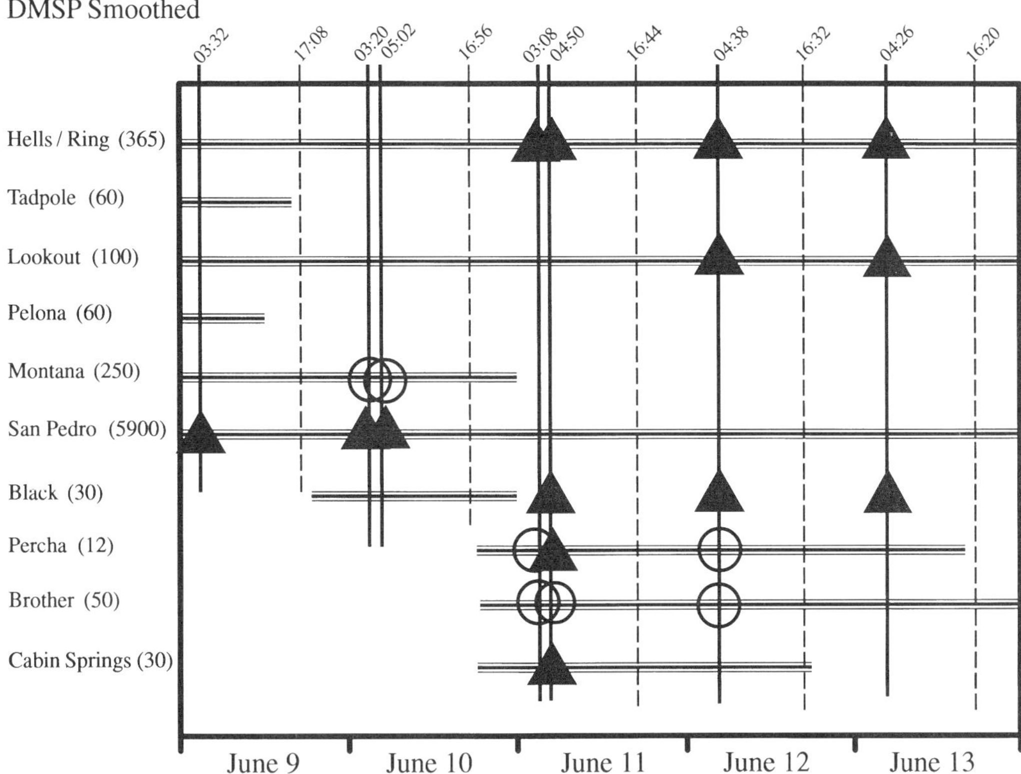

Figure 7.6. Fire detection results from DMSP-OLS smoothed data during the June 9 to 13, 1996 period.

4.4 DMSP Direct Readout

Due to constraints in the level of effort afforded to the processing of the DMSP direct readout data, only a small set of the total number of scenes recorded were actually analyzed. The focus of the analysis was to examine the potential use of fine resolution thermal data for fire detection and to compare visible band direct readout fire detection against results obtained by NGDC with the smoothed data.

Fine resolution (0.55 km) thermal band OLS data was examined for several fires which were detected with the nighttime visible band data. The results indicate that the OLS thermal band data, even in the fine spatial resolution mode, is of limited value in fire detection. As an example of the results found in New Mexico, Figure 7.7 shows the observed visible and thermal band values for a transect over a fire detected on June 5, 1996 at the White Sands Missile Range. Visible band counts outside the fire are in the 5–10 count range, but saturate at values of 63 over the fire. There is a slight thermal spike in the area of the fire. However, many parts of the scene, including areas on the right-hand side of the transect, have comparable or higher thermal band readings.

In comparing the visible band fire detection results between the smoothed and direct readout sources, one might expect that the direct readout data with the 0.55 by 2.7 km pixels would detect substantially smaller fires than the smoothed data with 2.7 km pixels. In examining simultaneously acquired smoothed and direct readout OLS data in New Mexico from June 5,

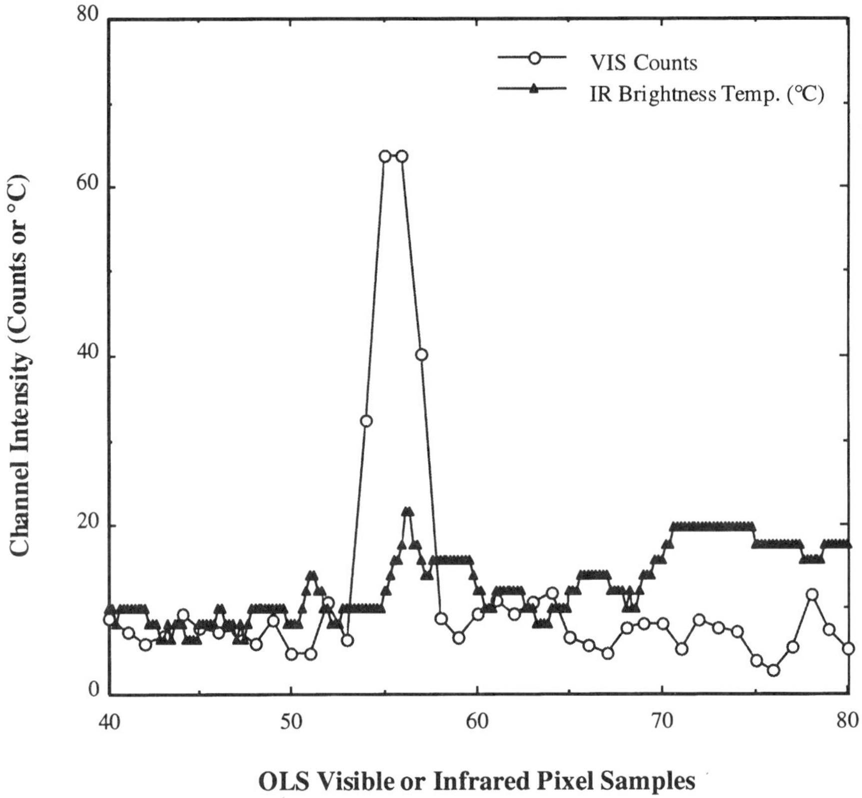

Figure 7.7. Transect of visible and thermal band values observed for the June 5, 1996 fire shown in Figure 7.2.

11, and 18 it was found that the smoothed data detected the same fires as the direct readout data. However, the appearance of the fires was distinctly different (Figure 7.8), with the direct readout data exhibiting a "boatdock" effect produced from the shape of the pixels and the alternating directions of the individual scans.

5.0 DISCUSSION

Each of the three systems has a documented capability to detect wildfires. Below is a discussion of the advantages and limitations noted for each of the meteorological satellite systems:

5.1 GOES

The GOES sensor has the advantage of producing large numbers of observations on a daily basis (90 +). Fire detection with GOES-8 can be accomplished with data acquired at any time of day or night, although in New Mexico the utility of the data was limited in some regions due to surface heating and solar contamination in the 4 mm band throughout the afternoon hours. In the New Mexico experiment, diurnal GOES-8 data detected the Lookout fire the same number of times as the AVHRR instrument and provided many more detections of the San Pedro and Black fires than the AVHRR and DMSP. The GOES-8 instrument did not detect the Hells/

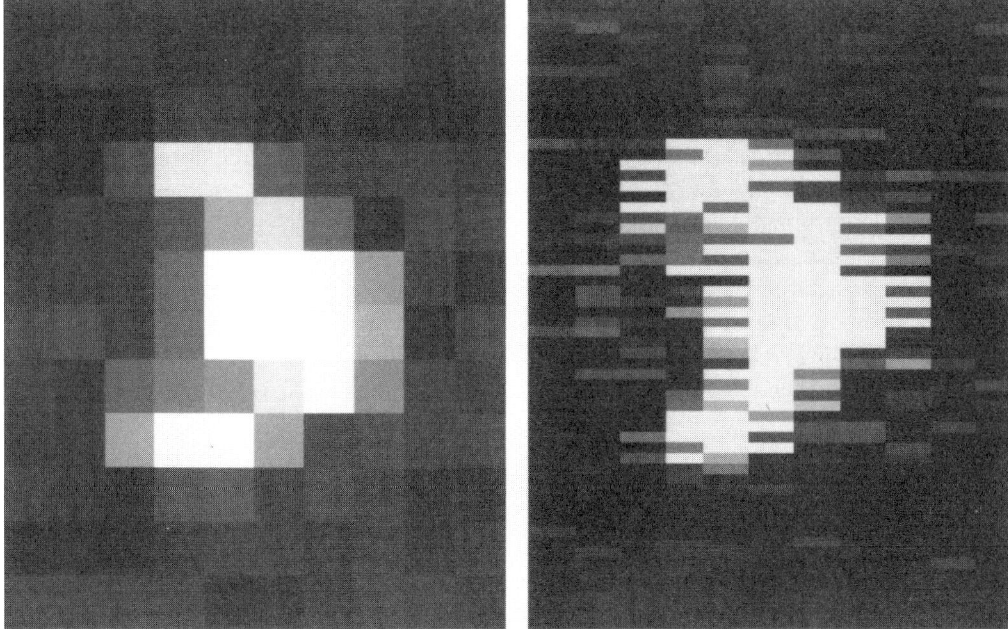

Figure 7.8. Comparison of DMSP-OLS smoothed (2.7 km) versus direct readout (2.7 x 0.55 km) nighttime visible band data from the June 5, 1996 fire on White Sands Missile Range.

Ring, Tadpole, Percha and Cabin Springs fires which were observed by the DMSP and/or AVHRR. The minimum detectable fire size for the GOES-8 in North America is unclear. In this experiment the smallest fire detected with the GOES-8 was the Saddle fire (5 acres), but it was only detected once. Validation studies in South America have verified GOES-8 ABBA detected fires on the order of 1 to 2 acres in size.

5.2 AVHRR

AVHRR has proved to be a very successful system for fire detection and has the advantage of providing four or more usable observations per day. Fire detection with AVHRR can be accomplished with either daytime or nighttime imagery. One known problem with AVHRR is the TIR band saturation at 320 K. This has two adverse effects on fire detection. First, there are many daytime passes over arid or semiarid regions where the TIR band data will be saturated, making the data unusable for fire detection. The second problem is that the TIR bands tend to saturate on fires. The saturated fire detects are useful for indicating the presence of fires, but cannot be used to model the burn size or temperature.

5.3 DMSP-OLS

DMSP-OLS data yielded the higher fire detection success rating (66 percent) than GOES or AVHRR. Despite the high level of performance, there are some serious restrictions to the fire detection capabilities of DMSP. Only very large hot fires will be detected with the TIR band on the OLS. As a result, fire detection with OLS data is generally only effective with nighttime visible band data which has been intensified with the PMT. Thus only one or two passes per 24 hours will provide usable data. The PMT data will not be used for nighttime data acquisitions at high latitudes near the summer solstice due to high levels of solar illumination. This is a particular problem for locations like Alaska, but the effect is present in the extreme western portions of orbital passes acquired of the northern tier of U.S. states during the summer solstice period. Another phenomenon which is unique to the OLS are the irregular-shaped areas of saturated visible band data known as "glare" which result from sunlight scattering off parts of the DMSP platform and entering the OLS telescope. The glare precesses on an annual cycle and will impact fire detection where present. Finally, the gain on the nighttime visible band is adjusted to track the monthly cycle of lunar illumination. The gain is turned up during new moon nights, permitting smaller fires to be detected. The gain is turned down under full moon conditions, resulting in lower sensitivity to fires and greater clutter problems associated with moonlit clouds and reflections of moonlight off features such as lakes or reservoirs.

6.0 SUMMARY

We report on the first known intercomparison of the fire detection capabilities of NOAA-GOES, NOAA-AVHRR, and DMSP-OLS data. The differences between the three systems in terms of number of daily passes, relative capabilities for fire detection, and daily or seasonal restrictions in fire detection capabilities indicate that there is little or no redundancy in the three systems. That is to say, a program dedicated to either early fire detection or the complete inventory of the daily pattern of fires would achieve its best results by having the capability to include data from all three systems.

For the multisensor intercomparison times the GOES data had the lowest fire detection success rating (22.2 percent), AVHRR was intermediate with a fire detection success rating of 37.5 percent. DMSP-OLS data yielded a fire detection success rating of 66.7 percent. Of the three systems, only DMSP detected a fire prior to its reported occurrence by the U.S. Forest Service. Although GOES had a lower fire detection success index, the frequent observations made by GOES provides a fire detection capability that the polar orbiting AVHRR and DMSP-OLS are unable to match. For instance, the half-hourly GOES data provided the same number of observations of the Lookout fire as the AVHRR and over 7 times as many observations of the San Pedro and Black fires than either AVHRR or DMSP/OLS. The GOES was able to detect the largest (San Pedro) and smallest fire (Saddle) detected during the experiment, but did not detect several of the intermediate fires which were detected by the other sensors. While DMSP-OLS data scored the highest fire detection success rating, this system's fire detection capabilities are restricted to nighttime passes and other phenomena such as the persistence of sunlight into early evening at high latitudes at or near the summer solstice detract from the fire detection capabilities of this source.

The study provided a unique opportunity to compare the fire detection capabilities of smoothed (2.7 km) DMSP-OLS data and fine resolution data (0.55 km TIR band and 2.7 × 0.55 km visible band) from the direct broadcast of the DMSP data stream. The poor capabilities of the smoothed TIR data to detect fires had been previously noted by Elvidge et al. (1996). The results obtained with the fine resolution TIR data in New Mexico indicate that even with 25 times better spatial resolution, the OLS thermal band has limited value for fire detection. Only very large hot fires will result in distinctive TIR band anomalies in the OLS TIR band data. In comparing the visible band data in the two spatial resolution modes, there were no fires detected with the 2.7 × 0.55 km data that were not also detected with the 2.7 km smoothed data. This still leaves open the possibility that 0.55 × 0.55 km nighttime visible band data may be superior to the 2.7 km resolution data. This should be the subject of a future study.

While the experiment was conducted to provide data for the intercomparison of the fire detection capabilities of the three meteorological satellite systems, the experimental data are not suitable for establishing the detection limits of the systems in terms of the area of burning required to result in a positive fire identification. We believe this issue could be addressed for the GOES system using airborne surveys of fires starting from ignition. The airborne surveys could be timed to coincide with the anticipated GOES data acquisitions (every 15 minutes). The observational problems associated with determining detection limits are more difficult for the polar orbiting systems (AVHRR and DMSP) due to the smaller number of daily data acquisitions. Perhaps detection limits for these systems could be bracketed by examining large numbers of fires which had small final burn sizes. An example of this would be the detection of the Saddle and Four Five fires using OLS data. These fires are reported to have burn sizes of 5 acres, establishing an upper bound for the OLS's fire detection limits.

ACKNOWLEDGMENTS

The authors gratefully acknowledge the NOAA-NESDIS Special Projects Office for providing support for NOAA-NGDC and NOAA-CIMSS. The Aerospace Sponsored Research Program provided support for the analysis of the direct readout DMSP data. The authors also acknowledge their data sources, including the Defense Meteorological Satellite Program, Air Force Global Weather Central, NOAA-NESDIS, and the U.S. Department of Agriculture Forest Service.

REFERENCES

Arino, O., J.M. Melinotte, and G. Calabresi. Fire, cloud, land, water: the 'Ionia' AVHRR CD-Browser of ESRIN. *EOQ*, 41, July 1993, European Space Agency—ESTEC, Noordwijk, 1993.

Bywaters, K.W. and E.M. Prins. An interactive WWW tool for coupling satellite and meteorological data in real time. Presented as a paper at the *12th International Conference On Interactive Information and Processing Systems (IIPS) for Meteorology, Oceanography, and Hydrology*, Atlanta, GA, Jan. 28–Feb. 2, 1996, pp. 382–384.

Cahoon, D.R., Jr., B.J. Stocks, J.S. Levine, W.S. Cofer, III, and K.P. O'Neill. Seasonal distribution of African savanna fires. *Nature*, 359, 812–815, 1992.

Croft, T.A. Burning waste gas in oil fields. *Nature*, 245, 375–376, 1973.

Croft, T.A. The brightness of lights on Earth at night, digitally recorded by the DMSP satellite. *Stanford Research Institute*, final report prepared for the U.S. Geological Survey, Palo Alto, CA, 1979.

Crutzen, P.J. and M.O. Andreae. Biomass burning in the tropics: impact on atmospheric chemistry and biogeochemical cycles. *Science*, 250, 1669–1673, 1990.

Dowty, P. The simulation of AVHRR data for the evaluation of fire-detection techniques. In: *Biomass Burning and Global Change*, J.S. Levine, Ed., The MIT Press, Cambridge, MA, 1, 40–50, 1996.

Dozier, J.A. A method for satellite identification of surface temperature fields at subpixel resolution. *Remote Sensing of Environ.*, 11, 221–229, 1981.

Elvidge, C.D., H.W. Kroehl, E.A. Kihn, K.E. Baugh, E.R. Davis, and W.M. Hao. Algorithm for the retrieval of fire pixels from DMSP Operational Linescan System data. In *Biomass Burning and Global Change*, J.S. Levine, Ed., The MIT Press, Cambridge, MA, 1, 73–85, 1996.

Flannigan, M.D. and T.H. Vander Haar. Forest fire monitoring using NOAA satellite AVHRR. *Canadian Journal of Forestry Research*, 16, 975–982, 1986.

Giglio, L., J.D. Kendall, and C.O. Justice. Evaluation of global fire detection algorithms using simulated AVHRR infrared data. *International Journal of Remote Sensing*, submitted, 1997.

Justice, C. and P. Dowty. Technical Report of the IGBP-DIS Satellite Fire Detection. *IGBP-DIS Working Paper 9*, Paris, France, 1994.

Justice, C.O. and J.P. Malingreau. The IGBP-DIS Fire Algorithm Workshop 2. *IGBP-DIS Working Paper 14*, Paris, France, 1996.

Justice, C.O., J.P. Malingreau, and A.W. Setzer. Satellite remote sensing of fires: Potentials and limitations. In *Fire in the Environment*, P.J. Crutzen and J.G. Goldammer, Eds., John Wiley and Sons Ltd., pp. 77–88, 1993.

Justice, C.O., J.D. Kendall, P.R. Dowty, and R.J. Scholes. Satellite remote sensing of fires during the SAFARI campaign using NOAA-AVHRR data. *Journal of Geophysical Research*, 23, 851–863, 1996.

Kennedy, P.J., A.S. Belward, and J.M. Gregoire. An improved approach to fire monitoring in West Africa using AVHRR data. *International Journal of Remote Sensing*, 15, 2235–2255, 1994.

Kihn, E.A. Forest fire detection from DMSP Operational Linescan System (OLS) imagery. In *Biomass Burning and Global Change*, J.S. Levine, Eds., The MIT Press, Cambridge, MA, 1, 86–91, 1996.

Langaas, S. Nighttime observations of West African bushfires from space: Studies on methods and applications of thermal NOAA/AVHRR satellite data from Senegal and Gambia. *Ph.D. Dissertation*, Department of Geography, University of Oslo, Norway, 1995.

Lee, T.F. and P.M. Tag. Improved detection of hotspots using the AVHRR 3.7 μm channel. Bulletin of the American Meteorological Society, 71, 1722–1730, 1990.

Levine, J.S., D.R. Cahoon, J.A. Costulis, R.H. Couch, R.E. Davis, P.A. Garn, A. Jalink, Jr., J.A. McAdoo, D.M. Robinson, W.A. Roettker, A.S. Washito, R.T. Sherrill, and K.D. Smith. FireSat and the global monitoring of biomass burning. In *Biomass Burning and Global Change*, J.S. Levine, Ed., The MIT Press, Cambridge, MA, 1, 107–132, 1996.

Lipsett, M., K. Waller, D. Shusterman, S. Thollaug, and W. Brunner. The respiratory health impact of a large urban fire. *American Journal of Public Health*, 84, 434, 1994.

Malingreau, J.P. and A.S. Belward. Recent activities in European Community for creation and analysis of global AVHRR data sets. *International Journal of Remote Sensing*, 15, 3397–3416, 1994.

Matson, M. and J. Dozier. Identification of subresolution high temperature sources using a thermal IR sensor. *Photogrammetric Engineering and Remote Sensing*, 47, 1311–1318, 1981.

Matson, M. and B. Holben. Satellite detection of tropical burning in Brazil. *International Journal of Remote Sensing*, 8, 509–516, 1987.

Matson, M., S.R. Schneider, B. Aldridge, and B. Satchwell. Fire detection using the NOAA series satellites. *NOAA Technical Report NESDIS 7*, 1984.

Menzel, W.P. and E.M. Prins. Monitoring biomass burning with the new generation of geostationary satellites. In: *Biomass Burning and Global Change*, J.S. Levine, Ed., The MIT Press, Cambridge, MA, 1, 56–64, 1996.

Muirhead, K. and A.P. Cracknell. Straw burning over Great Britain detected by AVHRR. *International Journal of Remote Sensing*, 6, 827–833, 1985.

Overby, S.T. and H.M. Perry. Direct effects of prescribed fire on available nitrogen and phosphorus in an Arizona chaparral watershed. *Arid Soil Research and Rehabilitation*, 10, 347, 1996.

Prevendel, D.A. Project Sparkey: A strategic wildfire monitoring package using AVHRR satellite data and GIS. *Photogrammetric Engineering and Remote Sensing*, 61, 271, 1995.

Prins, E.M. and W.P. Menzel. Geostationary satellite detection of biomass burning in South America. *International Journal of Remote Sensing*, 13, 2783–2799, 1992.

Prins, E.M. and W.P. Menzel. Trends in South American biomass burning detected with the GOES VAS from 1983–1991. *Journal of Geophysical Research*, 99(D8), 16719–16735, 1994.

Prins, E.M. and W.P. Menzel. Monitoring fire activity in the western hemisphere with the new generation of geostationary satellites. *22nd Conference on Agricultural and Forest Meteorology with Symposium on Fire and Forest Meteorology*, Atlanta, GA, Jan. 28–Feb. 2, 1996, pp. 272–275.

Prins, E.M., W.P. Menzel, and D.E. Ward. GOES-8 ABBA Diurnal Fire Monitoring During SCAR-B. In *SCAR-B Proceedings*, V.W.J.H. Kirchhoff, Ed., Trastec Editorial, Sao Jose dos Campos, Brazil, pp. 153–157, 1997.

Rauste, Y. Forest fire detection with satellites for fire control. *International Archives of Photogrammetry and Remote Sensing*, 31, 584, 1996.

Robinson, J.M. Fire from space: Global evaluation using infrared remote sensing. *International Journal of Remote Sensing*, 12, 3–24, 1991.

Scott, D.F. and D.B. Van Wyk. The effects of fire on soil water repellency, catchment sediment yields and streamflow. *Ecological Studies: Analysis and Synthesis*, 93, 216–239, 1992.

Setzer, A.W. and M.C. Pereira. Amazonia biomass burning in 1987 and an estimate of their tropospheric emissions. *Ambio*, 20, 19–22, 1991.

Setzer, A.W. and M.M. Verstraete. Fire and glint in AVHRR's channel 3: A possible reason for the non-saturation mystery. *International Journal of Remote Sensing*, 15, 711–718, 1994.

Weaver, J.F. and J.F.W. Purdom. Observing forest fires with the GOES-8, 3.9-μm imaging channel. *Weather and Forecasting*, 10, 803–808, 1995.

Whelan, R.J. *The Ecology of Fire.* Cambridge University Press, New York, 346 p., 1995.

CHAPTER 8

Detection of Fires and Power Outages Using DMSP-OLS Data

Christopher D. Elvidge, Kimberly E. Baugh, Vinita Ruth Hobson, Eric A. Kihn, and Herbert W. Kroehl

1.0 INTRODUCTION

Recently developed algorithms for processing nighttime data acquired by the Defense Meteorological Satellite Program (DMSP) Operational Linescan System (OLS) make it possible to detect fires and power outages worldwide. The DMSP-OLS has a unique capability to observe faint sources of visible-near infrared emissions present at night on the Earth's surface (Figure 8.1), including cities, towns, villages, gas flares, and fires (Croft 1973, 1978, 1979). By analyzing a time series of DMSP-OLS images, it is possible to define a reference set of "stable" lights that are present in the same location on a consistent basis. Fires are identified as lights detected on the land surface outside the reference set of stable lights. Power outages are detected based on the absence of lights in locations within the reference set of stable lights.

Since the early 1970s the U.S. Air Force has operated DMSP polar orbiting platforms carrying cloud imaging satellite sensors capable of detecting clouds using two broad spectral bands: visible-near infrared (visible-near infrared) and thermal infrared (TIR). The program began with the SAP (Sensor Aerospace Vehicle Electronics Package) which were flown from 1970–1976. The current generation of OLS sensors began flying in 1976 and are expected to continue flying until ~2010. At night the visible band is intensified with a photomultiplier tube to permit detection of clouds illuminated by moonlight. While DMSP data were not officially classified, digital data were not available or preserved during the first 20 years of the program. An OLS film archive at the University of Colorado, National Snow and Ice Data Center provided the scientific community with access to a subset of OLS data in analog form. A digital archive for the DMSP-OLS data was established in mid-1992 at the NOAA National Geophysical Data Center.

The potential use of DMSP-OLS data for the inventory of human settlements and energy consumption patterns has been noted since the 1980s (Welch, 1980; Foster, 1983). Sullivan (1989) produced a 10 km resolution global image of OLS-observed visible-near infrared emission sources using film data. These early studies with OLS data relied on the analysis of film strips, which limited the studies on population and energy to a small number of sites. The first global map of nighttime lights observed with DMSP-OLS data, published by Sullivan (1989), was also derived from film, selected based on the presence of large number cloud-free visible-near infrared emission sources and mosaicked into a global product. As a result, many of the features presented in areas such as Africa are ephemeral visible-near infrared emissions from

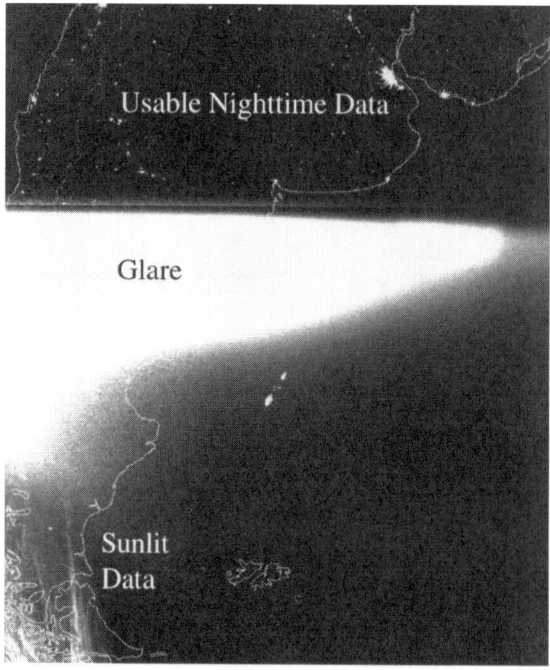

Figure 8.1. Nighttime visible band DMSP-OLS data of Chile and Argentina acquired January 27, 1995 showing city lights, sunlight contamination and solar glare.

fires. The only systematic use of film data for the analysis of fires was the inventory of African fires reported by Cahoon et al. (1992).

Recent results obtained at NOAA-NGDC with the digital DMSP-OLS data (Elvidge et al., 1996 and 1997) suggest that nightly broad area inventories of fires and power outages could be readily produced. Algorithms have been developed to identify and geolocate visible-near infrared emission sources in nighttime OLS imagery. This chapter describes the procedures which have been developed for detection of ephemeral events (fires and power outages) in DMSP-OLS data. The method relies on identifying a reference set of stable lights (cities, towns, villages, gas flares) database using a time series of OLS observations. Once the stable lights data set has been derived, it is possible to overlay the lights from an individual orbit to detect new visible-near infrared emission sources (e.g., fires) or locations where visible-near infrared emission was expected, but not observed (power outages).

2.0 THE DMSP OPERATIONAL LINESCAN SYSTEM

The DMSP generally operates two satellites in sun-synchronous orbits, one in a dawn-dusk orbit, the other in a day-night orbit. The Operational Linescan System (OLS) is an oscillating scan radiometer designed for cloud imaging with two spectral bands (visible and TIR) and a swath of ~3000 km. The "visible" bandpass straddles the visible and near-infrared (visible-near infrared) portion of the spectrum (0.5 to 0.9 μm). Satellite attitude is stabilized using four gyroscopes (three axis stabilization), a star mapper, Earth limb sensor, and a solar detector. The OLS visible band signal is intensified at night using a photomultiplier tube (PMT), for the detection of moonlit clouds (Figure 8.1). The low light sensing capabilities of the OLS at night

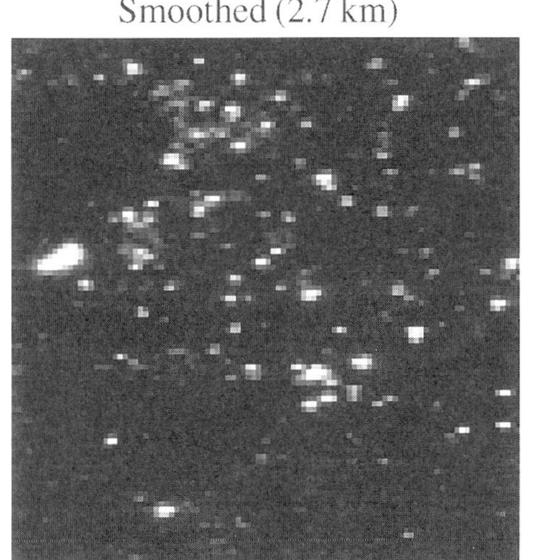

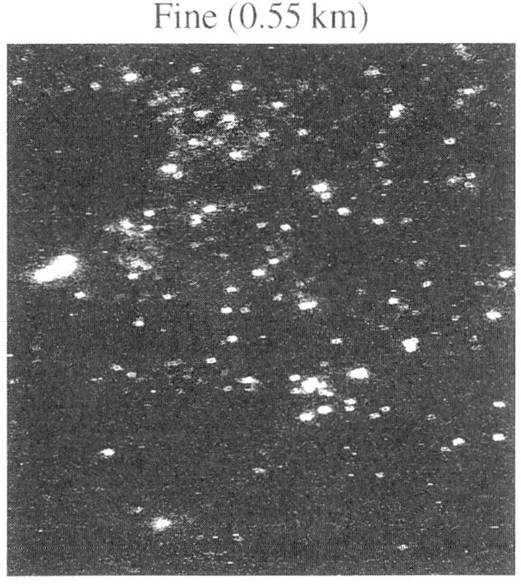

Figure 8.2. Simultaneously acquired smoothed versus fine resolution nighttime visible band DMSP-OLS of fires in Para, Brazil.

permit the measurement of radiances down to 10^{-9} watts/cm^2/sr/μm. This is more than four orders of magnitude lower than the OLS daytime visible band or the visible-near infrared bands of other sensors, such as the NOAA AVHRR or the Landsat Thematic Mapper.

There are two spatial resolution modes in which OLS data can be acquired. The full resolution data, having nominal spatial resolution of 0.56 km, is referred to as "fine." Onboard averaging of five by five blocks of fine data produces "smoothed" data with a nominal spatial resolution of 2.7 km. Most of the data received by NOAA-NGDC is in the smoothed spatial resolution mode. Acquisitions of the fine resolution data can be made (on a noninterference basis) through request to the Air Force. There are spatial details present in the fine resolution data that are absent in the smoothed data. Figure 8.2 shows examples of smoothed versus fine nighttime visible band data of fires in the state of Para, Brazil on August 22, 1995.

The OLS uses several methods to constrain the enlargement of pixel dimensions which normally occur as a result of cross track scanning (Lieske, 1981). The OLS features a sinusoidal scan motion, which maintains a nearly constant along track pixel-to-pixel ground sampling distance (GSD) of 0.56 km at all scan angles in the fine resolution data. The along scan GSD for fine resolution data starts at 0.31 km at nadir and slowly rises to 0.55 km at 1200 km surface distance from nadir, decreasing again to 0.5 km by the end of scan. The detector image rotates as a function of the scan angle and the shape, size, and orientation of the OLS detectors were designed to take advantage of this rotation to reduce the expansion of the EIFOV (effective instantaneous field of view) at off-nadir scan angles. In addition, the PMT electron aperture is magnetically switched (deflected) during the outer quarter of each scan, again reducing the size of the detector image on the ground surface. The EIFOV of the nighttime visible band fine resolution data starts at 2.2 km at the nadir and expands to 4.3 km at 766 km out from the nadir. After the aperture is switched, the IFOV is reduced to 3.0 km and expands to 5.4 km at the far edges of the scan. Thus the EIFOV is substantially larger than the GSD in both the along track and along scan directions.

The gain for the OLS is programmed to track changes in illumination associated with the lunar cycle. The lowest gain setting coincide with the full moon, at which time clouds and the outline of land surface features such as coastlines and major drainage patterns can be readily observed in visible band data. High visible band gain settings are used during the darkest 10–12 nights of each lunar cycle, when lunar illumination at the time of the DMSP overpass is less than half of the maximum possible. Even with the high gain settings, clouds and most of the earth's surface remain dark in OLS data acquired during the darkest nights of the lunar cycle, leaving visible-near infrared emission sources as the primary features observable in the night-time visible band data. As a result, fainter areas of visible-near infrared emission are detectable in OLS data acquired under new moon conditions than under full moon conditions.

3.0 METHODS AND RESULTS

3.1 Stable Lights

The procedures used to assemble stable lights databases are outlined in Figure 8.3. As an example, we will follow the steps used to generate a stable lights product of South America. The structure of the OLS data includes a byte for data quality which is used to tag pixels based on the presence or absence of a visible-near infrared emission source, cloud, or data quality problem. These tags are subsequently used in the time series analysis.

Scene Selection

DMSP-OLS data from 246 smoothed resolution orbits of South America acquired with less than half lunar illumination conditions between October 27, 1994 and April 30, 1995 were extracted from the digital archive. We use data acquired with less than 50 percent lunar illumination for two reasons: (1) During these nights the visible-near infrared band gain on the OLS is set to its highest monthly level, permitting detection of smaller light sources present on the Earth's surface. And (2) With low levels of lunar illumination it is possible to avoid inclusion of moonlight reflectance off clouds and water, which can be confused with visible-near infrared emissions from anthropogenic activity.

Establishment of a Reference Grid

We have used the one-kilometer grid developed for the NASA-USGS Global 1 km AVHRR project (Eidenshink and Faundeen, 1995) interrupted Goode Homolosine Projection (Goode, 1925; Steinwand, 1993). This projection is optimized to provide a uniform grid cell size at all latitudes and contiguous land masses (except Antarctica).

For South America, a 7700 line by 5300 sample reference grid of 1 km pixels was extracted from the global reference grid established for the 1 km AVHRR project (Eidenshink and Faundeen, 1995). Our reference grid begins (northwest corner) with line 10900, sample 7200 of the larger global grid. To fill geolocated OLS data into the reference grid we find the grid cell closest to the latitude and longitude of the OLS pixel center and fill in the surrounding three by three block of pixels.

Geolocation

Our geolocation algorithm operates in the forward mode, projecting the center point of each pixel onto the Earth's surface. The geolocation algorithm estimates the latitude and longitude

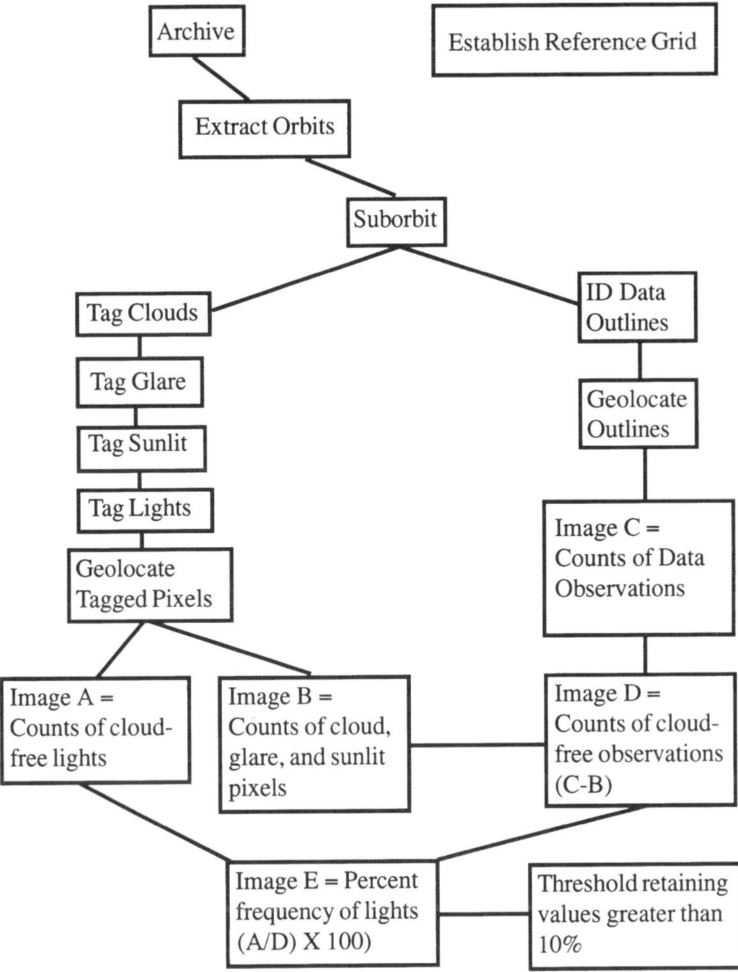

Figure 8.3. Flow chart indicating sequence of steps used to derive the nighttime lights from a time series of DMSP-OLS orbits.

of pixel centers based on the geodetic subtrack of the satellite orbit, satellite altitude, OLS scan angle equations, an Earth sea level model, and digital terrain data. The geodetic subtrack of each orbit is modeled using daily radar bevel vector sightings of the satellite (provided by Naval Space Command) as input into an Air Force orbital mechanics model that calculates the satellite position every 0.4208 seconds. The satellite heading is estimated by computing the tangent to the orbital subtrack. We have used an oblate ellipsoid model of sea level and have used Terrain Base (Row and Hastings, 1995) as a source of digital terrain elevations.

Tracking of Orbital Coverages

The number OLS data coverages for each of the 1 km cells in the reference grid was determined by geolocating the outer perimeters of the OLS data segments from each orbit to create coverage polygons. Areas inside the polygons were assigned values of 1. The total number of coverages for each grid cell can then be determined by adding the values of all the coverage polygons in the time series (Figure 8.4). Total number of coverages over the majority of the continent range from 70 to 100.

Glare and Sunlit Data Removal

Data of southern portions of the continent were contaminated by sunlight (Figure 8.1) during dates near the southern hemisphere's summer solstice (December 21). Data was screened for sunlight contamination based on solar elevations reported for the nadir of each scanline. Pixels in scanlines with solar elevations exceeding 14 degrees were tagged as sunlit.

One adverse effect of the light intensification is that the OLS is quite sensitive to scattered sunlight. Under certain geometric conditions, portions of the OLS are illuminated by sunlight. Scattering of sunlight into the optical path results in visible band detector saturation (Figure 8.1), a condition referred to as glare. The exact shape and orbital position of the glare changes through the year. An automated algorithm has been developed to detect and remove glare from OLS images. The algorithm searches the OLS data to detect the occurrence of 100+ adjacent pixels of saturated data (DN=63). Once such a block of data is encountered, all adjacent pixels having DN values greater than 40 are tagged as glare.

Cloud Screening

Because of the low level of lunar illumination present in the suborbits, it was not possible to use the visible band to assist in the identification of clouds. The cloud screening was based entirely on thresholds set on the TIR band. Clouds are generally colder than the earth's surface. However, the separation of cloud pixels from earth surface pixels using TIR thresholding is complicated by seasonal, latitudinal, and altitudinal variations in the background earth surface temperature. The separation of clouds from earth surface pixels is relatively easy at low latitudes where there is generally a large temperature difference between pixels of cloud tops and pixels containing land or ocean. Because of the strong latitudinal effects on the TIR threshold for cloud screening in our data, we segmented each of the orbital sections into a series of latitudinal bands for determination of a TIR threshold for discrimination and tagging of the cloud pixels.

Cloud Free Observation

Cloud, glare, and sunlit pixels were geolocated and the pixels were projected into the reference grid, where a counter tallied the number of times each grid cell contained unusable data. The resulting image of cloud, glare, and sunlit pixel counts is subtracted from the counts of data observations (Figure 8.4), yielding an image of the cloud-free observations of the earth's surface (Figure 8.5). The vast majority of the land areas had more than 25 cloud-free observations. The lowest numbers of cloud-free observations occurred in the Amazon basin and in high elevation areas in the Andes mountains, where snow pack was probably misidentified as cloud cover in the OLS TIR band data.

Identification of Visible-Near Infrared Emission Sources

Because of brightness variations which occur within and between orbits, it is not possible to set a single digital number (DN) threshold for identifying visible-near infrared emission sources.

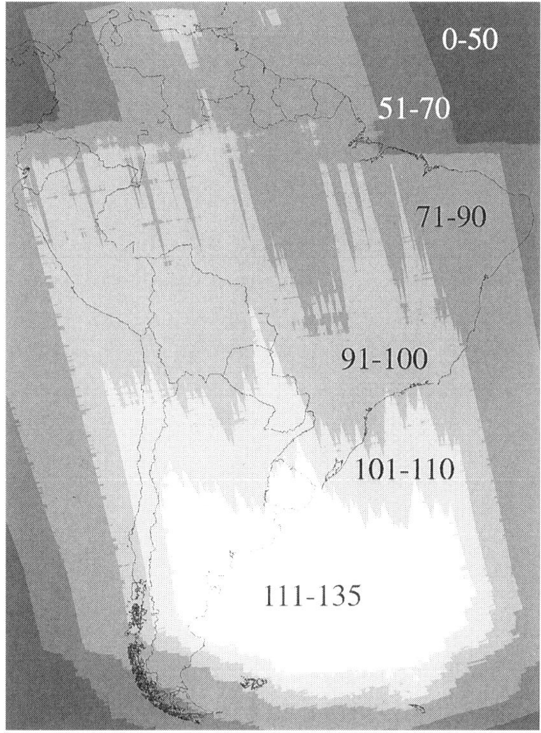

Figure 8.4. Number of nighttime OLS data coverages in the reference grid of the South America.

We have developed an algorithm for automatic detection of visible-near infrared emission sources (lights) in nighttime OLS data using thresholds established based on the local background. Lights are identified in 20 by 20 pixel blocks, with the local background being drawn from the surrounding 50 by 50 pixel block. This "light picking" algorithm excludes bright pixels in the selection of a background pixel set, which is then analyzed to establish a brightness threshold for the detection of visible-near infrared emission sources. The resulting threshold is applied to the central 20 by 20 pixel block inside the 50 by 50 pixel block. Processing of an image proceeds by tiling the results from adjacent 20 by 20 pixel blocks. Only pixels that were not previously tagged as cloud, glare, or sunlit are permitted to be tagged as visible-near infrared emission sources.

If an OLS line scan tracks across a cloud illuminated by lightning, a linear feature is produced which can be detected using the light picking algorithm described above. Typically the lighting is observed in one or two scan lines, but can occur in up to three or four scanlines if multiple lightning strikes continue to illuminate the same cloud mass. Because lightning is typically only detected in areas detected as cloud in the thermal band, the detects are generally excluded from the time series of observations used to generate the stable lights.

Identification of Stable Lights

Pixels tagged as visible-near infrared emission sources are geolocated and projected into the reference grid, where a counter tallied the number of times each grid cell contained a light source. The image of light counts is then divided by the number of cloud-free observations

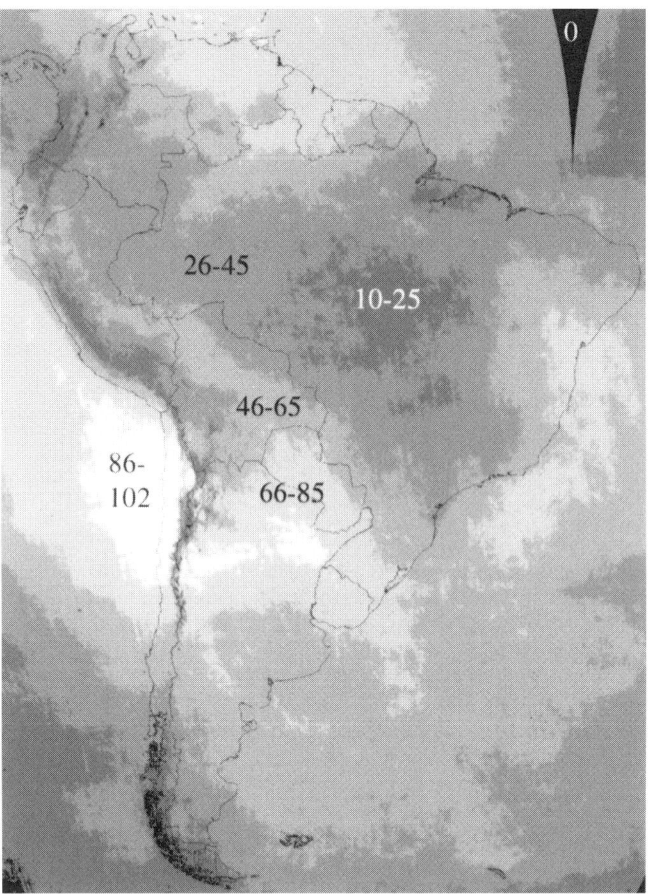

Figure 8.5. Number of cloud-free OLS data observations in the reference grid of South America.

(Figure 8.5) and the resulting real numbers are multiplied by 100 and truncated to byte values ranging from 0 to 100. These values represent the percent frequency with which visible-near infrared emission was detected in each grid cell based solely on cloud-free OLS observations. The locations of stable lights are defined as all areas which had DMSP-OLS detectable lights in at least 10 percent of the cloud-free observations. This 10 percent threshold was used to remove ephemeral visible-near infrared emission sources (fires, lightning, and random noise). In other regions of the world we were able to use a threshold of 6 percent, but random noise levels are higher over the region due to ionospheric disturbances over southern Brazil. The resulting georeferenced stable lights image of South America is shown in Figure 8.6. Larger cities were detected in 90 to 100 percent of the cloud-free observations. Smaller towns were detected with less frequency, in some cases only 10–20 percent of the time.

3.2 Identification of Fires

Fires are identified as visible band emission sources that are not associated with either stable lights or lightning. The procedure begins with the detection of lights and clouds in

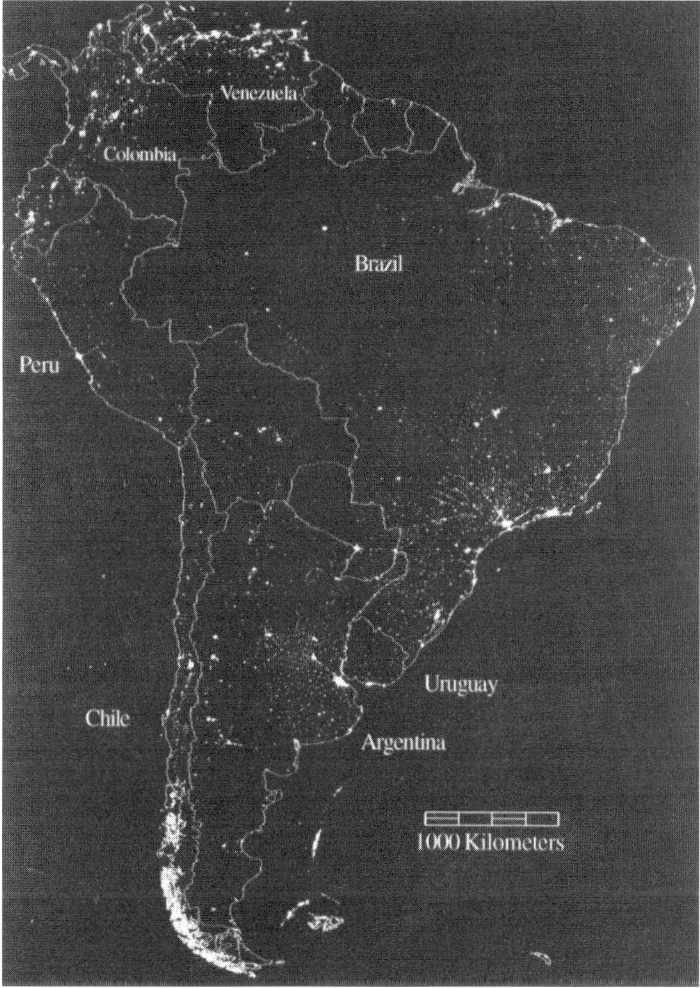

Figure 8.6. Nighttime lights of South America derived from cloud-free portions of 246 DMSP-OLS orbits acquired between October 1, 1994 and March 30, 1995.

incoming nighttime OLS using the algorithms described above. However, in the case of fire detection, lights detected in the presence of clouds are retained. In suborbits where visual inspection identifies the presence of lighting, the lightning illuminated pixels are manually removed. Lights and clouds are then geolocated and projected into the reference grid. Stable lights are then masked out with an algorithm that identifies all lit pixels which occur in or directly adjacent to the known stable light locations and sets their DN values to zero. The remaining pixels, which contain ephemeral visible band emission sources are taken to be fires. Figure 8.7 shows the fires which were detected from the September 8, 1996 nighttime OLS orbit over Brazil, during the peak of the burn season. By compositing burn observations from adjacent orbits over specific time intervals it is possible to assemble continental-scale depictions of the spatial and temporal patterns of biomass burning. Figure 8.8 shows the cumulative burn observed during a six month period over northern hemisphere Africa.

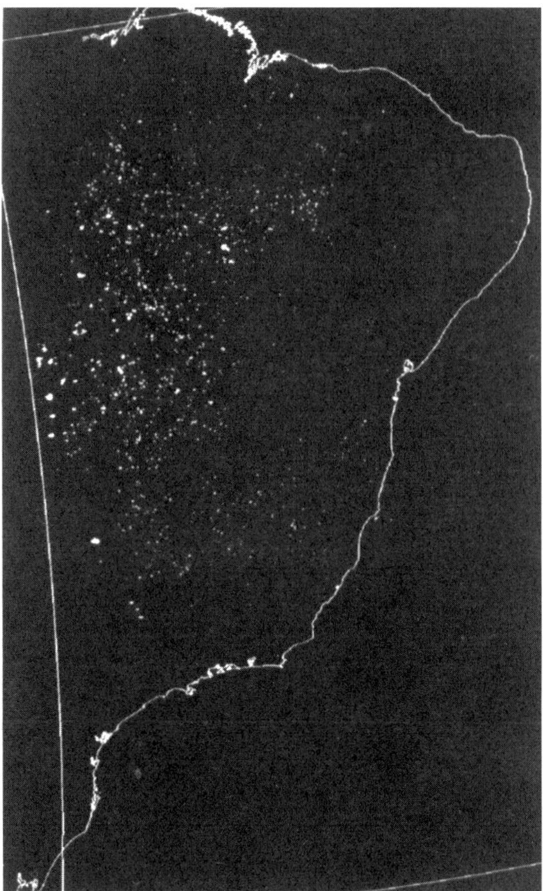

Figure 8.7. Nighttime fires in Brazil observed in a single DMSP-OLS orbit acquired September 8, 1996.

3.3 Identification of Power Outages

During the past 100 years electric power has become the principal source of nocturnal lighting in human settlements, roads, and industrial sites. Typically, electric production is centralized and the power is transmitted using a network of power lines and distribution wiring. Because many of the power lines and wires are aboveground, they are susceptible to damage, especially during severe weather events such as hurricanes, typhoons, thunderstorms, and in some instances snow or ice storms. Less frequent events, such as earthquakes, can also produce electric power outages. Another class of electric power outages, termed brownouts, results from an insufficient capacity to generate electric power or an inability to sustain the generation of electric power. In this case, the electric power supply becomes intermittent or is delivered at reduced levels.

Reporting on power outages is typically limited to eyewitness reports of individuals within the outage area. Because many types of communication (radio, television, telephone, facsimile) are also impacted by power outages, information on the extent of power outages during a disaster may be very incomplete. Having an observational basis for delineating the extent of power outages could provide valuable information to guide relief, repair, and cleanup efforts.

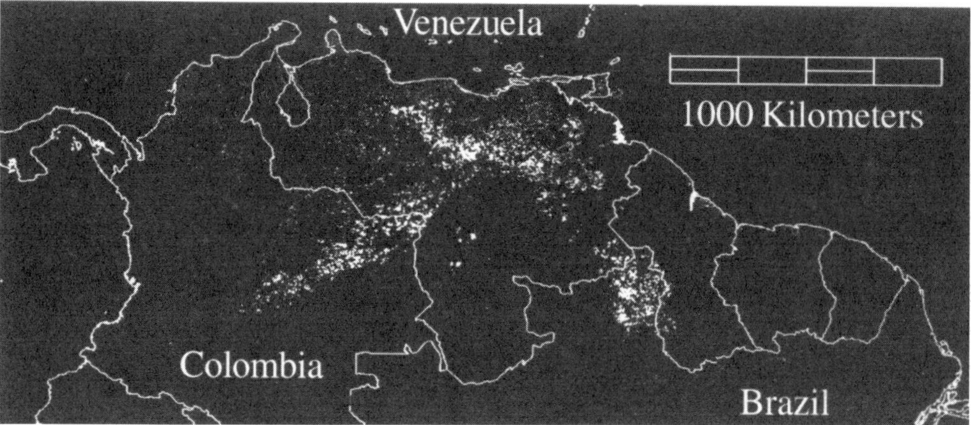

Figure 8.8. Cumulative DMSP-OLS fire detects in northern hemisphere Africa, October 1, 1994 through March 31, 1995.

Power outages are detected in nighttime DMSP-OLS data based on the absence of visible-near infrared emission in locations within the reference set of stable lights. As an example we present color Plate 15, which shows power outages detected September 7, 1996 in North Carolina following the September 6 passage of Hurricane Fran. The procedure for locating the power outages begins by overlying the lights detected on September 7 on top of the reference set of stable lights to form a color composite image. Because the September 7 nighttime OLS image was cloud-free, it was possible to examine the color composite and visually identify areas where power appeared to be on, areas with complete power outage, and areas that appeared to have partial power. Locations with power outage or partial power were manually outlined on the color composite. This outline was used to rearrange the color composite to generate the derived product shown in Plate 15. In areas where no power outage was detected, the reference set of stable lights are shown as white (Plate 15). In the outlined area containing power outages, detected visible-near infrared emissions are shown as white and power outage areas are shown as red. Thus towns shown as red had no power detected on September 7 and towns with partial power have a set of white pixels surrounded by red. For reference, the yellow line indicates the track of the hurricane center, as recorded by the National Hurricane Center. Note that most of the power outages occurred to the north of the hurricane landfall, encompassing the barrier islands of the central North Carolina coast. This is consistent with the fact that much of a hurricane's force, prior to landfall, is concentrated in its northeast quadrant. Moving inland, the zone of power outages tapers toward the overland track of the hurricane center. The OLS detected a diagonal corridor through the center of Raleigh, North Carolina where the power was knocked back harder than to either side. This corridor is parallel and adjacent to the hurricane path.

3.4 Factors Affecting the Detection of Fires and Power Outages

There are a number of factors or conditions which will impede the detection of fires and power outages with DMSP-OLS data, including: the OLS gain setting, cloud cover, and solar glare. The reference set of stable lights are derived using OLS data acquired during the darkest half of the lunar cycle, during which time the OLS visible band gain is set to its highest monthly levels. During the full moon, the OLS visible band gain is set at a much lower state. As a result, many of the small towns indicated in the reference set of stable lights will not be detected in OLS data acquired during the full moon. Thus the detection of both power outages and fires with OLS data works best during the dark half of each lunar cycle. Fewer fires will be detected during the brightest nights of lunar illumination cycle. Without factoring in the gain changes, it would not be possible to distinguish power outages from detection losses associated with lower gain settings during the brightest nights of the month.

Heavy cloud cover blocks the transmission of visible-near infrared light from the earth's surface, blocking the observation of fires and power outages. Light cloud cover tends to diffuse lights present on the Earth's surface, making them appear larger than their actual size. Many fires are detected under light cloud cover; however, their outlines appear diffused. The diffusion caused by light cloud cover could cause partial power outages to be indistinguishable from the normal power conditions.

Other factors impeding the detection of fires and power outages include data dropouts, sunlight, and solar glare. Data dropouts occur randomly. Thus there is always some chance that DMSP-OLS data of a particular event will not transmitted to NGDC. Solar glare results in unusable nighttime visible band data. This phenomenon precesses, and thus will affect portions of an orbit and gradually shift to the north or south. In some cases it is possible to use data from the edge of scan in an adjacent orbit to observe visible-near infrared emissions that are blocked by solar glare in the orbit. Sunlight will impact the detection of fires and power outages at high latitudes at or near the summer solstice.

The digital values in the reference set of stable lights indicate the percent frequency with which the light was detected in the set of cloud-free OLS observations. If a light source is reported to occur in 93 percent of the cloud-free observations and it is missing, along with a large number of its neighbors on a night following a natural disaster, then there is little doubt that the power is out. For small towns that are near the detection limits of the OLS, which may be detected in 15–20 percent of the cloud-free observations, the interpretation of the results gets trickier. Here again, the interpretation must rely on the local pattern of visible-near infrared emission detects. For instance, if town "A" has a detection frequency of 17 percent and no emission detected on a particular night and is surrounded by power outages in neighboring towns with higher detection frequencies, while nearby towns with detection frequencies in the 15–20 percent range had emissions the same night, then it would be reasonable to conclude that there is a high probability that town "A" has a power outage along with its neighbors.

4.0 SUMMARY

We have outlined the algorithms for nighttime fire and power outage detection with data from the DMSP-OLS and provided examples. While other sensors, such as GOES and NOAA-AVHRR have demonstrated capabilities for fire detection, the DMSP-OLS is unique in its capability to detect power outages. No other known sensor has this capability.

Over the next year we expect to complete a global map of stable visible-near infrared emission sources which will be available to the scientific community for the analysis of social, environmental, and energy issues. By combining the map of stable lights with observations from single nights, it will be possible to detect both fires and electric power blackouts worldwide, within the observational constraints noted in Section 3.3. In the longer term, we expect it will be possible to periodically update the global city light map to detect the expansion of urban areas. The DMSP program is expected to continue to operate OLS sensors continuously until the later part of the next decade and perhaps until the year 2010. The NOAA-DoD converged system of meteorological sensors will preserve the low light sensing capability initiated with the OLS. Thus the mapping of stable visible-near infrared emission sources using nighttime satellite data can be expected to be a continuing source of information for the coming decades.

ACKNOWLEDGMENTS

The authors appreciate the DMSP program office and the U.S. Air Force Global Weather Central for providing NOAA-NGDC with DMSP data used in this research.

REFERENCES

Cahoon, D.R., Jr., B.J. Stocks, J.S. Levine, W.S. Cofer, III, and K.P. O'Neill. Seasonal distribution of African savanna fires. *Nature*, 359, 812–815, 1992.

Croft, T.A. Burning waste gas in oil fields. *Nature*, 245, 375–376, 1973.

Croft, T.A. Nighttime images of the earth from space. *Scientific American*, 239, 68–79, 1978.

Croft, T.A. The brightness of lights on Earth at night, digitally recorded by DMSP satellite. *Stanford Research Institute Final Report,* prepared for the U.S. Geological Survey, 1979.

Eidenshink, J.C. and J.L. Faundeen. The 1-km AVHRR global land data set: First stages in implementation. *International Journal of Remote Sensing*, 15, 3443–3462, 1994.

Elvidge, C.D., H.W. Kroehl, E.A. Kihn, K.E. Baugh, E.R. Davis, and W.M. Hao. Algorithm for the retrieval of fire pixels from DMSP Operational Linescan System. In: *Global Biomass Burning,* J.S. Levine, Ed., MIT Press, 1996, pp. 73–85.

Elvidge, C.D., K.B. Baugh, E.A. Kihn, H.W. Kroehl, and E.R. Davis. Mapping city lights with nighttime data from the DMSP Operational Linescan System, *Photogrammetric Engineering and Remote Sensing*, 63, 727–734, 1997.

Foster, J.L. Observations of the Earth using nighttime visible imagery. *International Journal of Remote Sensing*, 4, 785–791, 1983.

Goode, J.P. The Homolosine projection: A new device for portraying the Earth's surface entire. *Association of American Geographers, Annals,* 115, 119–125, 1925.

Lieske, R.W. DMSP primary sensor data acquisition. *Proceedings of the International Telemetering Conference,* 17, 1013–1020, 1981.

Row, L.W., III and D.A. Hastings. *TerrainBase Worldwide Digital Terrain Data*, Documentation Manual and CD-ROM. NOAA National Geophysical Data Center, Boulder, CO, NGDC Publication KGRD 30, 1995.

Steinwand, D.R. Mapping raster imagery into the interrupted Goode Homolosine Projection. *International Journal of Remote Sensing*, 15, 3463–3472, 1993.

Sullivan, W.T., III. A 10 km resolution image of the entire night-time Earth based on cloud-free satellite photographs in the 400–1100 nm band. *International Journal of Remote Sensing*, 10, 1–5, 1989.

Welch, R. Monitoring urban population and energy utilization patterns from satellite data. *Remote Sensing Environ.*, 9, 1–9, 1980.

CHAPTER 9

Change Identification Using Multitemporal Spectral Mixture Analysis: Applications in Eastern Amazonia

Dar A. Roberts, Getulio T. Batista, Jorge L.G. Pereira,
Eric K. Waller, and Bruce W. Nelson

1.0 INTRODUCTION

Land cover change can be the most significant regional anthropogenic disturbance to the environment (Vitousek, 1992). It scales upward to impact global climate (Lean and Warilow, 1989; Shukla et al., 1990), biogeochemical cycles (e.g., carbon, Detwiler and Hall, 1988; Lugo and Brown, 1992) and biodiversity through habitat fragmentation (Lovejoy and Bierregarrd, 1990; Skole and Tucker, 1993). In many regions of the globe, records of historic land cover are either poor or nonexistent. Furthermore, the rapid pace of change and global scope requires tools that provide repeated, large-scale coverage. In these regions, historic satellite data offer, in many cases, the only practical method for determining what changes have occurred and when they have occurred (Detwiler and Hall, 1988; Hall et al., 1991a).

The importance of remote sensing in monitoring global deforestation and mapping changes in land cover is widely recognized (e.g., Detwiller and Hall, 1988; Brown and Lugo, 1992; Moran et al., 1994). Initial analysis in regions such as the Amazon Basin focused on the problem of mapping forest conversion and estimating rates of deforestation (e.g., Fearnside and Salati, 1985; Stone and Woodwell, 1988; Myers, 1988; Tucker and Skole, 1993). More recent studies have focused on more detailed patterns of change, such as mapping the extent of pastures and regenerating forest (Lucas et al., 1993; Moran et al., 1994; Brondizio et al., 1994; Adams et al., 1995; Foody et al., 1996). Although a knowledge of rates of forest conversion is of critical importance to an understanding of global carbon stores, a growing awareness of the potential importance of secondary forest as a carbon sink (e.g., Lugo and Brown, 1992), the significance of different successional patterns associated with variations in land use (Uhl et al., 1988; Foody et al., 1996), and local to regional variation in land use patterns following forest conversion (Moran et al., 1994) commonly justify analysis in greater detail at finer scales.

In this chapter we describe a general approach for identifying the physical nature of changes in land cover from remotely sensed data using spectral mixture analysis (Adams et al., 1993). We describe a multistage process, starting with reflectance retrieval, followed by single-date spectral mixture analysis (SMA) using reference endmembers, relative radiometric calibration of multitemporal data sets, then multitemporal application of SMA. Once a standard frame of reference is established, changes in land cover are identified as either changes in spectral fractions or through changes in classified images. In Section 2 we provide background information on SMA. In Section 3 we provide an example application using Landsat TM time series covering a region in Eastern Amazonia from 1984 to 1994.

2.0 BACKGROUND

2.1 Spectral Mixture Analysis

At scales ranging from a few meters to kilometers, as sampled by remote sensing, most vegetated ecosystems are comprised of a mixture of canopy components, architecturally derived shadow, and potentially exposed soil. For example, within the 30 meter instantaneous field of view (IFOV) of Landsat Thematic Mapper (TM), a shrubland may consist of a mixture of shrubs consisting of green leaves and nonphotosynthetic vegetation (litter, stems, and branches), shadows cast by the shrubs, and exposed soil (Figure 9.1). The spectrum measured by the sensor is a mixture of the spectra of each of the components within the scene. If photons predominantly interact with a single component (i.e., undergo minimal multiple scattering), the mixed spectrum can be modeled as the sum of the pure spectra within the IFOV, weighted by the areal proportion of each material. The "pure" spectra are called endmembers, while the process of solving for endmember fractions is called linear Spectral Mixture Analysis (SMA) (Adams et al., 1993). When scattered photons interact with multiple components, the mixture has the potential of becoming nonlinear (Johnson et al., 1983; Shipman and Adams, 1987; Roberts et al., 1993a). For most applications, multiple scattering is assumed to be negligible, although nonlinear mixing can become significant for some types of vegetation (Huete, 1986; Roberts et al., 1993a; Borel and Gerstl, 1994; Ray and Murray, 1996). Results presented in this chapter are based on an assumption of linear mixing.

The process of solving for endmember fractions can be readily demonstrated with a NIR to red scatterplot (Figure 9.2). In this figure, reflected red and NIR light are plotted along the x and y axis, respectively. Endmembers, represented by NIR and red values for "pure" spectra are shown, labeled as Green Leaf (high NIR, low red), Soil (NIR roughly equal to red), and Shade/Shadow (NIR greater than red, both low). Each endmember is connected by a line in spectral space that represents linear mixtures between two of the components. All spectral mixtures are confined to the triangular region bounded by the endmembers (labeled mixtures on the figure). A mixed spectrum, $P_{i\lambda}'$, is modeled as the sum of **N** endmembers, P_{kl}, weighted by the fraction of the endmember, f_{ki}, within the field view of view at pixel **I** (Equation 1):

$$P_{i\lambda}' = \sum_{k=1}^{N} f_{ki} * P_{k\lambda} + \varepsilon_{i\lambda} \tag{1}$$

Unmodeled portions of the spectrum are expressed as a residual term, $\varepsilon_{i\lambda}$, at wavelength λ. Fractions are derived by inverting this equation and solving for the fractions that provide the best fit between the measurement and the model. Model fit is assessed as either an error in the fractions (negative fractions or fractions exceeding 100 percent), residuals, $\varepsilon_{i\lambda}$, at each wavelength or as a root mean squared (RMS) error across all bands (Equation 2):

$$\text{RMS} = \text{sqrt}\left(\sum_{k=1}^{N} \left(\varepsilon_{i\lambda}\right)^2 \right) / (N) \tag{2}$$

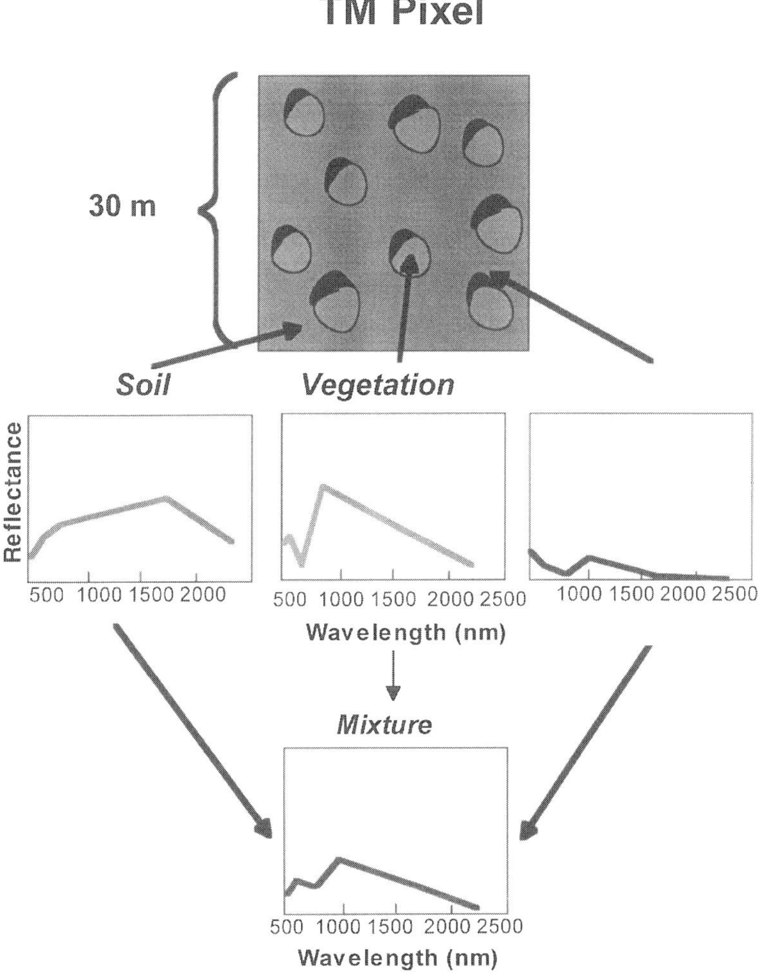

Figure 9.1. Example of a 30-m TM pixel consisting of shrubs, shadows, and bare soil. Spectra (endmembers) for each pure component in the scene are shown below the pixel. The sensor measures the mixed spectrum at the bottom of the figure.

A large number of solution methods exist for this type of linear problem. Common methods used in remote sensing include least squares estimation (Smith et al., 1990), singular value decomposition (Boardman, 1989), and Gram-Schmidt orthogonolization (Adams et al., 1993). Constrained and unconstrained solutions, in which fractions sum to unity or are constrained to remain positive, have been employed (e.g., Shimabukuro and Smith, 1991). In this chapter we employ the unconstrained Modified Gram-Schmidt least squares method (Golub and Van Loan, 1983). The fractions are constrained to sum to one, while individual fractions are allowed to be negative or superpositive (greater than 100 percent).

Fraction errors deserve further comment because they represent an additional measure of error in the model. They originate when a spectrum in the image is poorly represented by the endmembers (e.g., the target is darker or brighter than any of the endmembers), as a product of atmospheric contamination or instrumental noise, or if the dimensionality of the data differs from the number of endmembers used in the model (Sabol et al., 1992; Roberts et al., 1993a).

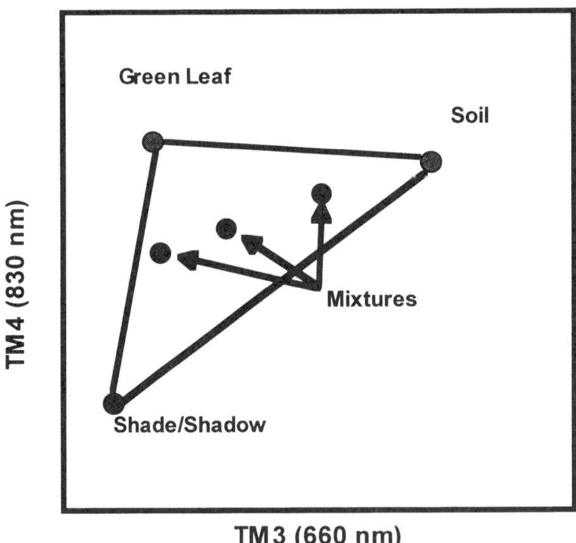

Figure 9.2. Three endmember mixture consisting of green leaf, soil, and shadows plotted on a NIR to red scatterplot.

For example, in the humid tropics, typical endmembers might include green leaves, nonphotosynthetic vegetation (NPV), soil, and vegetation shade (Roberts et al., 1993b; Adams et al., 1995). Clouds and muddy water excluded from the model will produce fraction errors where they occur in the image. Inclusion of an improper number of endmembers may have a similar effect. For example, a closed canopy forest might be best described as a combination of two or three endmembers, green leaf, shade and potentially, NPV. Addition of soil where no soil is present would generate a fraction error. In this case, natural variability in the spectra of canopy components, uncorrected atmospheric contamination or instrumental noise would be partitioned into the soil fraction. In this chapter, we present results from a simple model, in which a fixed set of four endmembers is applied to each image and thus fraction errors occur.

2.2 Image Endmembers vs Reference Endmembers

Endmembers can be derived from the image, or from field or laboratory spectra of known materials. In the terminology of Gillespie et al. (1990) and Adams et al. (1993), the former are known as image endmembers while the latter are called reference endmembers. The most common applications of SMA use image endmembers (e.g., Boardmen, 1989; Shimabukuro et al., 1994). The advantage of image endmembers is that they are easily obtained, represent spectra measured at the same scale as the data, and provide a simple measure of abundance. For example, a scene consisting of primary forest and pasture could be readily modeled by image endmembers for these two land cover types, providing areal estimates of each land cover at subpixel scales. However, these same image endmembers may not be portable across time, space, or sensor platforms. For example, changes in illumination, atmospheric contamination, radiometric calibration, spectral band pass, and pixel resolution may result in spectral changes in an image endmember that mimic real changes on the ground.

A key component of change identification using SMA is the use of reference endmembers. Reference endmembers provide a means for identifying what is present, how much is present, and how much change has occurred. For example, Adams et al. (1995), utilized reference endmembers to monitor changes in disturbed Amazonian vegetation over a four-year period in the vicinity of Manaus, documenting processes such as deforestation, regeneration, and pasture abandonment. While such physically based, standardized measures of change are of obvious value, application of reference endmembers is considerably more difficult than image endmembers, and comes with its own unique set of problems. For example, the approach requires the use of an appropriate spectral library collected at an appropriate scale. In addition, reference endmember selection can be subjective and flawed by the fact that different materials may have similar spectral properties (Price, 1994). Finally, the approach requires an intermediate step of calibration to link retrieved surface reflectance to a reference library. In order to distinguish areal cover from SMA estimates, it is important to remember that endmember fractions are spectral fractions, which may differ from true areal abundance depending on the degree of nonlinear mixing and the relationship between the reference endmembers and scene components. Through careful selection of reference endmembers, confidence in areal estimates can be greatly improved. We discuss one strategy for selecting reference endmembers in the following section.

2.3 Selecting Reference Endmembers

As described by Adams et al. (1993), reference endmembers are selected using an iterative approach starting with image endmembers. In this example, we will assume that our image has already been converted to apparent reflectance (see next section). The first step is to locate a set of image endmembers that represents relatively pure materials within the scene. In a tropical region, typical representatives might include crops or early successional forest for green vegetation, recently cleared forest or roads for soil, slash, or senesced grass as NPV, and shadows from vegetation, or water for shade. These endmembers are then applied, producing images of fractions, RMS error and, potentially, residuals for each band. Fraction and error images are then evaluated and the process repeated until the best set of image endmembers is located. As the least mixed representatives in the image, the image endmembers should form a mixing space that bounds a majority of spectra within the image (labeled Em_i on Figure 9.3).

Reference endmembers are then selected from a spectral library using the image endmembers and a series of constraints to locate the sets of spectra that best fit the image endmembers, while providing the most reasonable fractions. Because the image endmembers cannot be more pure than the reference endmembers, the most fundamental constraint is that all reference endmembers plot outside of the spectral space bounded by the image endmembers (potential candidate regions of Figure 9.3). As can be observed, this simple constraint greatly reduces the number of potential solutions. A secondary constraint is that the reference endmembers provide a reasonable fit, as represented by the magnitude of the RMS error and band residuals. For example, a three-endmember model defines a triangular plane in spectral space. All candidate reference endmembers would be required to lie in a similar plane beyond the vertices defined by the image endmembers.

The first two constraints greatly reduce the number of candidate spectra. However, the remaining spectral space may still represent an unacceptably large pool of candidates and does not account for the fact that different reference endmembers may produce significantly different fractions. In order to select the best possible candidates, it is important to remember that we rarely analyze images without some a priori knowledge of the physical characteristics of the scene. For example, based on fieldwork, it may be possible to place constraints on likely fractions for a given image endmember; a flat, bare soil surface might be expected to

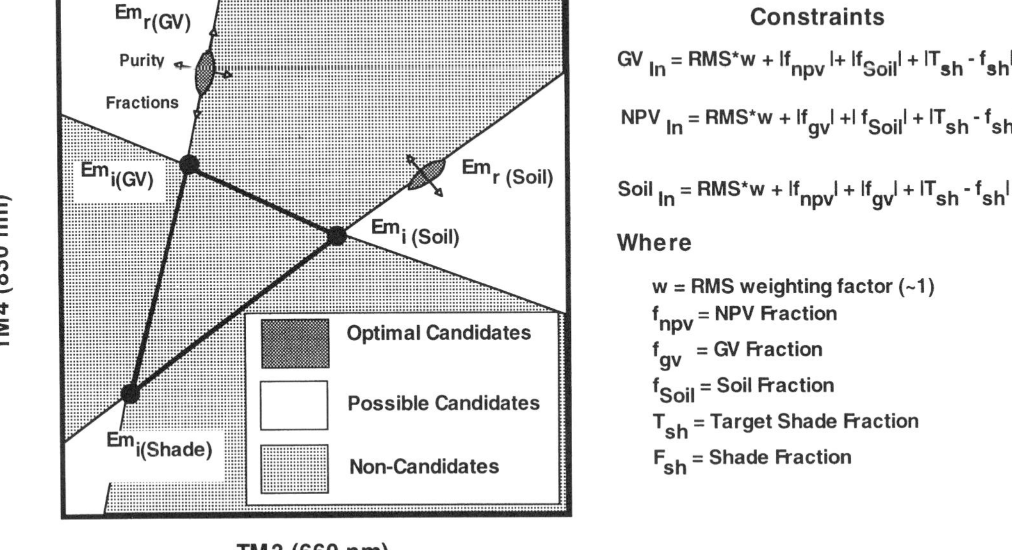

Figure 9.3. NIR to red scatterplot showing the spectral regions for optimal and possible candidates. Most of the spectral space is occupied by noncandidates. Example indices for GV, NPV, and soil are shown to the right. Although a single shade endmember is shown, multiple shades may occur (Roberts et al., 1993a).

have little or no shade, while second-growth forest may have moderate amounts cast by the crown or produced through canopy gaps. Aerial photography, if available, can also be used to estimate areal fractions. Through intelligent use of collateral information, the spectral space of potential candidates may be greatly reduced to a relatively small area (optimal candidates in Figure 9.3).

In practice, it is relatively easy to develop a set of indices and order reference endmember solutions on the basis of RMS error and fraction constraints. Three example indices are shown in the figure. Fractions are generated by reversing the process of SMA and unmixing a reference library using a set of four-image endmembers. Reference endmembers for each image endmember are ranked on the basis of a weighted RMS error and a measure of departure from the expected fractions. For example, candidates for GV are ordered on the basis of a weighted RMS error, the sum of the absolute value for the NPV and soil fractions, and the absolute value for the difference between the expected and estimated shade fraction. An index is calculated for each reference endmember in the library, sorted, then a candidate is selected from the pool with the lowest index.

2.4 Reflectance Retrieval and Relative Radiometric Calibration

Reflectance retrieval can be defined as the process in which sensor radiance is converted to apparent surface reflectance (Gao et al., 1993). Recent advances in reflectance retrieval using imaging spectrometry (Gao et al., 1993; Green et al., 1993) have greatly simplified the process of selecting reference endmembers by providing apparent reflectance that accounts for seasonally and spatially varying atmospheres (Roberts et al., 1997). Unfortunately, a similar approach is not possible for band-limited data such as Landsat TM, where radiometric calibration may be uncertain (e.g., Markham and Barker, 1987; Che and Price, 1992) and full characterization of atmospheric properties through radiative transfer not possible. Regardless, a large body of literature exists that describes methods that can provide estimates of surface reflectance at sufficient accuracy for analysis of historical satellite data. These techniques can be roughly divided into absolute calibration (Kaufman, 1989; Moran et al., 1992; Gilabert et al., 1994) and relative reflectance retrieval (e.g., Roberts et al., 1985; Conel, 1990; Elvidge and Portigal, 1990).

Each set of techniques has its own merits. For example, absolute calibration techniques, as described by Kaufman (1989) and Gilabert et al. (1994) can be automated to adjust for heterogeneous atmospheres, and retrieve atmospheric properties in addition to reflected surface radiance. However, these techniques require highly accurate radiometric calibration and atmospheric inputs that may not be available. Relative reflectance techniques, in which field targets are used to develop linear equations relating encoded radiance (DN) to apparent surface reflectance, can provide accurate estimates of apparent reflectance with little knowledge of sensor or atmospheric characteristics. However, these techniques typically require large, well characterized homogeneous field calibration targets and assume a uniform atmosphere, neither of which may be possible in areas of high relief or areas dominated by closed canopy vegetation. In this chapter we employ a relative approach called the modified empirical line, in which the requirement of homogeneous, well characterized calibration targets is adjusted to permit the use of spectral analogs derived from a reflectance library (Roberts, 1991). Using this approach, image spectra of identifiable targets (e.g., bright soil, water, etc.) are regressed against reflectance spectra of similar materials to produce a series of linear equations relating image DNs to measured reflectance. Models are evaluated based on the correlation coefficient, the shape of the intercept (which should resemble Rayleigh scattering) and retrieved reflectance for independent calibration targets, such as dark water or vegetation. Although field spectra are not a requirement, when available they should be used. For example, spectra used in this study included tropical soil and water spectra from the Amazon Basin.

Relative radiometric calibration involves taking a series of images and placing them within a common frame of reference. In this manner, relative differences between images due to illumination, atmospheric contamination, and radiometric calibration are removed. Examples include the use of psuedo-invariant features (PIF) (Schott et al., 1988) and the relative radiometric calibration as described by Hall et al. (1991b). While this approach does not provide surface reflectance, it represents a powerful means for normalizing sets of image data collected by different sensors under different atmospheric conditions.

The combination of apparent surface reflectance for a single date, followed by relative radiometric calibration for the remaining dates, provides an opportunity to link a reference library to a multitemporal data set. In this manner, a single set of reference endmembers can be selected, then applied to all of the data sets. In this chapter we employ a variant of the PIF technique described by Schott et al. (1988), in which sets of temporally invariant targets covering a range in brightness are located in the image and are regressed against the reference image to generate a linear equation for each date and band. Although more user-intensive than semiautomated approaches described by others (e.g., Schott et al., 1988; Hall et al., 1991b), this approach works well for relatively small numbers of images.

2.5 A General Strategy for Change Identification

Using reference endmembers, change can be identified as explicit changes in the endmember fractions, or as a change in class as defined by endmember fractions (Adams et al., 1995). The former approach has the advantage in that it provides a direct measure of a physical change. For example, the process of deforestation might be characterized by an abrupt drop in green vegetation (GV) and shade typical of primary forest, followed by an increase in NPV (slash, trunks) and soil. Regeneration might be characterized by a gradual increase in GV and shade at the expense of NPV and soil. Change between up to three dates can be readily shown and identified using a standard false color composite and loading fractions for GV, soil, NPV, or shade as red, green, or blue for different dates (e.g., Adams et al., 1995; Roberts et al., 1997). Classification, on the other hand, provides a tool for extending analysis well beyond the limitations of three dates, while providing a direct link to established land cover classes. As will be shown, each technique has advantages and disadvantages.

The complete approach described is summarized in Figure 9.4. Analysis begins with a single date and the selection of image endmembers, followed by reflectance retrieval and selection of reference endmembers. Relative radiometric calibration techniques are then employed to extend the reference endmembers through time, followed by direct analysis of the fractions or classification. In the following section we provide an example derived from 10 years of Landsat TM from an area in eastern Amazonia.

3.0 EXAMPLES FROM EASTERN AMAZONIA

3.1 Study Site

Preliminary analysis of land cover change from 1984 to 1994 is presented for an area in Eastern Amazonia. The study represents a small part of larger NASA-sponsored research on Amazonian hydrology and biogeochemistry. The study site consists of a 8,100 km^2 area located approximately 50 kilometers south of Maraba at a latitude of 6° S and longitude of 49° W (Figure 9.5). This region is located along the extreme fringe of the Amazon Basin in a transitional zone between tropical rainforest and tropical seasonal forest. Dominant forested vegetation consists of upland forests with abundant vines ("vine forests") and lowland forests dominated by palms such as babaçu (*Attalea speciosa Mart. ex Spreng*; Anderson, 1990). Seasonal temperatures average 25°C with mean daily minima between 18° and 21°C and maxima from 29° to 32°C over forested sites (Culf et al., 1996). Total annual precipitation averages 1,800 mm, falling primarily between December and May with a pronounced dry season between June and August (Gash et al., 1996).

The study area contains some of the oldest converted forest in the state of Para (Gash et al., 1996). Land cover consists primarily of at least two primary forest types, large pastures, small farms/plantations, regenerating forest, barren soil, and open water. Land use includes large cattle ranches, clear-cut timber harvesting, selective logging, shifting cultivation, and mining. Starting in the 1980s, the region has experienced considerable pressure for large-scale development as a part of the Programma Grande Carajas (PGC), one of the most ambitious developments in the humid tropics. The PGC covers an area of 895,000 km^2 designed to develop mineral reserves (including the world's largest reserves of high-grade iron ore), and promote supporting infrastructure and agricultural projects (Anderson, 1990). Pasture expansion and timber extraction for the production of charcoal for pig iron manufacture have greatly increased pressures on existing primary forest. As of 1985, the Western Zone of the PGC, which includes

Multitemporal Spectral Mixture Analysis

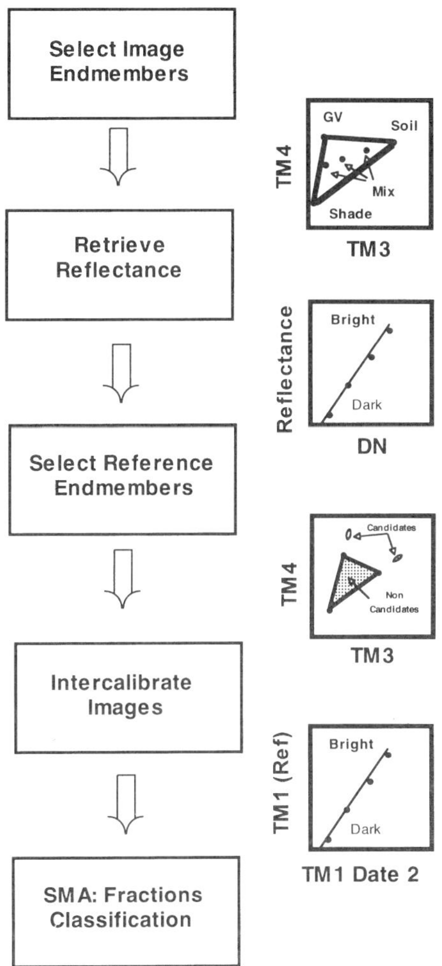

Figure 9.4. Flow chart for multitemporal SMA using reference endmembers.

Maraba, was reported to consist of 52.6 percent primary forest, 10.7 percent logged forest, 23.3 percent deforested land, 2.2 percent cerrado, and 11.1 percent other, prior to much of the current activity associated with the development (Anderson, 1990). Local variability in land cover may result in significantly different areal percentages (as will be shown for Maraba), but a general trend of decreasing primary forest should be true for much of the area.

The long-term objectives of the study are to evaluate land-cover change and its effect on biogeochemical and hydrologic processes at the mesoscale. The area around Maraba was selected as representative of land-cover change over a large area in Eastern Amazonia and along much of the fringe of the Amazon Basin. Specific objectives were to use multitemporal Landsat TM to capture dynamic changes in land cover in the region of relevance for hydrological or biogeochemical modeling. Example questions we are attempting to address include: (1) What is the rate of deforestation, and how does that change over time? (2) What is the areal extent and age structure of second growth forest and how long does second growth persist before it is cut

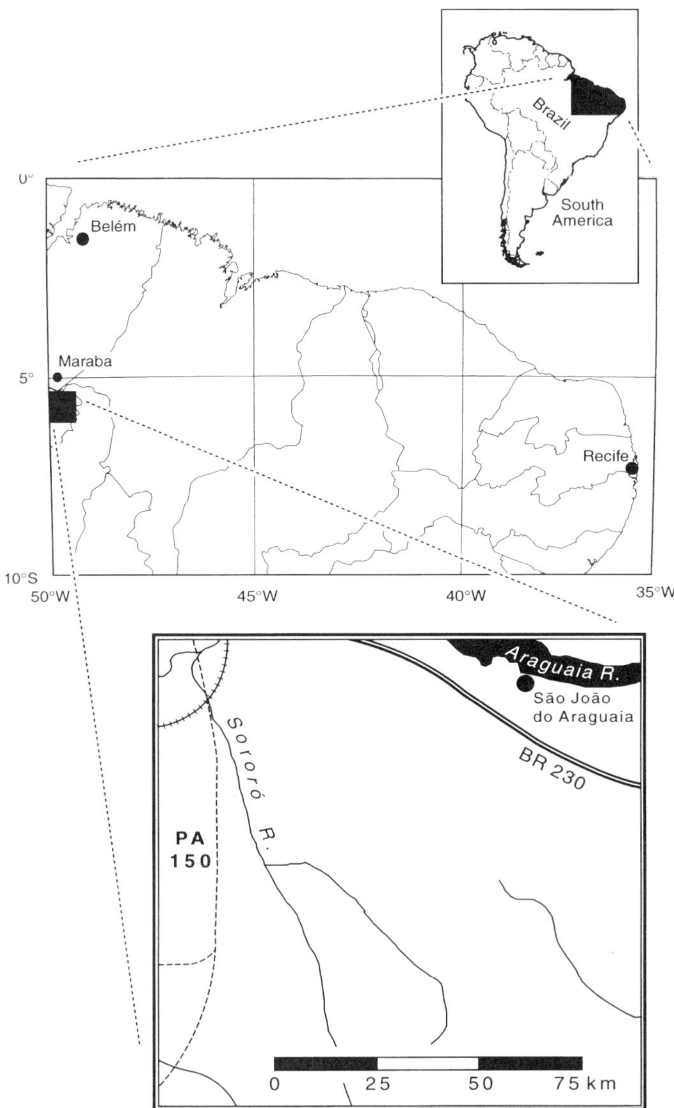

Figure 9.5. Index map of the study region.

down? and (3) What is the areal extent and age structure of pasture in the area and how does that change with respect to second growth and primary forest?

Supporting fieldwork was conducted over a one-month period starting in mid-September, 1994. Fieldwork included ground surveys, discussions with local farmers and ranchers, an overflight in a small plane over the northwestern portion of the study area, and destructive biomass measurements from several stands of second growth forest.

3.2 Image Preprocessing

Five Landsat TM quarter scenes were obtained for the southwest quadrant of Path 223, Row 64. Dates included 1984, 1989, 1992, 1993, and 1994. Data were restricted to a window from early July to mid-August, after the dry season had commenced but prior to the burn season which peaks in September in Maraba. As will be shown, even within this restricted window, variation in forest and pasture phenology between July and August and interannual variability in precipitation can significantly affect land cover characteristics and classification.

All image data were standardized to the 1993 quarter scene. As a first step, all of the image data were coregistered using nearest neighbor resampling to preserve spectral values. Reflectance retrieval was performed using the modified empirical line technique previously described. Reflectance spectra were selected from a library consisting of 485 spectra of soils, leaves, NPV, and water. Canopy level spectra were derived from apparent surface reflectance derived from AVIRIS over Jasper Ridge, California convolved to Landsat TM (Roberts et al., 1997). Soil and water spectra included water spectra covering a range in sediment levels from the low-sediment Rio Negro to the sediment-laden Rio Solimoes, and soil spectra collected in the vicinity of Manaus.

Once a set of linear equations have been developed that relate at sensor encoded radiance (DN) to reflectance, these equations can be used interchangeably to convert the image to apparent reflectance, or to convert a reflectance library to radiance as measured by the sensor. In order to convert DN to reflectance in 1993 (R_{93}) we used:

$$R_{93} = slope_{93}*DN_{93} + int_{93} \tag{3}$$

which can be rearranged to solve for DN_{93} as:

$$DN_{93} = R_{93}/slope_{93} - int_{93}/slope_{93} \tag{4}$$

Slope and intercept terms developed for this study are listed in Table 9.1. Note, the intercept term in Equation 3 is expressed in reflectance while the intercept term in Equation 4 is expressed in DN and thus resembles a path radiance term as might be estimated using a dark object in the image. The slope term incorporates all multiplicative factors including instrumental gains, the solar curve, and variable atmospheric transmittance, while the intercept term combines instrumental offsets and path radiance. In practice, because hard disk space can be limiting, the reflectance library is converted to encoded radiance for mixture analysis, leaving the image data in its original form.

The remaining images were intercalibrated using the 1993 image as a reference and 11 temporally invariant targets located across the image. Targets included two second-growth forests, one primary forest, dark water, two pastures, and five barren areas. Once linear equations have been developed, the reference year equations (3 and 4) can be modified for other image dates. For example, an equation converting 1992 data to 1993 is:

$$R_{93} = slope_{93}*(slope_{9293}*DN_{92} + int_{9293}) + int_{93} \tag{5}$$

Where all multiplicative and additive constants can be combined into two constants:

$$slope_{92} = slope_{93}*slope_{9293}$$

$$int_{92} = slope_{93}*int_{9293} + int_{93}$$

Equations intercalibrating each date to 1993 are provided in Table 9.2.

Table 9.1. Slope and Intercepts for Each Band.

	Radiance to Reflectance		Reflectance to Radiance	
Band	Slope	Intercept	Slope	Intercept
TM1	3.481×10^{-3}	–0.1457	287.279	41.847
TM2	6.696×10^{-3}	–0.0902	149.346	13.474
TM3	6.189×10^{-3}	–0.0646	161.567	10.440
TM4	4.493×10^{-3}	-9.696×10^{-3}	222.566	2.158
TM5	3.287×10^{-3}	–0.0171	304.203	5.211
TM7	8.063×10^{-3}	-4.305×10^{-3}	124.017	0.5339

3.3 Analysis of Fractions

Endmembers for GV, NPV, and soil were selected using the approach described previously (Figure 9.6). Endmembers were selected from a library of 485 reflectance spectra that included temperate and tropical soils, green leaves, and leaf stacks from temperate vegetation (no full range tropical leaves were available) and NPV from both temperate and tropical areas. The shade endmember was the same as the one used by Adams et al. (1995) called vegetation shade to account for NIR scattering by vegetation (Roberts et al., 1993a).

These endmembers were applied to the five intercalibrated data sets to generate fraction images for each endmember and an RMS error image. Fraction and RMS error images are shown for a 30x30 km subset of the northwest corner of the study area (Figure 9.7). Land cover in the area can be divided into six major land-cover classes: primary forest (PRF), regeneration (Regen), pasture, open water, burn, and construction (con) (Table 9.3). As is shown in Table 9.3, each of these classes actually consists of several land cover types that are difficult to separate spectrally. For example, two dominant types of primary forest are actually present in the area, upland vine forest and lowland forests dominated by palms. These two forest types can be separated on the basis of a higher shade content in the lowland forest, but, because of a significant overlap in spectral characteristics, would be difficult to classify separately without large error. For this reason, in this chapter we treat them as a single cover type. Regeneration, as discussed here, includes plantations and shifting agricultural, while construction includes both urban areas and bare soil.

Analysis of the fractions provides some measure of the physical properties of the dominant land cover classes. Primary forest can be characterized as consisting primarily of green leaves and shade with a minor amount of NPV, consistent with the presence of large forest gaps and emergents in primary forest. Interestingly, the primary forest in Maraba tends to have a lower shade content than those in the vicinity of Manaus, potentially due to the presence of vines (see Adams et al., 1995). Regeneration, in contrast, has a lower shade content and higher green leaf, soil, and NPV fractions (Figure 9.7), consistent with low shrubs and small trees with few gaps or exposed soil and branches where gaps occur. Pastures can be characterized as dominated by NPV and soil with lesser amounts of GV, while construction has a higher soil fraction. Burned areas consist of high shade and high NPV, most likely due to the presence of low reflectance charcoal with partially burned trunks and branches. Water can be characterized as having a high shade content due to its low reflectance, with a high soil fraction due to sediments. Analysis of the RMS error shows that vegetated areas were well characterized by the endmembers (low RMS error), but burned areas, pasture, and construction were poorly characterized (high RMS). Analysis of the variance of fractions for each land-cover class shows high variance in the shade and NPV fractions for burn, pasture, and construction. The high variance in water is

Table 9.2. Intercalibration Equations.

Band	1984 to 1993			1989 to 1993			1992 to 1993			1994 to 1993		
	Slope	Int	R^2	Slope	Int	R^2	Slope	Int	R^2	Slope	Int	R^2
1	0.722	12.64	0.992	0.632	18.259	0.966	0.416	30.995	0.8749	1.038	3.8	0.9979
2	0.802	3.03	0.9936	0.6636	6.658	0.9879	0.709	6.77	0.9515	1.05	1.36	0.996
3	0.897	1.679	0.9916	0.7652	5.276	0.9962	1.105	−0.772	0.9707	1.014	3.341	0.999
4	0.898	3.993	0.99	0.7869	5.359	0.9885	1.099	−5.62	0.9944	1.017	2.75	0.998
5	1.003	1.59	0.9988	0.8433	5.451	0.9941	1.225	−2.107	0.997	1.1	0.402	0.9998
7	0.9702	0.6204	0.997	0.8043	1.327	0.992	1.017	−0.48	0.998	1.108	0.385	0.9993

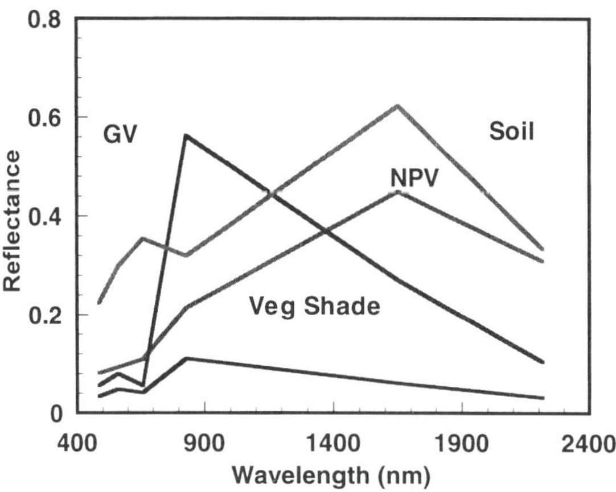

Figure 9.6. Endmembers used to model the scene.

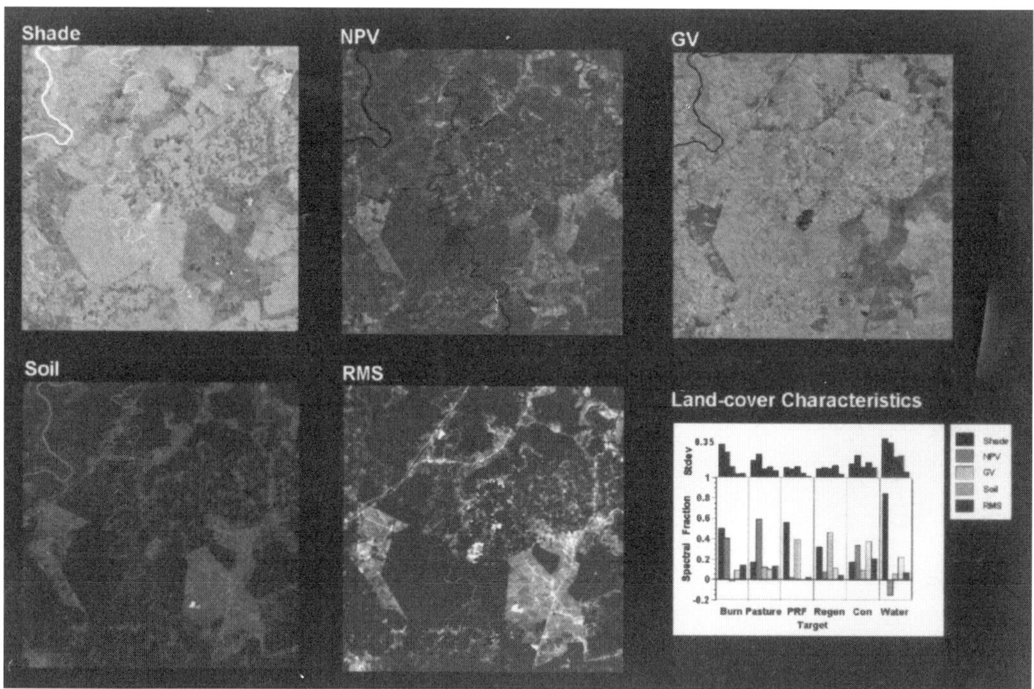

Figure 9.7. Fraction and RMS error images for a 30x30 km subset of the July 20, 1993 scene. Fractions for six land-cover classes are shown to the lower right. The standard deviation of fractions is plotted above the fractions.

due to variable sediment concentrations between rivers. The area shown consists primarily of two classes, primary forest (high shade, high GV, low soil) and pasture (low shade, high soil and NPV, high RMS error in Figure 9.7). Regeneration is most abundant in small cuts characterized by low shade but high GV. A recent burn is in the center of the image, shown as low GV, high shade and a high RMS.

Table 9.3. Land-Cover Classes.

Class	Cover Types	Characteristics
Water	Rivers, streams, lakes	High shade, high soil fractions High RMS for muddy water
Pasture	Pasture	High NPV, all other fractions below 10%, high RMS
Primary Forest	Vine forest, Babaçu Palm	Primarily shade and green leaves, some NPV
Regeneration	Regeneration, shifting agriculture, plantations	Primarily GV and lesser shade, some exposed soil and NPV.
Construction	Urban areas, bare soil, roads	High NPV and soil, high RMS.
Burn	Burned pasture, primary forest or regeneration	High shade, high NPV, high RMS

The simple processes of forest conversion and regeneration can be shown by comparing GV fraction images for 1992 to images for 1993 (see color Plate 16). In this image, the GV fraction for 1993 is loaded as green and blue, while the GV fraction for 1992 is loaded as red. We would expect forest conversion to display as red, due to a high GV fraction in 1992 followed by low values in 1993. Regeneration should be displayed as cyan, reflecting low GV fractions in 1992 and higher values in 1993. Areas of no change should grade between black (low GV both years) to white (high GV both years). As can be observed, a significant amount of the area is in one of these two states, demonstrating the dynamic nature of land cover change in the area. Comparison of GV for 1993 and 1992 on a scatterplot demonstrates how these patterns are reflected in spectral fractions. The large scatter around the 1:1 line shows that even those areas that remain in the same land-cover class actually show a significant amount of change from one year to the next.

Patterns of change over longer time spans can be shown on a simple ternary diagram (Sabol et al., 1995, Figure 9.8). In this figure, shade, GV, and combined NPV and soil are plotted at the corners of a ternary diagram. Fractions for four land-cover classes, burn, pasture, second growth, and primary forest are plotted as triangles, squares, diamonds, and circles. Arrows show expected patterns of change. Short-term transitions include four processes; primary forest to burn (forest conversion); second growth to burn (pasture maintenance); pasture to burn (pasture maintenance), and pasture to second growth (regeneration). Longer-term patterns, which cannot be observed in the 10-year time span here, include natural succession from pasture to second growth, and finally primary forest. Highly degraded soils might be expected to stall natural regeneration at a low biomass shrub land, while the pace of natural regeneration would be expected to vary depending on the length of time and intensity of pasture use (Uhl et al., 1988). Note, given a single observation between years, or a gap of several years between observations, some of these patterns would be expected to go unobserved. For example, most of the image data was acquired prior to the burning season. As a result, transitions to the burn land-cover class will be largely unobserved.

3.4 Multitemporal Classification

The fraction images were classified into the six classes listed in Table 9.3 using endmember fractions and RMS error as input variables into a decision-tree classifier implemented in S-Plus (Chambers and Hastie, 1992; Clark and Pregibon, 1992). A decision-tree classifier is similar to

Patterns of Change

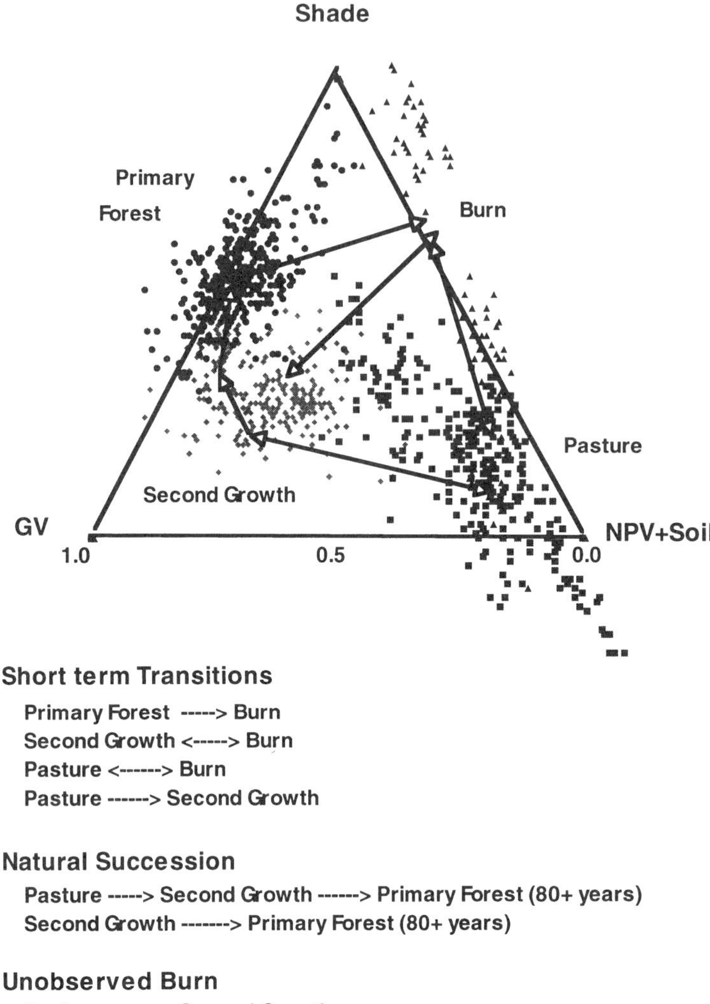

Figure 9.8. Ternary diagram for shade, GV, and Soil+NPV showing expected patterns of change.

a parallelepiped classifier but is hierarchical, developing a series of splitting rules that progressively divide heterogeneous training sets from several classes into more homogeneous groups of pixels within the same class. This technique was chosen because it requires a minimum number of assumptions about statistical properties of the classes, and has the potential of providing sets of rules that have a physical basis. Decision-tree classifiers have been used infrequently in remote sensing, most recently by Hess et al. (1995). In order to account for clouds and cloud shadows, which comprised up to five percent of some of the images, clouds were included as a separate class (number seven). The results presented here do not include a rigorous assessment of classification accuracy. They are provided primarily to illustrate the technique and provide a measure of the dynamic nature of land cover change in the area. A total of

2,189 data points were extracted from training areas to develop the decision tree. Although, in theory, training areas could be located in a single date and used for all of the images, we found that many of the classes required inputs from more than a single date to adequately account for the variation in the class. We determined this to be particularly true for clouds and cloud shadows, which were highly variable and thus difficult to characterize based on a single image. Some degree of temporal variability was observed in all of the classes.

The binary decision tree generated for this study consisted of 33 terminal nodes with an overall classification accuracy better than 75 percent for the training areas. Splitting rules selected by S-plus generally followed patterns that would be expected for the different land cover types. For example, primary forest (PF in Figure 9.9) is classified through a path of high shade (>36 percent), high GV (28 percent), low RMS error (< 5.2), and low soil (< 5 percent). Multiple paths to primary forest reflect natural variability in canopy composition as described by spectral mixtures. Pasture, on the other hand, is classified through low shade (primarily < 48 percent), and high NPV (>34 percent). Cloud obscured areas could be classified as either high or low shade depending on whether they were a shadow or a cloud. Once the tree was established, each pixel was classified hierarchically on the basis of these rules.

In addition to a binary decision-tree classifier, spatial and temporal information were incorporated to improve classification accuracy. Spatial filtering was accomplished using a median filter, which was only applied to a pixel if none or only one of its neighbors fell into the same class (converting it to the most common class within the 3x3 pixel block). A temporal filter was developed to remove physically unreasonable transitions due to classification error and undesirable transitions such as those due to clouds. For example, a direct transition from pasture or regeneration to primary forest is not possible given the short time frame of our study. This type of transition is relatively easy to locate, but not as easy to correct. In this chapter we explored the use of a simple reclassification rule in which pixels that undergo a disallowed transition are reclassified to the classes that bracket the year, provided they are in the same class. For example, a pixel that was primary forest in 1984, became cloud obscured in 1989, then was classified as primary forest again in 1992 would be reclassified as stable primary forest. If two transitions had occurred, such as from primary forest to cloud, followed by pasture, the pixel was not reclassified and stayed cloud. This approach does not solve all problems in classification. For example, misclassified pixels at the start or end date cannot be corrected. Nor does it account for errors in classification that result in allowed transitions.

Disallowed transitions evaluated in this study included:

- Cloud Transitions—Any transition from one land cover class to a cloud or cloud shadow.
- Water Transitions—Any transition from one land cover class to water. Primarily originates because cloud shadows and forest gaps were occasionally misclassified as water.
- Regeneration—Pasture or burn transitions to primary forest. This error can be a result of classification error or the natural process of regeneration where second growth forest takes on the characteristics of primary forest as it ages.

Classified images were generated for the five dates, then spatially and temporally filtered as described. As an example of land-cover transitions in the study region, a 30x30 km subsection is shown for the northwest corner of the quarter scene that is representative of general patterns over much of the quarter scene (see color Plate 17). Overall, the temporal patterns can be characterized by a general increase in pasture and loss of primary forest from 1984 to 1992, followed by continued forest conversion and forest regeneration from 1993 to 1994. In 1984, the region consisted largely of primary forest, with a few large pastures concentrated in the

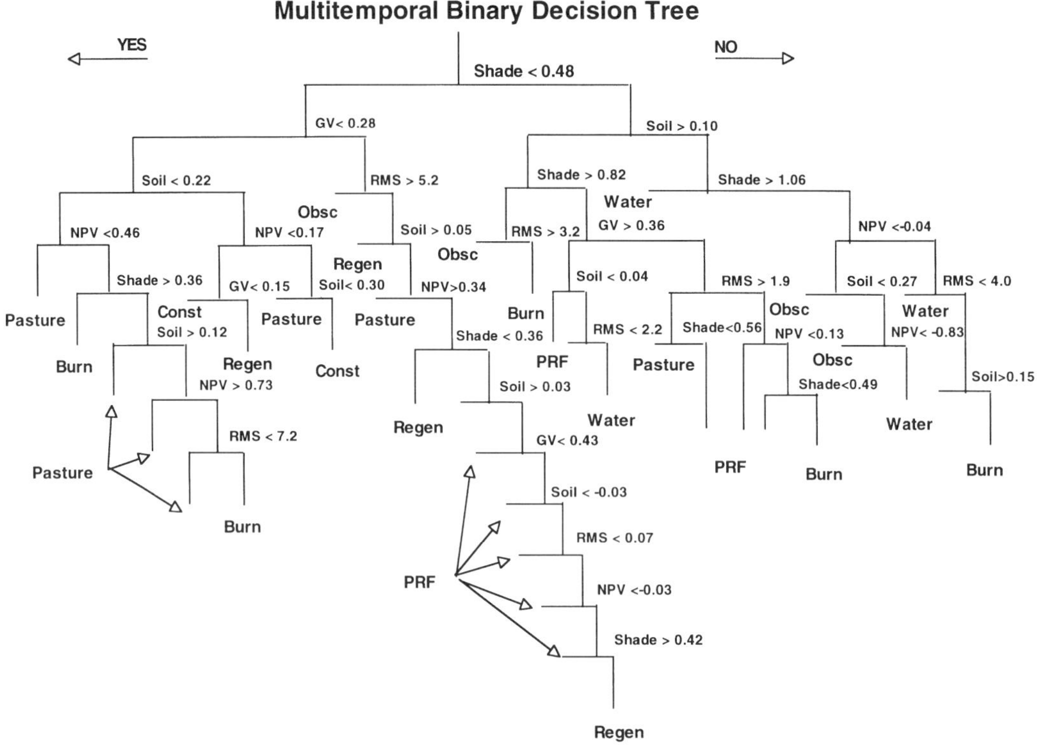

Figure 9.9. Binary decision tree used to classify the images based on spectral fractions and RMS.

eastern portion of the image (Plate 17). Minor amounts of regeneration occurred, primarily within blocks adjacent to larger pastures or in small forest patches in the western part of the image. Between 1984 and 1989 pasture increased significantly at the expense of primary forest, concentrated primarily in a few large cattle ranches expanding in the southeast and west. Second growth remained minor, concentrated primarily in recently cut primary forest rather than within older, established pastures. Between 1989 and 1992, rapid pasture growth continued. However, unlike the earlier time period, the largest growth occurred in small irregular patches within larger blocks of primary forest such as those observed in the central portion of the image. Second-growth remained a minor land-cover class.

After 1992, the area classified as second-growth forest increased dramatically. Initially, in 1993, second growth was concentrated primarily in the smaller patches and probably represented plantations or crops that are spectrally difficult to separate from second growth (Plate 17). Small blocks of second growth also occurred within larger pastures but remained a small portion of the total area. By 1994, second growth was the second most prevalent class, dominating much of the area previously mapped as pasture and expanding into primary forest along the margins of smaller irregular farms.

Differences in spatial patterns between small farms and large pastures represent different styles of land use with very different social and ecological consequences. Large cattle ranches are the predominant style of land use in Amazonia (Fearnside, 1988), primarily resulting from Brazilian fiscal policies that provided favorable subsidies and tax benefits throughout much of the 1980s (Moran, 1993). They support few people, concentrating land and wealth with a few landowners (Hecht and Cockburn, 1989). Ecologically, large cattle ranches concentrate deforestation into a few large patches, resulting in less favorable conditions for regeneration yet also less fragmentation of primary forest. In comparison, small farms support a greater number of people for a given area of forest conversion. Ecologically they divide up the deforested area into many more, small patches resulting in conditions that may be more favorable to regeneration but also lead to greater fragmentation of primary forest.

A quantitative description of the dynamic nature and magnitude of change can be shown by a set of transition matrices for each pair of years for the entire study area (Table 9.4). In this table, transitions are reported as percentages of a total of 8,100,000 pixels (2,700x3,000/image). Rows represent the number of pixels in classes in the first year, while columns represent the second year. Disallowed transitions are italicized, while transitions that represent pasture maintenance, deforestation, and regeneration or shown in bold. Areas that did not change class are underlined.

The dynamic nature of land-cover change is well illustrated in the transition matrices. For this discussion, we will divide the study into two periods, 1984 to 1992 and 1992 to 1994. During the period from 1984 to 1992 the general pattern for the entire area is forest conversion, pasture growth, and pasture maintenance. Based on total percentage area for each year, primary forest declined from 86.6 percent in 1984, to 76.9 percent in 1989, and 65.07 percent in 1992. Annual deforestation rates, by these estimates increased from 1.95 percent between 1984 and 1989 to 3.9 percent between 1989 to 1992. Much of the converted primary forest became pasture, which increased from 8.3 percent in 1984 to 16.8 percent in 1989 and 24.8 percent in 1992. Regeneration, unlike previous observations for Manaus (Lucas et al, 1993; Adams et al., 1995) and Altmira (Moran et al., 1994) remained relatively stable, increasing from 4.4 percent in 1984 to 5.2 percent in 1989, and 5.6 percent in 1992. Much of this increase came at the expense of primary forest, with regeneration losses due to pasture maintenance balanced by conversion of primary forest and subsequent regeneration. For example, between 1984 and 1989, 1.8 percent of the area converted from regeneration to pasture, while 0.5 percent underwent the reverse process, resulting in a net loss of 1.4 percent regeneration to pasture. Between 1989 and 1992 regeneration losses were even greater, at 2.0 percent.

Some measure of the magnitude of classification error can be determined based on the percentage of disallowed transitions in this time period. Disallowed transitions due to clouds and cloud shadows totaled slightly over 1.0 percent of the image between 1984 and 1989 and 2.9 percent between 1989 and 1992, attesting to the difficulty of screening clouds over a changing landscape. Disallowed transitions to primary forest totaled 2.8 percent between 1984 and 1989 and 1.9 percent between 1989 and 1992. Although relatively small, these still may be significant in terms of biogeochemical models. In addition, they represent a potentially significant underestimate of the total error.

The period between 1992 and 1994 showed a very different pattern. Based on total percentage area for each year, primary forest declined from 65.1 percent in 1992, to 60.7 percent in 1993 and 50.8 percent in 1994. These results suggest very large deforestation rates of close to 5.0 percent and 10.0 percent for 1992 and 1993, respectively. Unlike the previous period, a majority of the conversion was to regeneration, not pasture. Patterns of pasture growth and maintenance differed as well. Pasture area decreased between 1992 and 1993 from 24.8 to 18.1 percent, and from 18.1 to 16.2 percent between 1993 and 1994. Regeneration, according to this model, increased dramatically from 5.6 percent in 1992 to 17.1 percent in 1993 and 28.9 percent in 1994.

Based on our analysis, two patterns emerge for the area. For the first eight years, we observed a pattern of deforestation, pasture maintenance, and slowly increasing regeneration. Over the last two years, we observed faster rates of regrowth and deforestation. Two hypoth-

Table 9.4. Transition Matrices: 1984 to 1994. (Disallowed transitions are italicized, while transitions that represent pasture maintenance, deforestation, and regeneration are shown in bold. Areas that did not change class are underlined.)

				1984 to 1989				
1984	**water**	**pasture**	**regen**	**Primary F**	**obscured**	**cons**	**burn**	**total 84**
water	0.13	***0.058***	***0.01***	***0.076***	***0.012***	***0.003***	***0.002***	**0.291**
pasture	***0.104***	6.34	**0.455**	***1.254***	***0.033***	0.045	**0.051**	**8.282**
regen	***0.044***	**1.816**	0.906	***1.499***	***0.049***	0.02	**0.015**	**4.349**
Primary F	***0.277***	**8.358**	**3.747**	73.872	***0.127***	0.041	**0.167**	**86.589**
obscured	***0.017***	0.051	0.016	0.131	0.029	0.008	0.001	**0.252**
cons	***0.004***	0.11	0.004	0.004	***0***	0.017	0.001	**0.141**
burn	***0.002***	0.057	0.008	***0.024***	***0***	0	0.004	**0.096**
Total 89	**0.578**	**16.79**	**5.146**	**76.86**	**0.25**	**0.134**	**0.241**	99.999

				1989 to 1992	1992			
1989	**water**	**pasture**	**regen**	**Primary F**	**obscured**	**cons**	**burn**	**Total 89**
water	0.158	***0.243***	***0.023***	***0.062***	***0.05***	***0.003***	***0.038***	**0.577**
pasture	***0.123***	13.329	**0.775**	***1.204***	***0.138***	0.073	**1.149**	**16.79**
regen	***0.035***	**2.795**	1.266	***0.616***	***0.157***	0.052	**0.227**	**5.147**
Primary F	***0.805***	**8.112**	**3.471**	63.112	***0.862***	0.076	**0.424**	**76.861**
obscured	***0.017***	***0.048***	***0.017***	***0.05***	0.101	***0.001***	***0.016***	**0.251**
cons	***0.001***	0.107	0.002	0.002	***0.006***	0.002	0.014	**0.134**
burn	***0.005***	**0.169**	**0.023**	***0.026***	***0.002***	0.002	0.014	**0.24**
Total 92	**1.144**	**24.803**	**5.577**	**65.072**	**1.316**	**0.209**	**1.882**	100.003

1992 to 1993

1992	water	pasture	regen	Primary F	1993 obscured	cons	burn	Total 92
water	0.165	0.314	0.115	0.5	0.037	0.006	0.004	1.143
pasture	0.417	13.615	7.798	1.18	0.655	1.091	0.046	24.803
regen	0.028	0.531	4.802	0.105	0.07	0.033	0.007	5.577
Primary F	0.14	2.582	3.751	58.228	0.263	0.038	0.071	65.072
obscured	0.094	0.225	0.116	0.632	0.223	0.021	0.005	1.315
cons	0.003	0.125	0.063	0.006	0.001	0.01	0	0.208
burn	0.174	0.682	0.493	0.058	0.272	0.196	0.007	1.882
Total 93	1.021	18.074	17.138	60.709	1.521	1.395	0.14	99.998

1993 to 1994

1993	water	pasture	regen	Primary F	1994 obscured	cons	burn	Total 93
water	0.306	0.185	0.39	0.069	0.04	0.031	0.001	1.022
pasture	0.25	9.831	6.733	0.962	0.15	0.134	0.013	18.075
regen	0.153	2.46	13.084	1.168	0.193	0.077	0.004	17.139
Primary F	0.841	2.38	7.522	48.417	1.29	0.215	0.044	60.709
obscured	0.111	0.297	0.826	0.111	0.12	0.054	0.002	1.521
cons	0.023	0.995	0.273	0.017	0.006	0.08	0	1.395
burn	0.003	0.039	0.069	0.02	0.003	0.001	0.003	0.139
Total 94	1.687	16.187	28.897	50.764	1.802	0.592	0.067	99.996

eses may account for the change we observed between these two time periods. One hypothesis would be that the deforestation rates and maps of second growth for the 1993 to 1994 time period are erroneous due to misclassification. This hypothesis is supported in part by fieldwork in 1994, which found some errors in large pastures where green grass was misclassified as regeneration, and errors in some areas of primary forest, which were also misclassified as regeneration. The errors may arise in part because our model fails to account for interannual variability in precipitation and the fact that the images were not collected at the same seasonal cycle. The 1992, 1993, and 1994 images were acquired on August 10, July 20, and July 13, respectively, forming a temporal transect that progresses toward the rainy season. The resulting shift toward a period when both pastures and vine forest are greener (Bohlman, 1997) would result in overclassification of regeneration at the expense of the other two classes. Potential solutions for this type of error include analysis of additional images collected later in the dry season for 1993 and 1994 or adjustment of class boundaries to account for phenological variability within a class (e.g., Hall et al., 1991a).

An alternative hypothesis is that known classification errors for 1993 and 1994 are relatively minor and that most of the deforestation and regeneration modeled were real. To test this hypothesis, subsets of the image were examined in detail and compared to the original, unclassified imagery. In most cases, visual analysis of the unclassified images supported an interpretation of second growth. In order to choose between these two hypotheses, a more detailed accuracy assessment is currently underway using aerial photographs collected in 1994.

Other transitions are also of some interest. Although the area of burns and construction are never large, they show clear patterns. The percentage of area burned varied between the years, with the highest area occurring in 1992 (1.9 percent), followed by 0.2 percent in 1989 and 0.1 percent in 1984. Both 1993 and 1994 had very low fractions. These results are probably more representative of the temporal proximity of the images to the peak season for burning, than real rates of burning. Areas classified as construction increase unidirectionally from 1984 to 1994, increasing from 0.1 percent in 1984 to 1.4 percent in 1994. This pattern most likely represents increases in bare soil, roads, and urban areas as development progresses.

4.0 SUMMARY

In this chapter we present a general strategy for using multitemporal spectral mixture analysis for identifying change. We believe the techniques are powerful, providing a direct measure of what is changing and by how much. The techniques are portable across sensors and across geographic space. The technique requires careful attention to atmospheric correction and careful selection of reference endmembers which become standards for identifying change. The general strategy described here represents a collection of techniques that have either been adapted from others or developed to provide complete analysis from reflectance retrieval, to intercalibration, and finally analysis of fractions.

In order to illustrate an application of the technique, we provide an example derived from research in the humid tropics of South America. We focus on an area that has undergone rapid changes in land cover over the past few decades, located along the eastern fringe of the Amazon Basin. Although the results are preliminary, they provide some measure of the potential of the technique. Changes in spectral fractions show an incredibly dynamic landscape in which change is the rule, rather than the exception over 10 years. We identify the nature of change in terms of changes in the physical properties of the land cover classes, then extend the analysis to classify the multitemporal data sets. Although seasonal effects and classification errors provide some confusion, a coherent story of deforestation, pasture maintenance, and slowly increasing regeneration emerges.

ACKNOWLEDGMENTS

Funding for this research was provided in part by NASA Grant NAGW-2652, University of Washington Subcontract 470097, and the University of California, which supplied start-up funds for computing facilities. The authors wish to thank David Lawson and Brad Allen for designing the index map to the study region.

REFERENCES

Adams, J.B., M.O. Smith, and A.R. Gillespie. Imaging spectroscopy: interpretation based on spectral mixture analysis. In *Remote Geochemical Analysis: Elemental and Mineralogical Composition*, C.M. Pieters and P. Englert, Eds., 7, Cambridge University Press, New York, 1993, pp. 145–166.

Adams, J.B., D.E. Sabol, V. Kapos, R. Almeida Filho, D.A. Roberts, M.O. Smith, and A.R. Gillespie. Classification of multispectral images based on fractions of endmembers: Application to land-cover change in the Brazilian Amazon. *Remote Sens. Environ.,* 52, 137–154, 1995.

Anderson, A.B. Smokestacks in the rainforest: industrial development and deforestation in the Amazon Basin. *World Development*, 18(9), 1191–1205, 1990.

Boardman, J.W. Inversion of imaging spectrometry data using singular value decomposition. *Proceedings IGARSS 1989*, Vancouver, BC, 1989, pp. 2069–2072.

Bohlman, S. Seasonal foliage changes in the eastern Amazon detected from Landsat Thematic Mapper satellite images. *Biotropica.*, in press, 1997.

Borel, C.G. and S.A.W. Gerstl. Nonlinear spectral mixing models for vegetative and soil surfaces. *Remote Sens. Environ.*, 47, 403–416, 1994.

Brondizio, E.S., E.F. Moran, P. Mausel, and Y. Wu. Land use change in the Amazon estuary: patterns of Caboclo settlement and landscape management. *Human Ecology*, 22(3), 249–278, 1994.

Chambers, J.M. and T.J. Hastie. *Statistical Models in S. Pacific Grove, CA.* Wadsworth and Brooks/Cole, 1992.

Che, N. and J.C. Price. Survey of radiometric calibration results and methods for visible and near infrared channels of NOAA-7, -9 and -11 AVHRRS. *Remote Sens. Environ.*, 41, 19–27, 1992.

Clark, L.A. and D. Pregibon. Tree-based models. In: *Statistical Models in S. Pacific Grove, CA,* J.M. Chambers and T.J. Hastie, Eds., Wadsworth and Brooks/Cole, 1992, pp. 377–420.

Conel, J.E. Determination of surface reflectance and estimates of atmospheric optical depth and single scattering albedo from Landsat Thematic Mapper data. *Int. J. Remote Sens.,* 11, 783–828, 1990.

Culf, A.D., J.L. Esteves, F.A.O. Marques, and H.R. da Rocha. Radiation, temperature and humidity over forest and pasture in Amazonia. In: *Amazonian Deforestation and Climate.* J.H.C. Gash, C.A. Nobre, J.M. Roberts, and R.L. Victoria, Eds., John Wiley & Sons, New York, 1996, pp. 175–191.

Detwiler, R.P. and C.A.S. Hall. Tropical forests and the global carbon cycle. *Science*, 239, 42–47, 1988.

Elvidge, C.D. and F.P. Portigal. Change detection in vegetation using 1989 AVIRIS data. *Proceedings SPIE Imaging Spectroscopy of the Terrestrial Environment*, G. Vane, Ed., Orlando, FL, April 16–17, 1990, pp. 178–189.

Fearnside, P.M. and E. Salati. Explosive deforestation in Rondonia, Brazil. *Environmental Conservation*, 12(4), 355–356, 1985.

Fearnside, P.M. An ecological analysis of predominant land uses in the Brazilian Amazon. *The Environmentalist*, 8(4), 281–300, 1988.

Foody, G.M., G. Palubinskas, R.M. Lucas, P.J. Curran, and M. Honzak. Identifying terrestrial carbon sinks: classification of successional stages in regenerating tropical forest from Landsat TM data. *Remote Sens. Environ.*, 55, 205–216, 1996.

Gao, B.C., K.B. Heidebrecht, and A.F.H. Goetz. Derivation of scaled surface reflectance from AVIRIS data. *Remote Sens. Environ.*, 44, 165–178, 1993.

Gash, J.H.C., C.A. Nobre, J.M. Roberts, and R.L. Victoria. An overview of ABRACOS. In: *Amazonian Deforestation and Climate*, J.H.C. Gash, C.A. Nobre, J.M. Roberts, and R.L. Victoria, Eds., John Wiley & Sons, New York, 1996, pp. 1–14.

Gilabert, M.A., C. Conese, and F. Maselli. An atmospheric correction method for the automatic retrieval of surface reflectance from TM images. *Int. J. Remote Sens.,* 15(10) 2065–2086, 1994.

Gillespie, A.R., M.O. Smith, J.B. Adams, S.C. Willis, A.F. Fischer, and D.E. Sabol. Interpretation of residual images: spectral mixture analysis of AVIRIS images, Owens Valley, California. *Proceedings 2nd Airborne Visible/Infrared Imaging Spectrometer (AVIRIS) Workshop*, R. Green, Ed., Pasadena, CA, June 4–5, JPL Publication No. 90-54, 1990, pp. 243–270.

Golub, G. and C. Van Loan. *Matrix Computations*. Johns Hopkins Press, Baltimore, MD, 1983.

Green, R.O., J.E. Conel, and D.A. Roberts. Estimation of aerosol optical depth and additional atmospheric parameters for the calculation of apparent surface reflectance from radiance measured by the Airborne Visible-Infrared Imaging Spectrometer (AVIRIS). *Summaries of the 4th Annual JPL Airborne Geoscience Workshop,* Washington DC, Oct 25–29, 1, 73–76, 1993.

Hall, F.G., D.B. Botkin, D.E. Strebel, K.D. Woods, and S.J. Goetz. Large-scale patterns of forest succession as determined by remote sensing. *Ecology,* 72, 628–640, 1991a.

Hall, F.G., D.E. Strebel, J.E. Nickeson, and S.J. Goetz. Radiometric rectification, toward a common radiometric response among multidate, multisensor images. *Remote Sens. Environ.*, 35, 11–27, 1991b.

Hecht, S. and A. Cockburn. *The Fate of the Forest*. Verso, London, 1989.

Hess, L.L., J.M. Melack, S. Filsos, and Y. Wang. Delineation of inundated area and vegetation along the Amazon floodplain with the SIR-C synthetic aperture radar. *IEEE Trans. Geosci. Remote Sens*., 33(4), 896–904, 1995.

Huete, A.R. Separation of soil-plant spectral mixtures by factor analysis. *Remote Sens. Environ.,* 19, 237–251, 1986.

Johnson, P.E., M.O. Smith, S. Taylor-George, and J.B. Adams. A semiempirical method for analysis of the reflectance spectra of binary mineral mixture. *J. Geophys. Res.*, 88, 3557–3561, 1983.

Kaufman, Y.J. *Theory and Applications of Optical Remote Sensing*. G. Asrar, Ed., John Wiley & Sons, New York, 1989.

Lean, J. and D. Warilow. Simulation of the regional climate impact of Amazon deforestation. *Nature*, 342, 411–412, 1989.

Lovejoy, T.E. and R.O. Bierregaard. Central Amazonian forests and the minimum critical size of ecosystems. In *Four Neotropical Rain Forests,* A. Gentry, Ed., Yale University Press, New Haven, CT, 1990, pp. 60–74.

Lucas, R.M., M. Honzak, G.M. Foody, P.J. Curran, and C. Corves. Characterising tropical secondary forests using multi-temporal Landsat sensor imagery. *Int. J. Remote Sens*., 14, 3061–3067, 1993.

Lugo, A.E. and S. Brown. Tropical forests as sinks of atmospheric carbon. *Forest Ecol. Manage.* 54(1), 239–255, 1992.

Markham, B.L. and J.L. Barker. Radiometric properties of U.S. processed Landsat MSS data. *Remote Sens. Environ*., 22, 39–71, 1987.

Moran, E.F. Deforestation and land use in the Brazilian Amazon. *Human Ecology*, 21(1), 1–21, 1993.

Moran, E.F., E. Brondizio, P. Mausel, and Y. Wu. Integrating Amazonian vegetation, land-use and satellite data. *BioScience*, 44(5), 329–338, 1994.

Moran, M.S., R.D. Jackson, P.N. Slater, and P.M. Teillet. Evaluation of simplified procedures for retrieval of surface reflectance factors from satellite sensor outputs. *Remote Sens. Environ.*, 41, 169–184, 1992.

Myers, N. Tropical deforestation and remote sensing. *For. Ecol. Manage*., 23, 125–225, 1988.

Price, J.C. How unique are spectral signatures? *Remote Sens. Environ*., 49, 181–186, 1994.

Ray, T.W. and B.C. Murray. Nonlinear spectral mixing in desert vegetation. *Remote Sens. Environ.*, 55, 59–64, 1996.

Roberts, D.A. Separating spectral mixtures of vegetation and soils. *Ph.D. Dissertation*, University of Washington, 1991.

Roberts, D.A., J.B. Adams, and M.O. Smith. Discriminating green vegetation, non-photosynthetic vegetation and soils in AVIRIS Data. *Remote Sens. Environ.*, 44(2/3), 255–270, 1993a.

Roberts, D.A., J.B. Adams, and D.E. Sabol. Remote sensing of vegetation in Amazonia, ecological implications of spectral mixtures. *Presented at the 78th Annual Meeting of the Ecological Society of America*, Madison, WI, July 31–Aug. 4, 1993b.

Roberts, D.A., Y. Yamagvch, and R.J.P. Lyon. Calibration of airborne imaging spectrometer data to percent reflectance using field spectral measurements. In *Proceedings 19th International Symposium on Remote Sensing of Environment*, Ann Arbor, MI, October 21–25, 1985, pp. 619–688.

Roberts, D.A., R.O. Green, and J.B. Adams. Temporal and spatial patterns in vegetative and atmospheric properties determined using AVIRIS. *Remote Sens. Environ.*, in press, 1997.

Sabol, D.E., J.B. Adams, and M.O. Smith. Quantitative sub-pixel spectral detection of targets in multispectral images. *J. Geophys. Res.*, 97, 2659–2672, 1992.

Sabol, D.E., M.O. Smith, J.B. Adams, J.H. Zukin, C.J. Tucker, D.A. Roberts, and A.R. Gillespie. AVIRIS spectral trajectories for forested areas of the Gifford Pinchot national forest. *Summaries of the Fifth Annual JPL Airborne Earth Science Workshop*, R.O. Green, Ed., Pasadena, CA, Jan. 23–26, 1, 133–136, 1995.

Schott, J., C. Salvaggio, and W. Volchok. Radiometric scene normalization using pseudoinvariant features. *Remote Sens. Environ.*, 26, 1–16, 1988.

Shimabukuro, Y.E., B.N. Holben, and C.J. Tucker. Fraction images derived from NOAA AVHRR data for studying the deforestation in the Brazilian Amazon. *Int. J. Remote Sens.*, 15, 517–520, 1994.

Shimabukuro, Y.E. and J.A. Smith. The least-squares mixing models to generate fraction images derived from remote sensing multispectral data. *IEEE Trans. Geosci. Remote Sens.*, N1, 16–20, 1991.

Shipman, H. and J.B. Adams. Detectability of minerals in desert alluvial fans using reflectance spectra. *J. Geophys. Res.*, 92(B10), 10391–10492, 1987.

Shukla, J., C. Nobre, and P. Sellers. Amazon deforestation and climate change, *Science*, 247, 1322–1325, 1990.

Skole, D. and C. Tucker. Tropical deforestation and habitat fragmentation in the Amazon: satellite data from 1978 to 1988. *Science*, 260, 1905–1910, 1993.

Smith, M.O., S.L. Ustin, J.B. Adams, and A.R. Gillespie. Vegetation in deserts: a regional measure of abundance from multispectral images. *Remote Sens. Environ.*, 31, 1–26, 1990.

Stone, T.A. and G.M. Woodwell. Shuttle imaging radar analysis of land use in Amazonia. *Int. J. Rem. Sensing*, 10, 1663–1672, 1988.

Uhl, C., R. Buschbacher, and E.A.S. Serrao. Abandoned pastures in eastern Amazonia: patterns of plant succession. *J. Ecol.*, 76, 663–681, 1988.

Vitousek, P.M. Global environmental change: an introduction. *Annual Review of Plant Ecology and Systematics*, 23, 1–14, 1992.

CHAPTER 10

Seasonal Vegetation Patterns in a California Coastal Savanna Derived from Advanced Visible/Infrared Imaging Spectrometer (AVIRIS) Data

Susan L. Ustin, Dar A. Roberts, and Quinn J. Hart

1.0 INTRODUCTION

There is a need to better understand seasonal and interannual variations in ecosystem characteristics and how these properties interact with the climate system, in order to predict long-term environmental consequences of climate and land use changes on ecosystem function and sustainability (USGCRP, 1996). The consequences of changes in the hydrologic regime, including changes in clouds and radiative feedbacks, precipitation, and evapotranspiration budgets, all of which vary at fine spatial scales and are highly dynamic in time, are among the most critical for understanding global change (IPCC, 1996). Watershed scale studies of biosphere functioning play an important role in this synthesis by contributing to the development and subsequent validation of more physically realistic models. The understanding of processes at scales where experimental control and data reliability are high can be used to extrapolate to continental and global scales, where model predictions cannot be readily tested. The watershed scale is particularly relevant for detailed remote sensing studies with regard to exchanges of gases and energy because land surfaces are heterogeneous at all scales and as such are difficult to measure or model using traditional biometeorological or eddy correlation methods.

The use of broad-band sensors to map vegetation and monitor changes in biophysical attributes using vegetation indices is well established (Justice et al., 1985; Holben, 1986; Tucker et al., 1986a; Malingreau et al., 1989; Sellers et al., 1989, 1992; Hall et al., 1991; Lucas et al., 1993; Skole and Tucker, 1993). Most of this research has focused on the use of vegetation indices (VIs) that use ratios of red to near-infrared (NIR) reflectance, measured from sensors such as Landsat Multispectral Scanner (MSS), Thematic Mapper (TM), and the Advanced Very High Resolution Radiometer (AVHRR), because of the availability of repeat sampling and long-term data sets.

Among the most common approaches for estimating remotely sensed vegetation properties has been to examine space or time dependent patterns in the Normalized Difference Vegetation Index (NDVI) with the assumption that fluxes are proportional to the amount of green leaf area or biomass (e.g., Tucker et al., 1986a). Vegetation types with different canopy architectures, plant densities, and phenology that have differing flux rates may not be well characterized by considering only the foliar greenness as estimated by a vegetation index. This concern is especially critical for evergreen communities that experience significant seasonal variation in biologic activities but often with little change in foliar biomass (Gamon et al., 1995). In central California, low light levels and cool temperatures may limit winter activity and drought may limit physiological activity in the summer and autumn despite canopy greenness.

For annual species, NDVI generally responds linearly over a range of leaf area up to a Leaf Area Index (LAI) of about three (Sellers, 1985; 1987). Many studies have supported the near-linear correlation between NDVI and fraction of photosynthetically active radiation absorbed by the canopy (FPAR), over a wide range of canopy types on both empirical and theoretical bases (Gallo et al., 1985; Sellers, 1985; Hall et al., 1992; Myneni and Williams, 1994). However, most studies have not examined relationships across a range of canopy types; instead, most have been of relatively uniform annual (herbaceous) vegetation under nonstressed conditions (Sellers, 1987; Gamon et al., 1993) or vegetation patterns at very large regional to global scales (Goward et al., 1985; Tucker et al., 1986b; Running and Hunt, 1993; Gash et al., 1994). The NDVI relationship becomes more nonlinear in vertically structured forest communities, and obvious problems arise during periods of physiological suppression when green foliage is still present. Gamon et al. (1995) examined the effect of reduced radiation-use efficiency on NDVI for woodland savanna plants under stressed conditions at Stanford University's Jasper Ridge Biological Preserve in the central coast range of California. They examined canopy NDVI of shrub and grassland species and found that low NDVI values (< 2.0) in the annual vegetation types, where Δ LAI is proportional to phenologic activity, provided a good indication of canopy structure and chemical composition. However, higher NDVI values were insensitive to canopy structure and chemical content, and were poor predictors of photosynthetic capacity and evapotranspiration rates in evergreen species. In a previous study of the grasslands, Gamon et al. (1993) showed spatial vegetation patterns in canopy greenness matched generalized community distributions when derived from the full reflected solar spectrum over the 400–2500 nm range rather than just NDVI alone. Smith et al. (1990b) also found that using a mixture analysis that included shortwave-infrared (SWIR) bands produced better green foliage fractions and a better estimate of evapotranspiration than did VIs alone. Thus developing methodologies using additional spectral regions in addition to the red and NIR bands may be useful for vegetation mapping and condition assessment studies.

In recent years, it has become possible to remotely measure changes in other canopy properties using imaging spectrometry, e.g., with NASA's Advanced Visible Infrared Imaging Spectrometer (AVIRIS). However, to date there are few examples of multitemporal imaging spectrometry data in the literature (e.g., Elvidge and Portigal, 1990; Roberts et al., 1997). Nonetheless, airborne and spaceborne imaging spectrometry can contribute significantly to an improved monitoring program for important environmental variables that are not accessible to broad-band sensors. The Lewis HyperSpectral Imager (HSI), launched in 1997, would have been the first spaceborne imaging system (DeLong et al., 1995; Staenz, 1995). However, more than 20 airborne HSI systems currently exist and several are planned or ready for launch. As more of these data sets become available, it is expected that HSI change detection applications will increase.

Imaging spectrometry is the acquisition of a contiguous spectrum across the visible and reflected infrared wavelengths for every pixel. Airborne and satellite imaging spectrometry offers the potential to identify a wide range of geologic and biogeochemical materials (Goetz et al., 1985; Kruse et al., 1993), soil properties (Palacios-Oreuta and Ustin, 1996), plant properties such as pigments (Gamon et al., 1993, 1995), liquid water thickness (Gao and Goetz, 1990, 1995; Green et al., 1991; Roberts et al., 1997), and leaf litter and woody stems (Roberts et al., 1993; Hart et al., 1992, 1993; Ustin et al., 1996). These sensors provide several significant characteristics needed for change detection, including better characterization of atmospheric and surface properties in terms of their spectral attributes, thus allowing better estimates of physical processes (Vane et al., 1993). Because better surface reflectance calibration can be obtained, imaging spectrometers can be used to calibrate other sensors, making it possible to use heterogeneous sensors for data fusing and multitemporal remote sensing. Because of the emphasis on spectral absorption features, radiometric calibration becomes important in order to retrieve surface reflectance. Use of this spectrally derived information should provide a basis for greater understanding of functional ecosystem processes.

To illustrate some of these features, we have examined the relationship between vegetation structure patterns, foliage distribution, and water availability for three dates in 1992 over a semiarid region in the central Coast Range of California. The plant communities at the site have significantly different phenological patterns that produce major functional differences in terms of canopy development, energy balance, gas exchange properties, and nutrient dynamics. If hypotheses about the biochemical determinants of leaf and canopy reflectances are correct, then the functional differences between these communities should be apparent in comparisons of canopy reflectance between communities and over time. We compare seasonal vegetation patterns when energy balance conditions are markedly different to explore the use of AVIRIS data for monitoring seasonal changes in surface and atmospheric properties.

2.0 BACKGROUND

2.1 Study Site

Stanford University's Jasper Ridge Biological Preserve (JRBP), is located approximately seven km west of Palo Alto, California, along the northeastern margin of the Santa Cruz Mountains (centered around 37°24′N, 122°13′30″W). The preserve is approximately 500 ha of natural and disturbed communities of the central California coast. Natural communities include redwood and mixed evergreen forests in the Santa Cruz Mountains to the west, evergreen woodlands on north and northeast facing slopes, deciduous oak woodlands at lower elevations, chaparral on south and west facing slopes, riparian forests in drainages, and introduced annual grasslands on southeast facing sites of low relief and unstable slopes. Figure 10.1 shows the distribution of the major vegetation types at Jasper Ridge with aspect. The preserve itself has less relief than the surrounding coastal mountains and so the topographic dependence of the vegetation types are less distinct than is characteristic of the entire area. The preserve is surrounded by dense urban use exhibiting a wide variety of land cover types. The area has two predominant soil types derived from the metamorphic Franciscan formation: serpentine and Franciscan greenstone (Page and Tabor, 1967). Serpentine endemic communities include serpentine grassland and serpentine chaparral. The contrasting nutrition and water-holding capacity of these soils produces markedly different productivity rates and species diversity over short distances. Elevation ranges from 600 meters in the Santa Cruz Mountains and to 20 meters in the vicinity of Palo Alto to the east. The area has a Mediterranean climate, characterized by mild winters with a winter precipitation regime and up to eight months of drought. Annual precipitation averages 48 cm, with nearly all rainfall between December and March (Thomas, 1961). Annual mean temperature is 15°C, ranging from 9°C in January to 20°C in July.

2.2 Imaging Spectrometry Data

Three AVIRIS data sets were acquired in 1992 along south-east/north-west trending flight lines on June 2 and September 4, and on an east-west trending flight line on October 6. AVIRIS

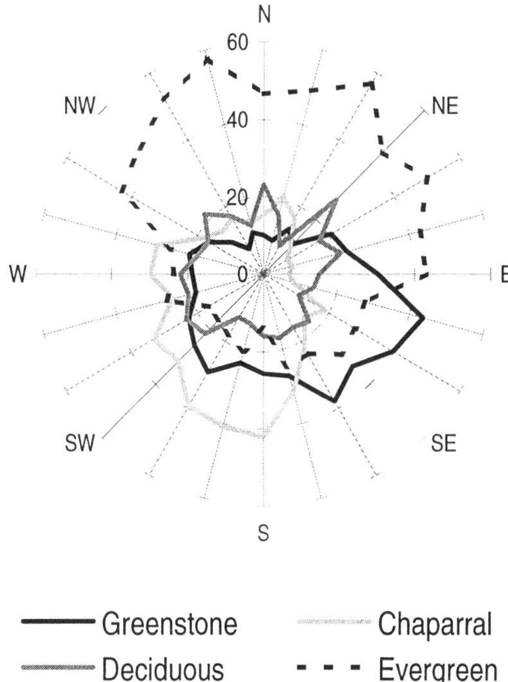

Figure 10.1. Aspect diagram showing the relative distribution of four major vegetation types, annual grasslands (identified as greenstone soil, the most abundant grassland type), deciduous oak woodlands, various chaparral communities, and mixed coastal evergreen forest) within the JRBP related to topographic position. The distance out from the origin indicates the relative abundance of each type on slopes of different aspect.

acquires continuous spectra in the 390 to 2500 nm spectral region in 224 bands with a nominal spectral response of 10 nm (Vane et al., 1993). The AVIRIS scenes are approximately 11 kilometers (long-track) by 8 kilometers (side-track) with a nominal instantaneous field-of-view of 20 meters on the ground. Sky conditions were clear and haze was slight during all of these acquisitions. At least 15 sites in the surrounding area were measured for surface reflectance in the field, in addition to sites within the JRBP. These sites represented a range in elevations from 40 to 560 meters and a range in cover types from disturbed and undisturbed grasslands to redwoods. Additional descriptions can be found in Gamon et al. (1993) and Roberts et al. (1997). Natural vegetation types of the region include mixed redwood and Douglas fir forests, evergreen broadleaf forests, forested wetlands, and chaparral. Disturbed areas included a golf course. A polo field was used as a temporally invariant ground target, and reflectance spectra were collected by the Jet Propulsion Laboratory on June 2, 1992, at the time of the first overflight.

2.3 GIS Database and Methods

A digitized version of the vegetation map of the preserve that identified six vegetation types was available for evaluation of image products (Plate 14a). The JRBP vegetation map was produced from aerial photographs and modified slightly by N. Chiariello based on field observations (personal communication). The vegetation map, while simplified, shows considerable spatial heterogeneity within the JRBP, largely due to topography and soil patterns. The map indicates six dominant vegetation types, three herbaceous and three woody tree/shrub types. The two annual grassland types are common in coastal prairies, one characteristic of the general soil type, Franciscan greenstone, the other restricted to serpentine soils. The former are dominated by introduced European annuals, including bromes, oat, and fescue, with California poppy, lupines, mustards, and clover common. The latter are generally sparse, low-growing endemic grasses and forbes that are tolerant of the low calcium soils. The wetlands are dominated by cattail, bulrush, water plantain, and other emergent aquatics. Chaparral includes several types of shrub communities, either mostly pure chamise stands or mixtures dominated by ceanothus or manzanita shrubs and various semiarid shrub species. The deciduous woodlands are dominated by blue or valley oak, California buckeye, and scrub oak species. The evergreen woodlands are dominated by coast live oak, California bay tanoak, madrone, and toyon. Douglas fir and redwood may be present.

A digital elevation model (DEM) of the USGS Menlo Park 7.5 minute quadrangle map was used to create slope and aspect maps in the public domain geographic information system (GIS) software package from the U.S. Army Corps of Engineers, GRASS. The DEM was resampled to 20 meter pixels. All output images were coregistered to the June 1992 data, then georeferenced to the DEM. To register the October data set to September and June, parts of two AVIRIS scenes had to be merged and rotated 180°. Georeferencing was accomplished using 68 tie points extracted from the June 1992 image and the DEM. The JRBP vegetation map was coregistered to the base map. From the DEM, solar irradiance was computed for all raster cells in the site on a monthly and annual scale (Plate 14b, Figure 10.2). The parameters used to compute the direct and diffuse radiation assumed a constant clear sky condition, equivalent to a standard optical depth and single scattering albedo, and did not account for potential cloud or fog effects. Therefore, Plate 14b indicates topographical differences in radiation for the scene. The figure shows the vegetation boundaries superimposed on the radiation field, which clearly illustrates delineations coincident with radiation differences among these communities. We calculated the mean monthly potential radiation for each vegetation type in the preserve. Because the radiation regime of grasslands and deciduous woodlands was similar, we combined them to simplify Figure 10.2 to deciduous, chaparral, and evergreen forest. The effect of the annual energy budget on these communities is clearly evident. Annual grasslands and deciduous woodlands are located on sites having significantly higher radiation budgets throughout the year. The higher radiation loads in the winter months contributes to winter germination of grasslands, and peak phenological activity shifted to spring months when other plant communities have low physiological activity.

2.4 Image Calibration Procedures

The approach we used to derive apparent surface reflectance was to calibrate atmospheric absorption and scattering based on a radiative transfer model, Modtran2 (Kneizys et al., 1988). The model uses look-up tables for path radiance and reflected radiance for water vapor for a range of values for specified date, time, location, and meteorological visibility and surface elevation. Modeled radiance is fitted to upwelling measured radiance using a nonlinear least squares fitting routine. In general, 20 bands centered over the 940-nm atmospheric water vapor band are used for the pixel-by-pixel water vapor correction (Green et al., 1991). Additionally, a simple surface reflectance model is used over the narrow spectral region of the liquid water and ice absorptions, which must be calculated to minimize confusion between

Figure 10.2. Summed monthly potential net radiation for the JRBP showing differences in radiation budget between plant communities. The category "Other" includes grasslands and deciduous forest types which exhibited the same radiation patterns. Differences in energy load help explain distribution differences and phenological patterns among communities.

different forms of water (Roberts et al., 1997). The model then generates water vapor, liquid water thickness, and ice images. These images are themselves of interest because they provide a representation of instantaneous atmospheric water vapor and surface liquid water content maps. These new data products, uniquely available from AVIRIS images, have not been sufficiently explored for their potential to contribute to ecosystem information and surface gas exchange processes.

The assumption in this model is that under clear sky summer conditions, detection of liquid water is from the terrestrial surface and largely originates from plant canopies. Rivers and lakes absorb most of the irradiance at all wavelengths, and little energy is reflected to the sensor; therefore, these sources of water are insensitive to this approach. The assumption that the liquid water measured is stored in the canopy is valid during the spring and summer drought months (such as in 1992) when soil surfaces are dry. Because this approach solves for path length, it is not possible to separate the effect of changes in liquid water thickness at the leaf scale from enhanced expression at the canopy scale that is due to multiple near-infrared (NIR) scattering. Consequently, the whole canopy reflectance will probably underestimate the liquid water content.

2.5 Spectral Mixture Analysis

AVIRIS images were modeled using Spectral Mixture Analysis (SMA) (Smith et al., 1990a; Adams et al., 1993) because the selection of endmembers allowed a consistent basis for change detection, based on changes in the proportional abundances of reference endmembers (Ustin et al., 1993; Roberts et al., 1998). SMA has been used at Jasper Ridge to estimate surface composition of the dominant scene components (Ustin et al., 1992, 1994; Gamon et al., 1993; Roberts et al., 1993). One potential advantage for choosing this image processing approach over NDVI is that the surface condition for shade/shadows, soils, and the nongreen (stem and litter) fractions of canopies can be estimated in addition to an estimate of the abundance of green foliage (GV). The shade fraction is functionally the inverse function of surface albedo. Adams et al. (1995) found this set of endmember types were appropriate to classify vegetation characteristics and changes in their abundances in the Brazilian Amazon using Landsat TM. This multitemporal analysis has been extended by Roberts et al. (1998). Thus, SMA potentially provides a benefit for improved plant community classifications based on distributions of continuously varying canopy categories (foliage, nonphotosynthetic canopy components, and bare soil) relevant to ecosystem processes (DeFries et al., 1995).

SMA is based on the assumption that the spectrum measured by a sensor is a linear combination of the spectra of all components within the pixel. For most remote sensing applications, a linear mixing model has been assumed (Graetz and Gentle, 1982; Pech et al., 1986; Smith et al., 1990a; Adams et al., 1993; Shimabukuro et al., 1994), although nonlinear mixing can be important in vegetation where significant multiple scattering of photons occurs (Allen and Richardson, 1968; Huete, 1986; Roberts et al., 1993; Borel and Gerstl, 1994; Myneni et al., 1995; Ray and Murray, 1996). If photons interact only with a single canopy component that has minimal multiple scattering, such as is the case under sparse single-story vegetation, the mixed spectrum can be modeled as the sum of the pure spectral endmembers, weighted by the areal proportion of each material (Smith et al., 1990a). Since most vegetation at JRBP consists of sclerophyllous chaparral shrubs, hardwoods, or conifers, vegetation types which have thicker, less transmittant foliage, or are dry herbaceous/grass litter, should produce relatively linear spectral mixtures (Roberts et al., 1992).

SMA was applied to each image date using a simple model of four basic endmembers. The model estimates the fractional abundance of each endmember in each pixel. Reference endmembers were chosen from a spectral library consisting of laboratory and field measured spectra that were applied to all image data sets. All fractions were constrained to sum to 1.0, although individual fractions are not constrained to lie between 0.0 and 1.0. Fraction errors, resulting in physically unrealistic fractions ($f < 0.0$ or $f > 1.0$) may occur because a limited set of endmembers is used to describe a scene in which variation exists within the four basic materials. The largest source of error is due to confusion between the soil and NPV components, which are only distinguishable by ligno-celloluse absorptions present in organic matter but lacking in soils (Roberts et al., 1993). The SMA was applied to the AVIRIS bands, avoiding noisy bands and those in the atmospheric water absorption regions.

2.6 Spectral Endmember Library

Multiple collections of leaf samples were made at JRBP in June, September, and December, 1992, and May, 1993. Approximately 10 replicate leaf samples were collected from the dominant plant species for spectral analysis (Grossman et al., 1996). Reflectance measurements were made with a NIRSystems Model 6500 spectrometer (NIRSystems Inc., Silver Spring, MD) from 400–2498 nm at 2.0 nm intervals with a full width/half maximum slit width of 10.0 nm. Means and standard deviations were computed for each sample. Surface (top cm) soil samples were collected from serpentine and greenstone derived soils (Gamon et al., 1993). Leaf litter was removed before collection. Dry grass litter was collected from several sample points. Additional field and laboratory spectra were measured on additional soils and plant samples from other sites within the flight lines. Furthermore, Roberts et al. (1997) has a large spectral library of plant and soil spectra from the region which was available for reference.

3.0 SEASONAL CHANGES IN ECOSYSTEM CHARACTERISTICS

3.1 Seasonal Spectral Characteristics of Different Vegetation Types

Figure 10.3 shows how mean AVIRIS reflectance for each vegetation type varies between vegetation types and changes over the growing season. The spectra were obtained in the GIS by averaging the reflectance for all pixels in each type. Grasslands were largely scenesced in June (earliest date) and remained dormant over the summer. Thus, they show little change during this period and AVIRIS reflectance spectra resemble dry plant litter. In contrast, the deciduous, evergreen and chaparral communities all show phenologic changes over the summer. Some of these seasonal changes can be attributed to decreasing solar radiance between June and October and others to changes in the contribution of different scene components (like soil, shade, or litter) to the spectra. However, Roberts et al. (1997) reported that the nearby polo field had less than a one percent difference in reflectance between these dates at most wavelengths, indicating that most changes are due to the plant canopies. The biochemical and physical properties that distinguish different vegetation types are clearly evident. Grasslands have lost the abosrptions resulting from photosynthetic pigments and water while the absorptions due to cellulose features in the 2.0–2.5 μm range are evident. Curran (1989) reports cellulose and lignin absorptions are centered near 1730, 2060, and 2270 nm in the SWIR. Differences in spectral contrast between the NIR and red are evident between chaparral, deciduous, and evergreen woodlands and are probably due to spectral differences between leaves, while differences in NIR between vegetation types are probably due to differences in scattering and shadows in canopies (Roberts et al., 1997). Canopy architectures vary markedly between these vegetation types.

Important products of the reflectance calibration include maps of equivalent liquid water thickness and column water vapor. A map of water vapor for the entire scene is shown for June (Figure 10.4a), September (Figure 10.4b), and October (Figure 10.4c), which can be compared to DEM of the area (Figure 10.4d). In these images, high values of water vapor are displayed as white, and low values as black. Atmospheric water vapor primarily varied with elevation and showed little relationship to the underlying surface cover (Plate 14a). However, even small canyons with less than 5 meters of relief could be located on the water vapor image and could be used to validate the registration. Jasper Ridge (150 meter relief) is observed within the scene as a localized region of low water vapor. Because variability is low within the JRBP, we display the entire scene in order to illustrate the correlation with topography. Lower elevation sites within the Santa Cruz Mountains, such as large canyons are readily apparent in the water vapor map as regions of high water vapor.

Roberts et al. (1997) demonstrated a near-linear negative relationship between water vapor and elevation over the 560-meter range in elevation (June, $r^2 = 0.994$). Their model found column water vapor decreased at 11.9 μm for every meter increase in elevation. An important application of this information is a spatially distributed instantaneous estimate of bulk atmospheric water vapor above the surface boundary layer. Provided elevation is known and estimates for barometric pressure and temperature are provided, it is possible to estimate some instantaneous properties of the bulk atmosphere using the relationship between elevation and water vapor. Roberts et al. (1997) estimated specific humidity from AVIRIS data to be 10.43 g kg^{-1}, which they found compared favorably to ground-measured point estimates of 14.51 g kg^{-1}. The September image showed significantly higher water vapor concentrations while the October image was considerably drier, corresponding to 5.42 g kg^{-1} versus 5.26 g kg^{-1} for September AVIRIS and ground-based estimates, respectively, and for October, 4.42 g kg^{-1}

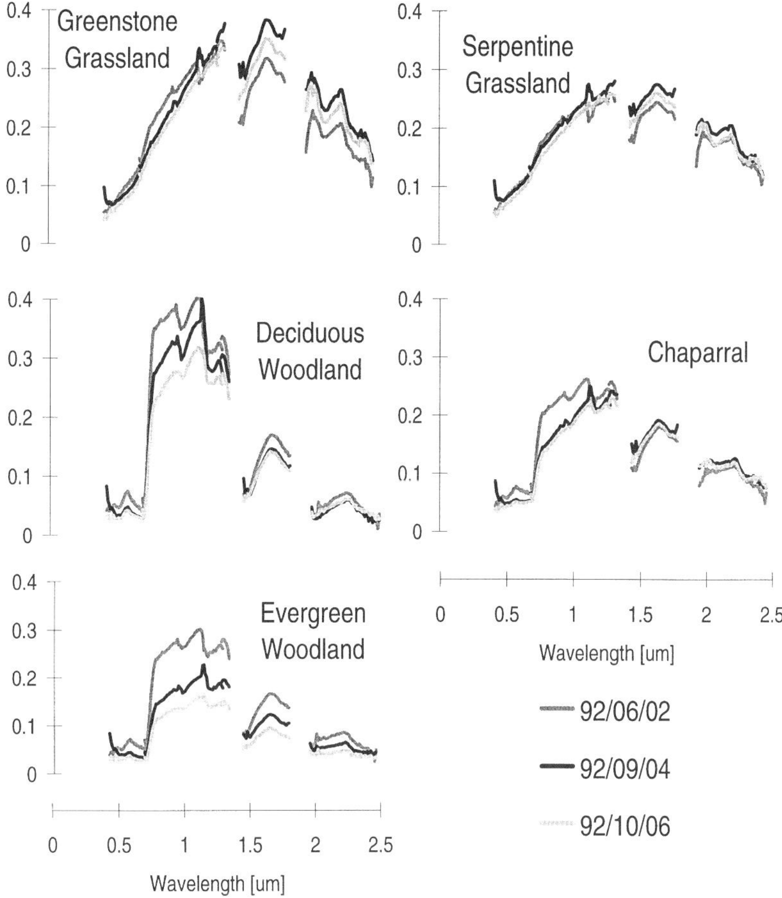

Figure 10.3. Comparison of mean reflectance for each vegetation type at JRBP derived from AVIRIS data for the three dates, from spring to autumn. Significant spectral differences exist between vegetation types and seasonal patterns differ.

versus 5.91 g kg^{-1}. Roberts et al. (1997) concluded that the AVIRIS estimates were remarkably similar, considering the potential complexity of temperature and water vapor vertical patterns within the surface boundary layer.

3.2 Temporal Patterns in Liquid Water

Some of the most interesting spatial patterns occur in the seasonal liquid water images (Figures 10.5a, b, c). As noted previously, the separation between vegetation types is most pronounced in the liquid water images. Increasing liquid water contents (brighter pixels) correspond with expected patterns for increasing leaf area. In our study, these range from low liquid water content surface types, like the senesced grasslands, to high water content sites like the forested wetlands and redwoods. This pattern suggests that because water absorption is less effective than chlorophyll, it does not saturate as quickly and is responsive to a wider range of leaf areas than VIs. Thus, the liquid water absorption should remain more sensitive to increasing foliage than NDVI or GV at high LAIs. This implies that the liquid water absorption band may be

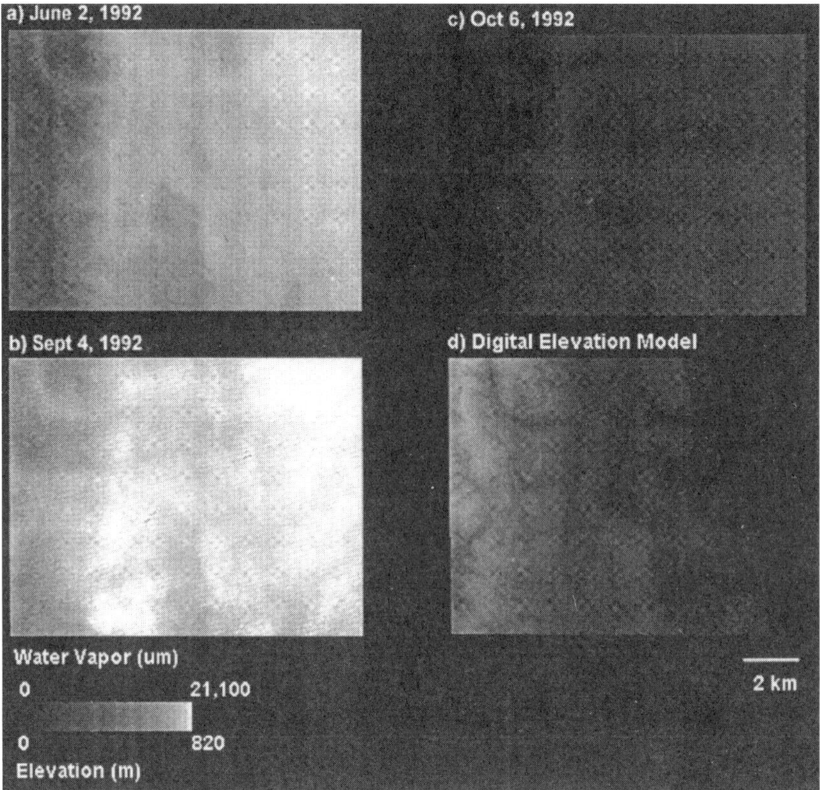

Figure 10.4. Estimated column water vapor for (a) June, (b) September and (c) October 1992, derived from the calibration of the AVIRIS images to surface reflectance. (d) DEM for the scene illustrating the comparison between topography and column water vapor.

better for estimating leaf area in forested areas. This assumption should be valid if the soil and canopy do not have free water on the surface. Such assumptions are reasonable in a Mediterranean climate system like this, which has no summer rain. The water content of leaves from functionally diverse species, such as conifers and deciduous broadleaf trees, are expected to vary considerably, as is demonstrated in the figure. Grossman et al. (1996) reported a mean water content of 183 g m^{-2} that ranges from 53 to 552 g m^{-2} for foliage from common species at Jasper Ridge on samples collected in the summer of 1992. Figure 10.5d shows mean liquid water thickness measured for the communities within the scene which also express an order of magnitude difference. The increased liquid water content in September may have been due to a small precipitation event that produced four mm of rain on August 30 and 31, five days before the overflight. However, it is more likely that the calibration algorithm is inconsistently separating liquid water and water vapor, causing an overestimate in liquid water and underestimate in water vapor at this time.

3.3 Spatial Trends in Surface Properties

The images were modeled as a simple linear spectral mixture of four reference endmembers. These endmembers accounted for over 98 percent of the spectral variability within the scene on

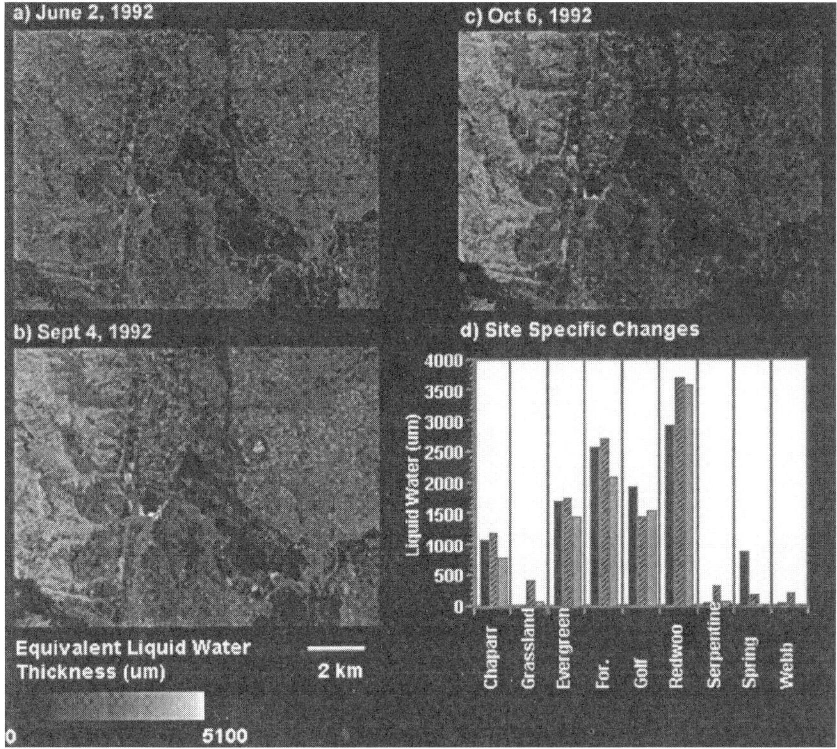

Figure 10.5. Apparent surface liquid water content is shown in (a) June, (b) September, and (c) October. Images show high values as white and low values as black. Differences in tone between dates indicates differences in liquid water contents, while tone within an image provides a measure of spatial distribution in water content. (d) shows liquid water thickness for several vegetation types within the scene on the three dates, June (black), September (hatched), and October (gray). Estimated water content varies over an order of magnitude between vegetation types.

all dates and were used to produce fraction images. In general, the scenes can be characterized as consisting primarily of high fractions of shade and GV, with areas of locally high nonphotosynthetic vegetation (NPV) and occasional patches of bare soil. Pixels with relatively pure soil fractions were restricted to some roads, and the vicinity of the Stanford Linear Accelerator (linear feature observed in Figure 10.5), and urban buildings where the fractions model both soil and materials with similar spectral shapes such as asphalt and roofing material. The spatial consequences for the fraction distributions are apparent when examined in the images. Plate 14c shows endmember changes in the fraction distributions for the site in June. Spectral patterns within the vegetation boundaries show that the actual community distribution is more complex and fragmented than the JRBP map.

In June, NPV fractions were highest (>50 percent) in annual grasslands at low elevation sites which had already undergone senescence (Figure 10.6). These sites also had high canopy shade/shadow fractions, with slightly less than 50 percent shade. NPV and shade fractions varied spatially, with the sites having highest shade contents along the spine of Jasper Ridge (the greenstone grassland identified in Plate 14a). In contrast, at high elevation grassland sites in the Santa Cruz Mountains, NPV fractions were lower and the GV fraction was higher. Such patterns are consistent with a milder climatic gradient delaying senescence at higher elevations.

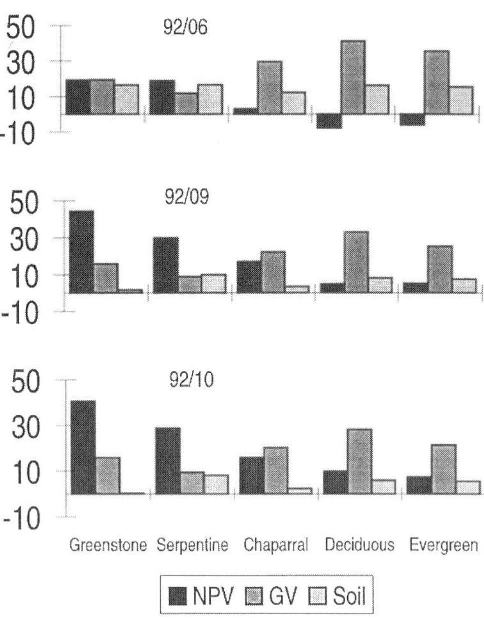

Figure 10.6. Histogram of mean endmember fractions by vegetation type for June, September, and October, with NPV (black), GV (medium gray), and soil (light gray).

The chamise chaparral communities were the only other communities with a NPV fraction greater than 15 percent, which was modeled primarily as equal proportions of NPV and GV. In contrast, GV fractions were highest in the forested wetland and on the nearby golf courses. The forested wetland was modeled primarily as GV (> 60 percent) and canopy shadows (<40 percent). The evergreen forest communities, which were either mixed conifer or coastal broadleaf forests, were modeled with higher shade and lower GV fractions. Higher shade fractions in these communities result from subpixel shadowing by leaves, branches, and trunks as well as shadowing due to topography. The broadleaf forest is found primarily on north facing slopes while the redwoods are predominantly found on northeast facing slopes. High shade fractions were also found in the serpentine grassland and chamise chaparral. Possibly, high shade fractions in these communities may indicate some inability to completely separate soil and shade fractions due to low spectral contrasts. In the serpentine grassland, the soil is much darker than the greenstone soil and has significantly lower reflectance, which with this model is identified as having high shade. Some chaparral species have narrow leathery leaves, such as chemise that typically has low NIR reflectance which tends to be confused with a high shade fraction.

Spatial patterns in endmembers were highly correlated between dates (Plates 14c–e). This temporal consistency allowed us to compare endmember fractions across images acquired on different dates to evaluate changes in ecosystem conditions. The frequency distribution of endmember fractions for these communities is remarkably stable between image dates over the growing season (Figure 10.7). The general trend observed is the fraction of green vegetation decreased over the growing season between May and October images, while fractions of soil and dry plant material increase. These patterns are consistent with expected phenological trends for a Mediterranean grassland savanna ecosystem. Another way to look at seasonal changes is to examine how each endmember fraction changes with community type and over the growing

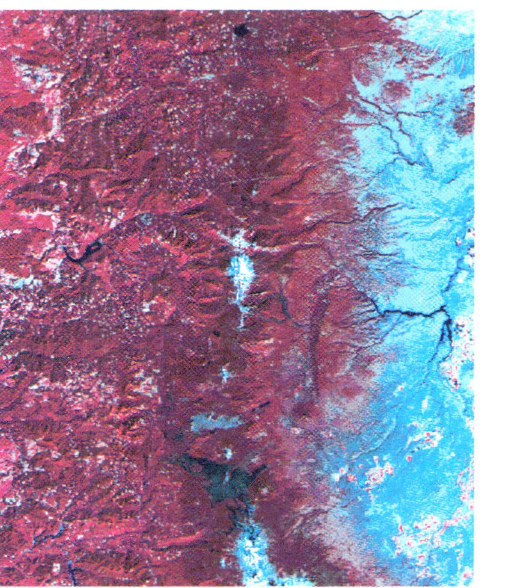

Plate 1. *Landsat MSS images of central Oregon from 1972 and 1992 which have been georeferenced and coregistered as part of the NALC project (Path 45, Row 29). Bands 4,2,1 are displayed as red, green, and blue, respectively. All three images have the same contrast stretch applied. The 1972 image is shown in its raw form on the right, and in the middle following ASCR radiometric normalization (Elvidge et al., 1995).*

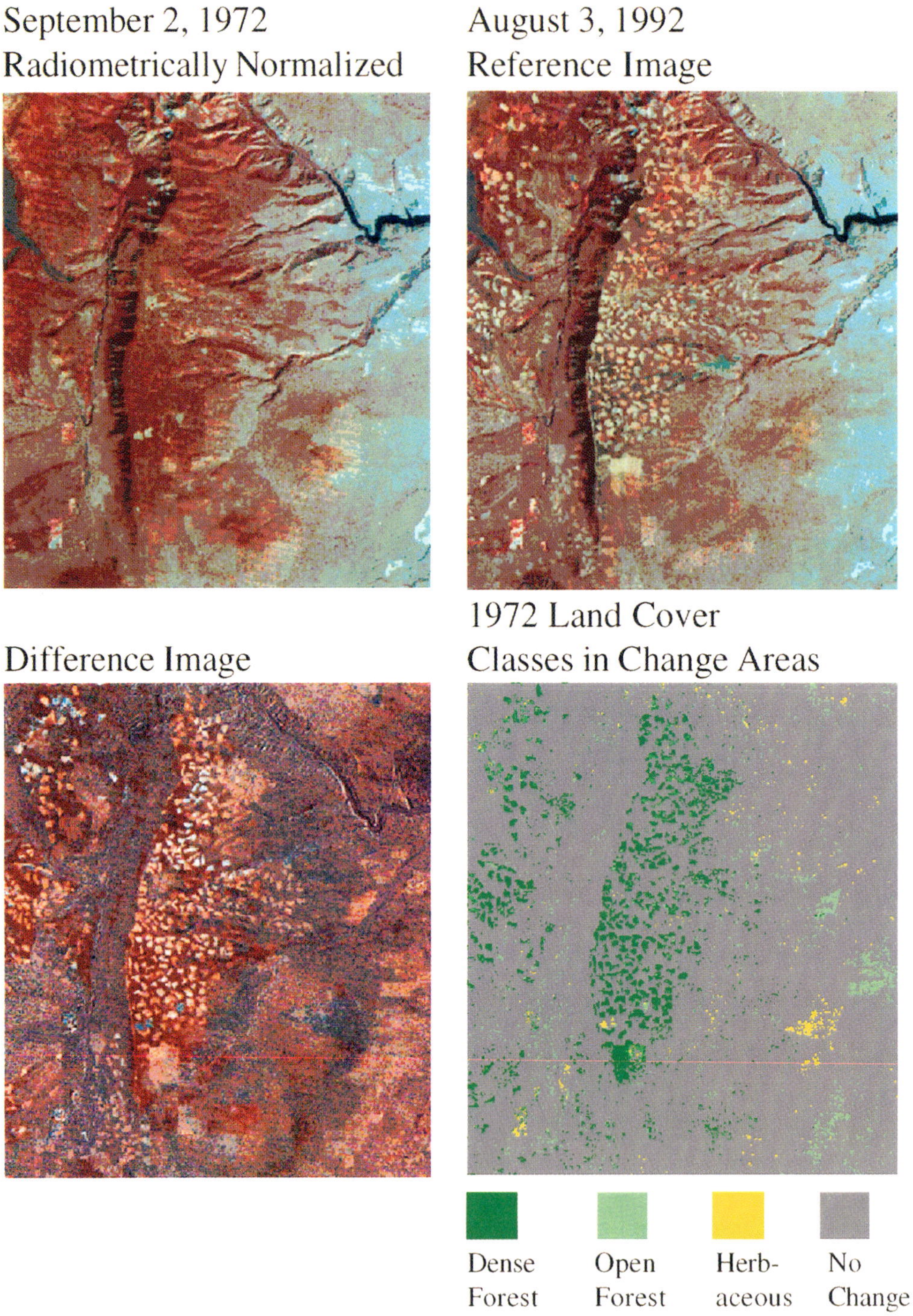

September 2, 1972
Radiometrically Normalized

August 3, 1992
Reference Image

Difference Image

1972 Land Cover
Classes in Change Areas

Dense Forest
Open Forest
Herb-aceous
No Change

Plate 2. *Outline of a hybrid spectral change identification/post-classification land cover change procedure.*

Plate 3. *NALC mosaic image of the Canadian and United States Laurentian Great lakes developed by the Canada Centre for Remote Sensing. A total of 82 Landsat MSS scenes, acquired over the 1985–1987 time window, were merged in order to cover the whole watershed. Note that the MSS data in the mosaic have been aggregated to a pixel resolution of 200 meters.*

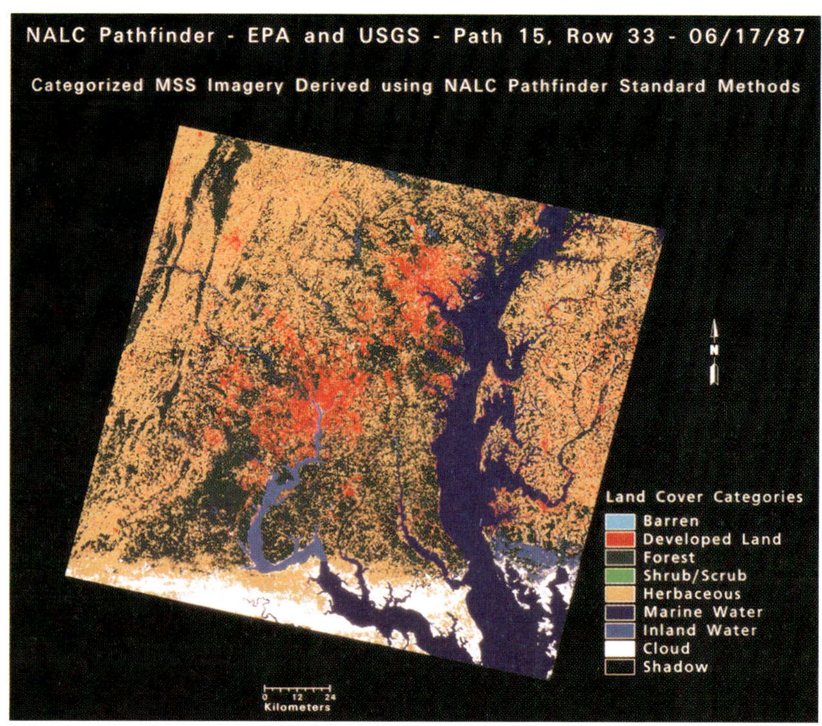

Plate 4. *Categorized June 17, 1987 NALC data from WRS-2 Path 15, Row 33.*

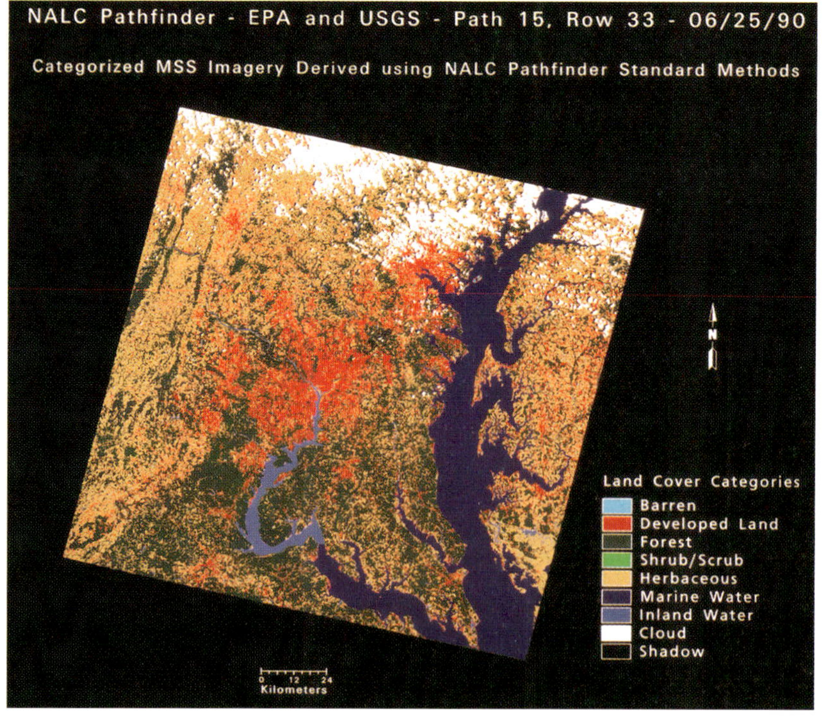

Plate 5. *Categorized June 25, 1990 NALC data from WRS-2 Path 15, Row 33.*

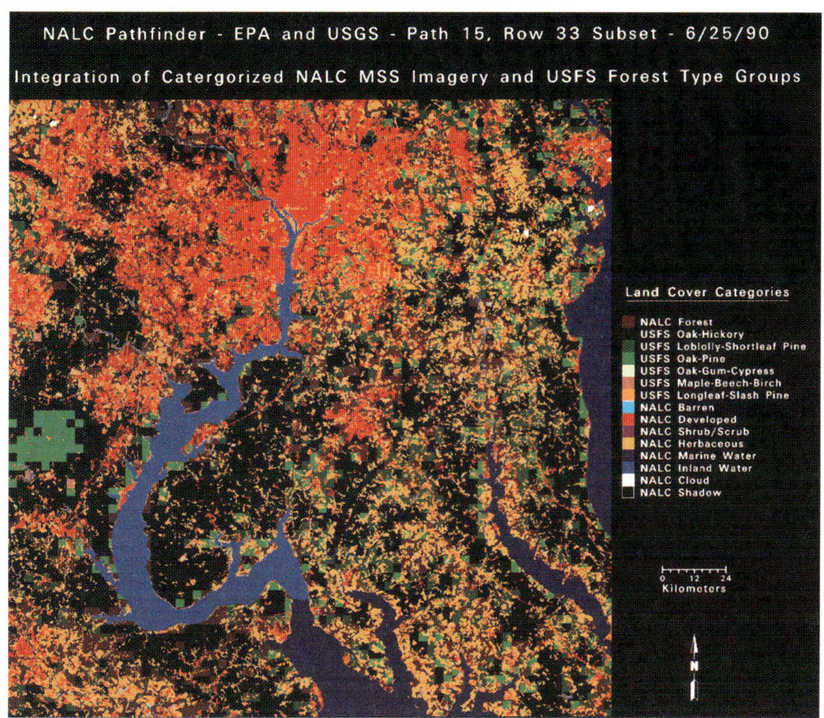

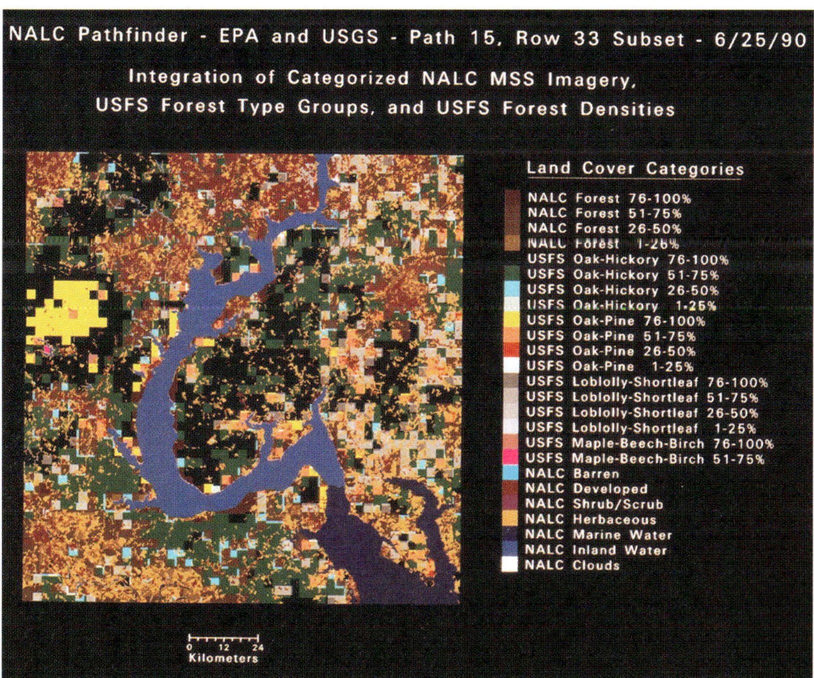

Plate 6. *Examples of "higher order" land cover products for potential application in the calculation of standing carbon stocks. Developed by integrating a NALC categorization product with the U.S. Forest Service forest type data (a) and forest type and forest density data sets (b).*

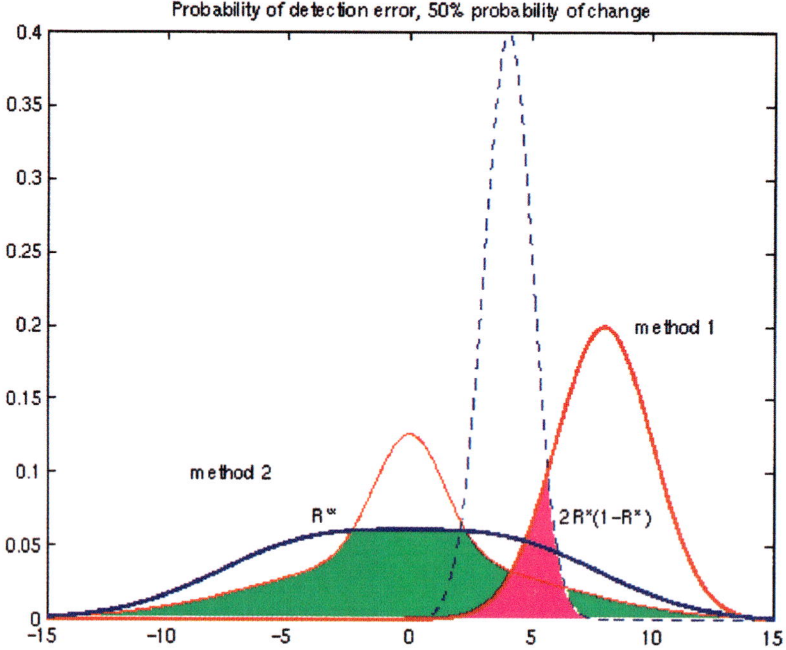

Plate 7. *Graphical representation of the probabilities of error of approaches 1 and 2.*

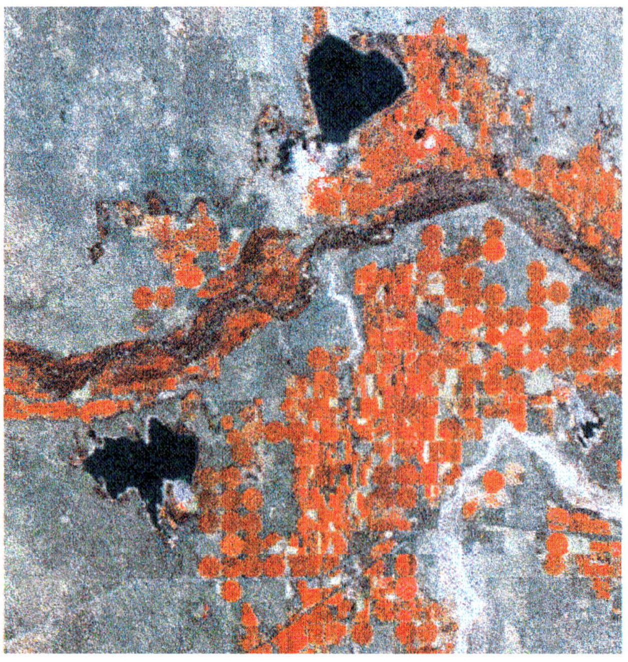

Plate 8. *The effects of NDVI on classification. From top to bottom, left to right: result of classification without NDVI, result of classification with NDVI. In the classified maps, green is agricultural, yellow is rangeland, blue is water, purple is wetland, gray is barren terrain.*

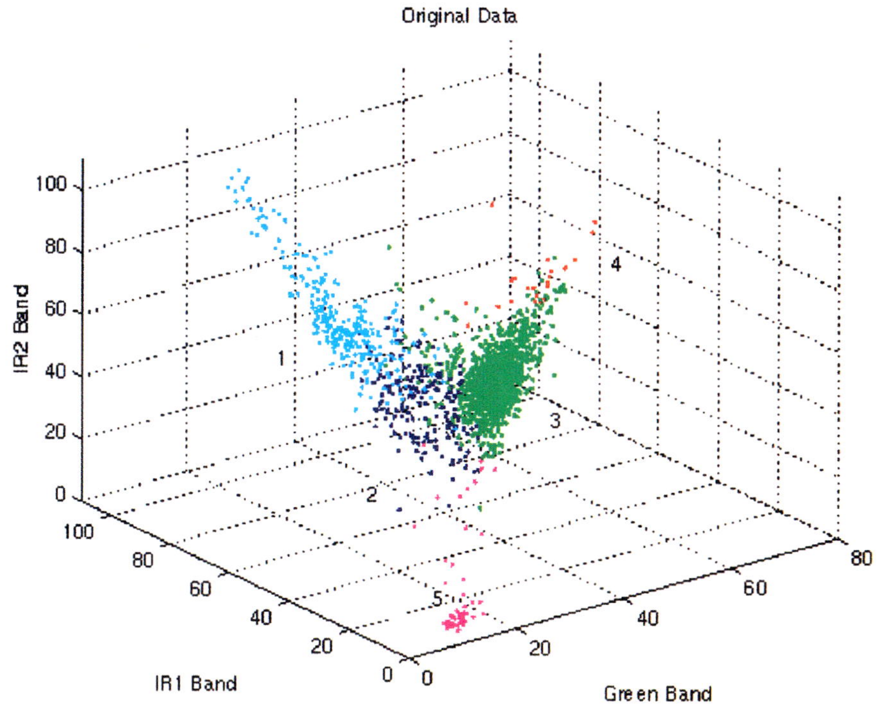

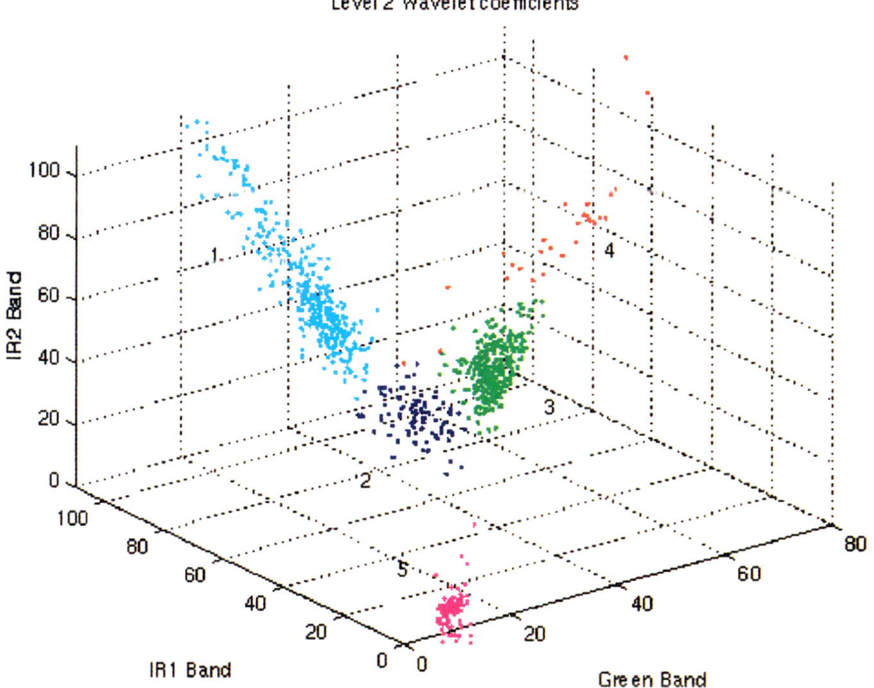

Plate 9. *The progressive classification scheme.*

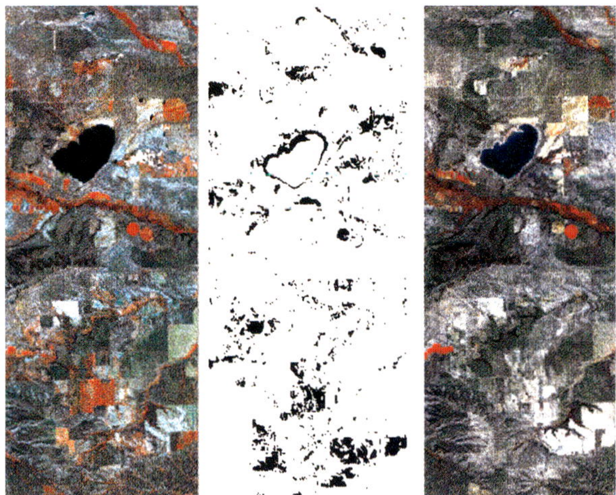

Plate 10. *Portion of a LULC map (left) and smoothed results of classification of the corresponding 72 NALC image. The classifier used to obtain the classified map on the right hand side was developed using standard techniques based on photointerpretation.*

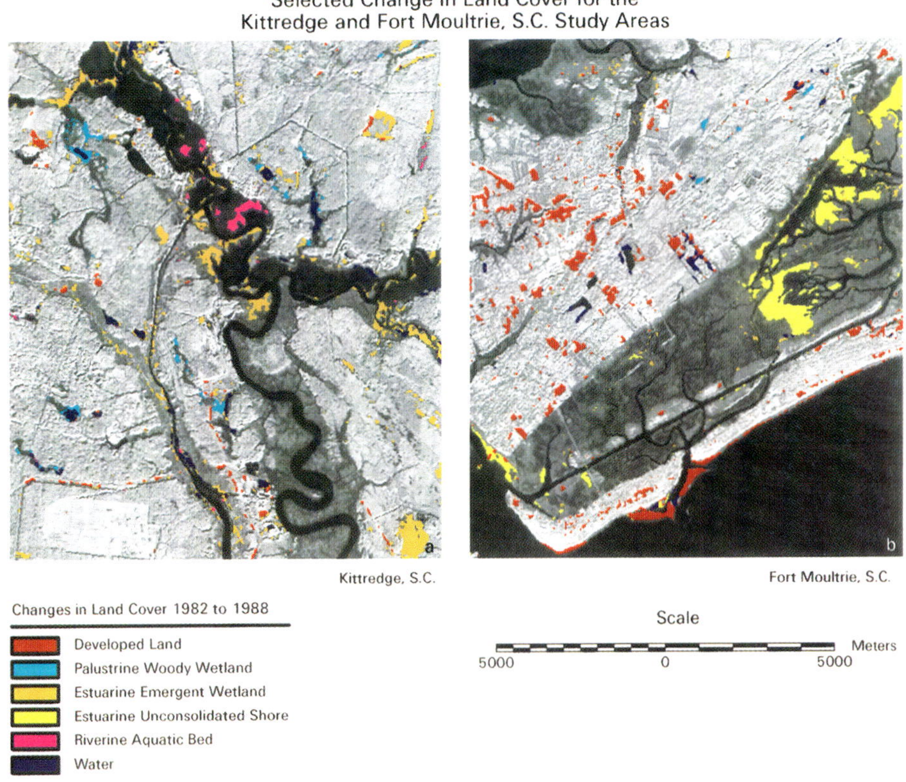

Plate 11. *Change detection maps of the Fort Moultrie and Kittredge, South Carolina study areas derived from analysis of 9 November 1982 and 19 December 1988 Landsat Thematic Mapper data. The nature of the change classes selected for display are summarized in Figure 5.1. The change information is overlaid onto the Landsat TM band 4 image of each date for orientation purposes.*

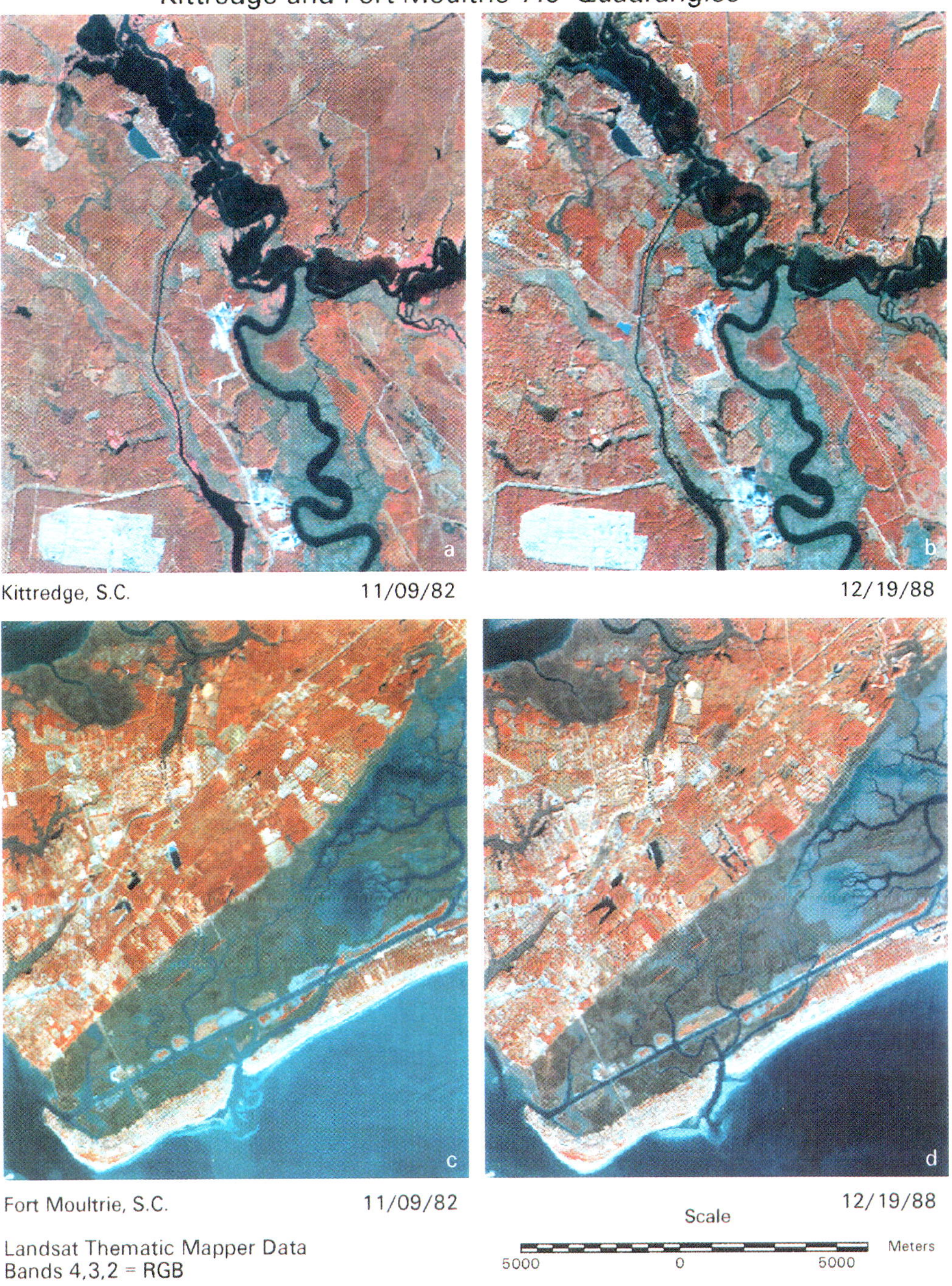

Plate 12. *Color-infrared color composites of the Landsat Thematic Mapper data bands (4, 3, 2 = RGB) of the Fort Moultrie, South Carolina and Kittredge, South Carolina study areas selected to test CoastWatch change detection protocols. Fort Moultrie and Kittredge contain extensive estuarine emergent wetland. Kittredge also exhibits palustrine forested wetland and riverine aquatic beds.*

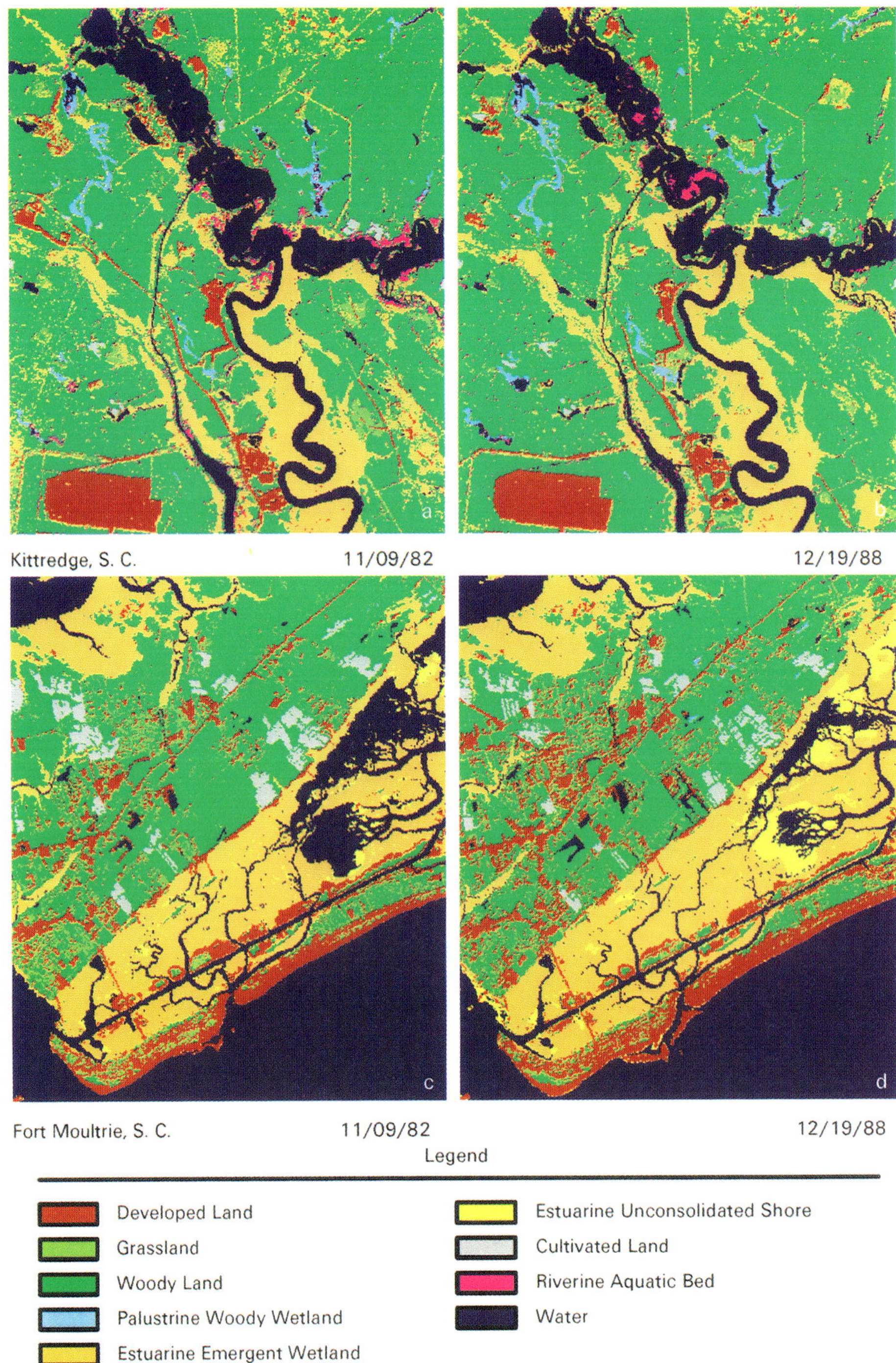

Plate 13. *Individual classification maps of the Fort Moultrie and Kittredge, South Carolina study areas in 1982 and 1988 based on the analysis of Landsat Thematic Mapper data. Classes are from the Tentative CoastWatch Classification Scheme.*

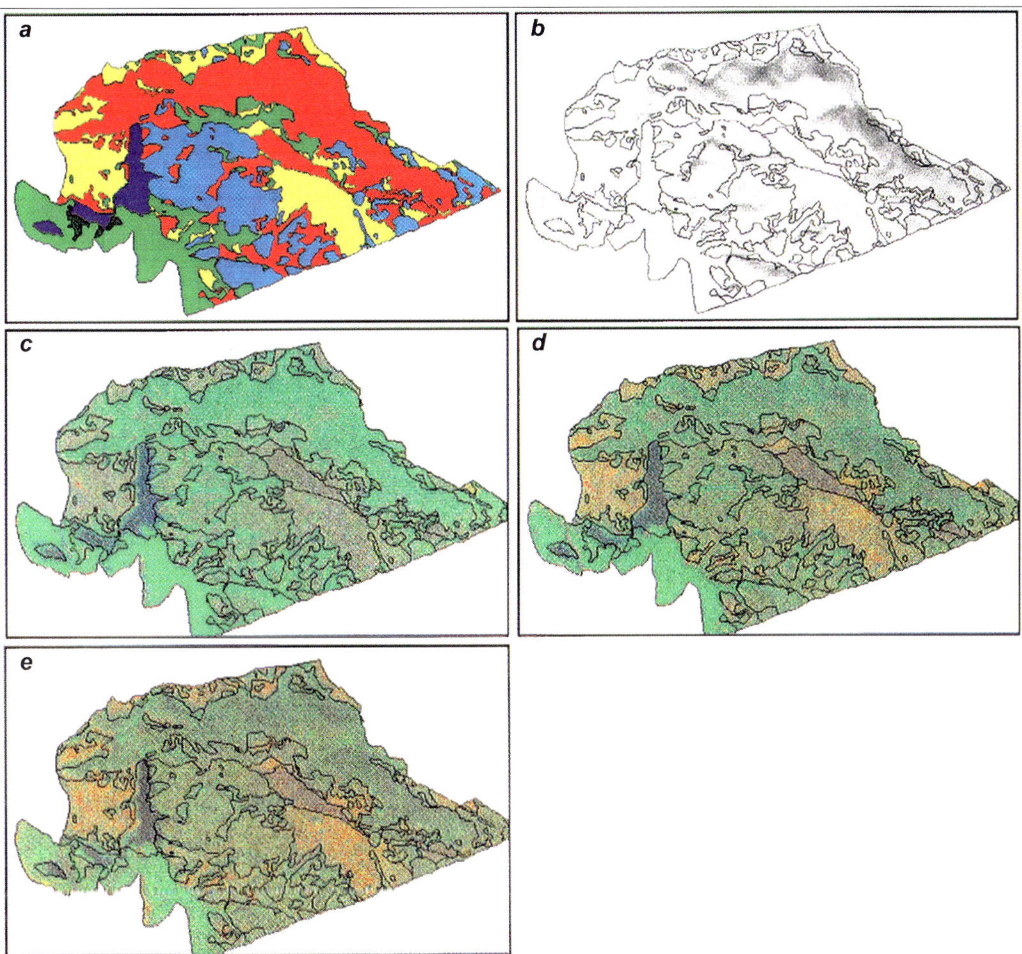

Plate 14. (a) *JRBP field-based vegetation map that was used to evaluate image predictions. The vegetation types are yellow (annual grasslands on greenstone derived soil), red (annual serpentine grasslands), light blue (chaparral communities), green (deciduous oak woodlands), pink (mixed evergreen coastal forest), black (freshwater forested wetlands), dark blue (lakes).* **(b)** *Annual potential (cloud and fog free) net radiation image for the JRBP. Image shows relative potential radiation regime for all sites within the JRBP, with white indicating maximum potential radiation and black indicating lowest potential radiation. The vegetation boundaries are superimposed on the figure to illustrate the relationship between vegetation type and irradiance conditions.* **(c-e)** *Change in spectral fractions displayed as a false color image for GV (displayed as green), NPV (displayed as red) and soil (displayed as blue) for the three dates between spring and autumn. (2c) June, (2d) September, and (2e) October. Color balance changes as spectral fractions change. Sites with nearly pure (i.e., 100 percent cover of one type) are nearest pure red, blue, or green while spectral mixtures are intermediate colors. A general loss of canopy greenness (green) is observed and increasing NPV fractions (red) over the season.*

Plate 15. *Power outages observed in North Carolina on September 7, 1996 following the September 6 passage of Hurricane Fran.*

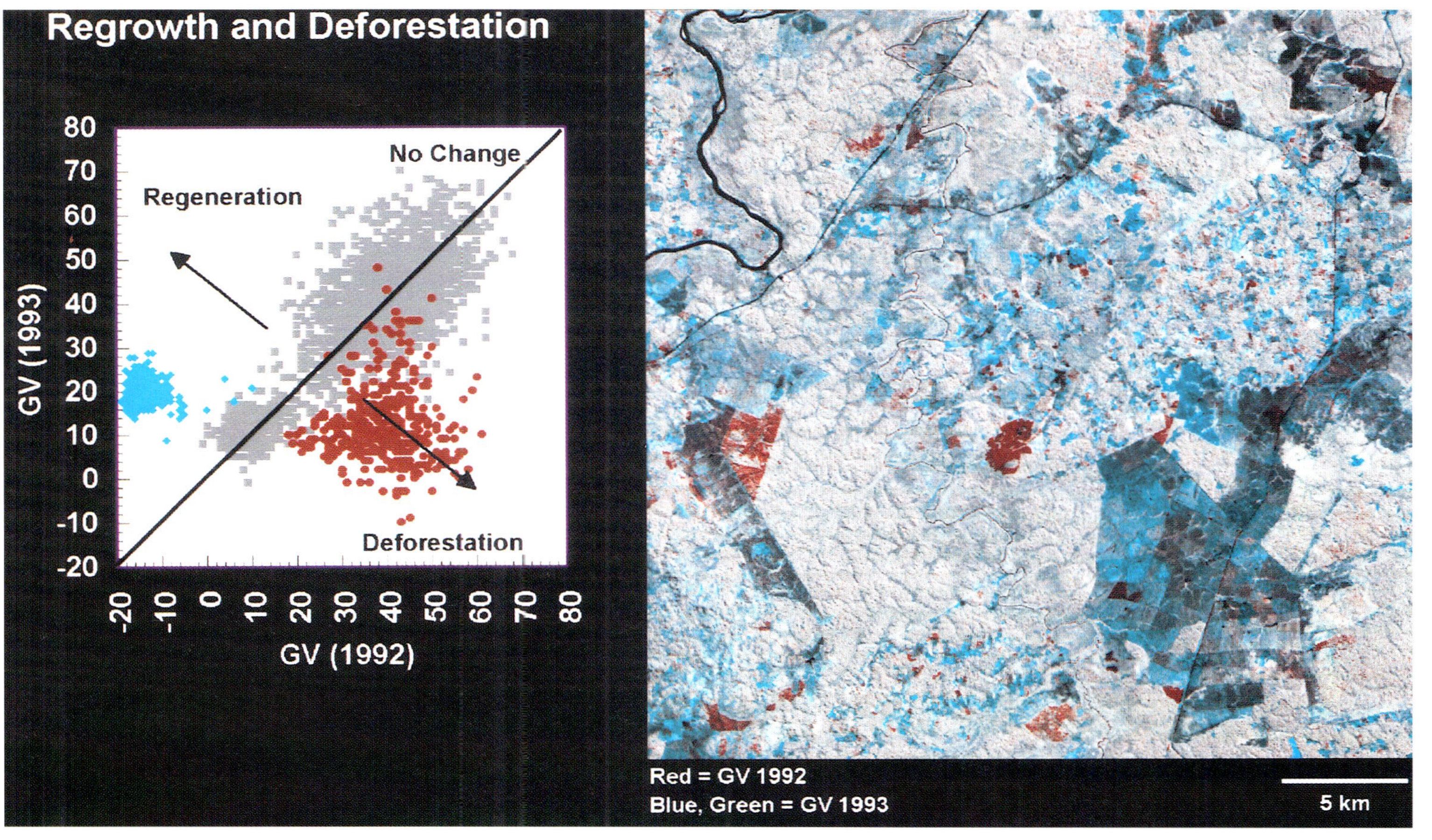

Plate 16. *GV fractions for 1992 (red) and 1993 (blue and green). Recently deforested areas are red while regeneration displays as cyan.*

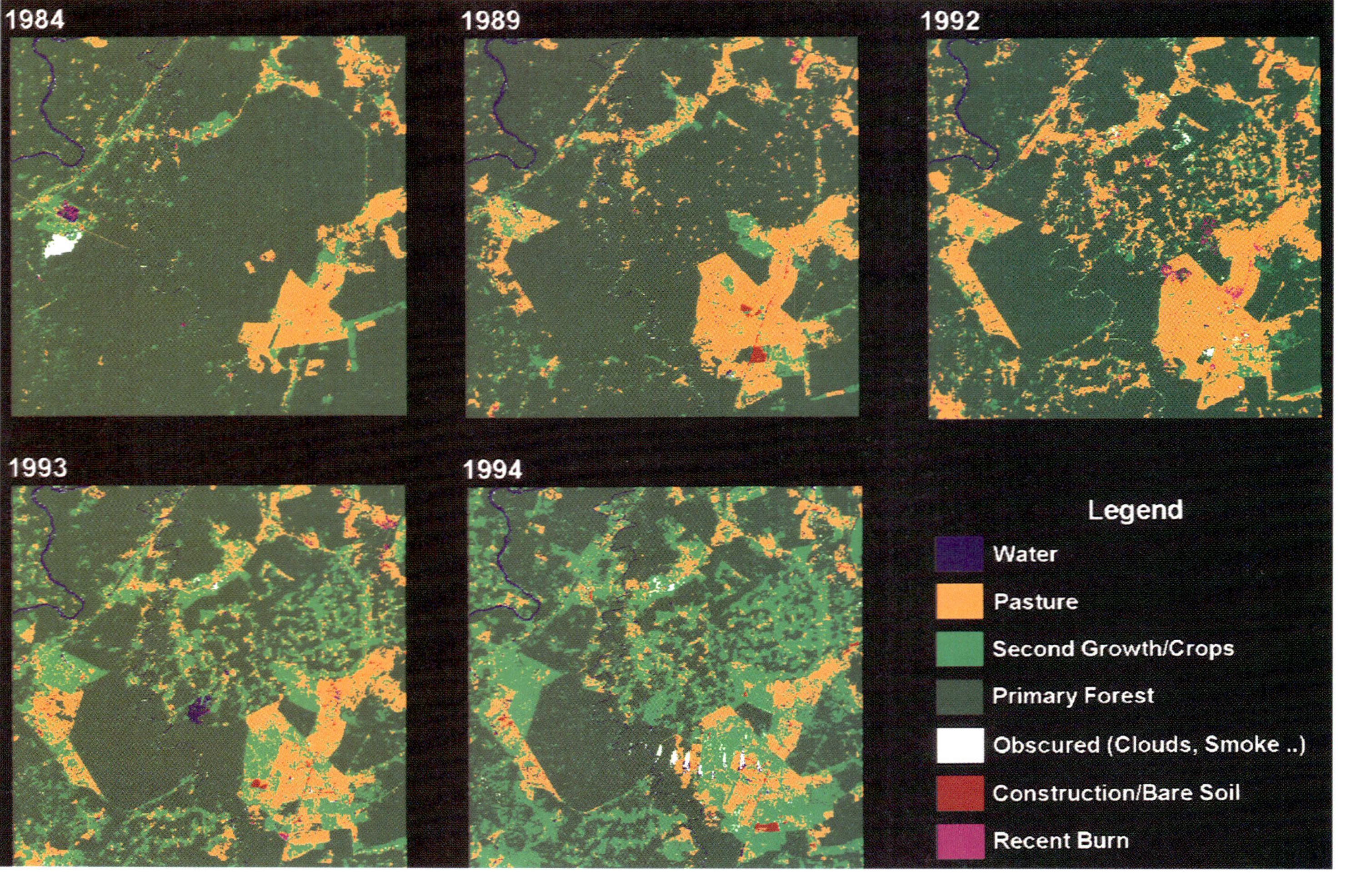

Plate 17. *Multitemporal classified images of a 30x30 km subset.*

Plate 18. *Site image - July 6, 1986 Landsat MSS showing the locations of the wetlands which were studied. Bands 1, 2, 4 as blue, green, and red.*

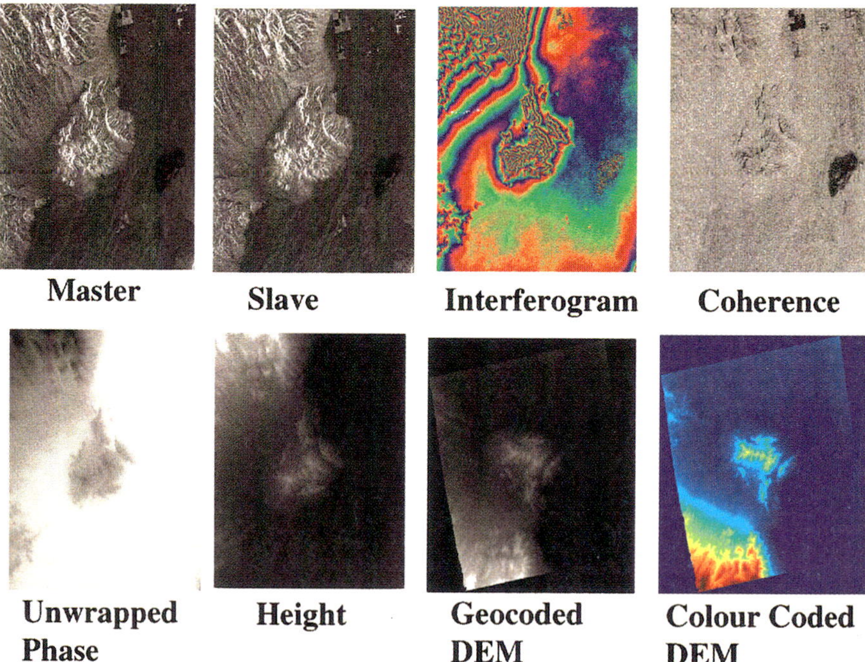

Plate 19. *This example shows the outputs of each step in the InSAR processing for an area in Death Valley, California. The data were acquired by the spaceborne SAR sensor RADARSAT using Fine Beam 2. The master image was acquired September 24, 1996, and the slave on October 18, 1996. The incidence angle at scene center is 41 degrees. The perpendicular baseline at scene center is* B_{perp}*=280 m. The height sensitivity is* $\Delta h/\Delta\phi=-63$ *m/cycle. (Signal data is © 1996 CSA. Radar imagery is courtesy of RADARSAT International. All other images are © 1997 Atlantis Scientific, Inc.)*

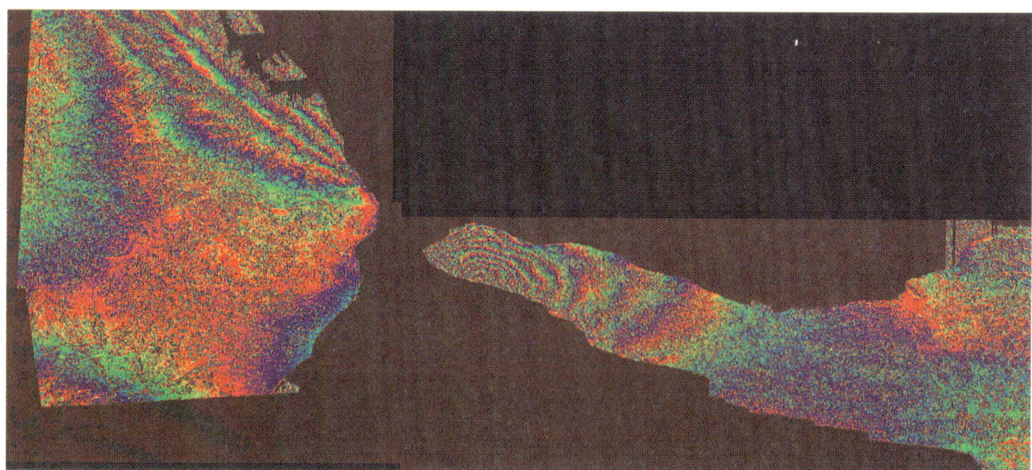

Plate 20. *Coseismic displacement mapped by differential InSAR (Ohkura, 1997a). These data were acquired by the JERS-1 spaceborne SAR on October 10, 1993 (Pass 1) and March 22, 1995 (Pass 2). The 1995 Hyogoken-Nanbu earthquake occurred near the city of Kobe, Japan on January 17, 1995. Each fringe represents a change in slant range distance from satellite antenna to the ground surface of 11.75 cm from the time of Pass 1 to the time of Pass 2. (© NIED, Japan. © 1997 Atlantis Scientific, Inc.)*

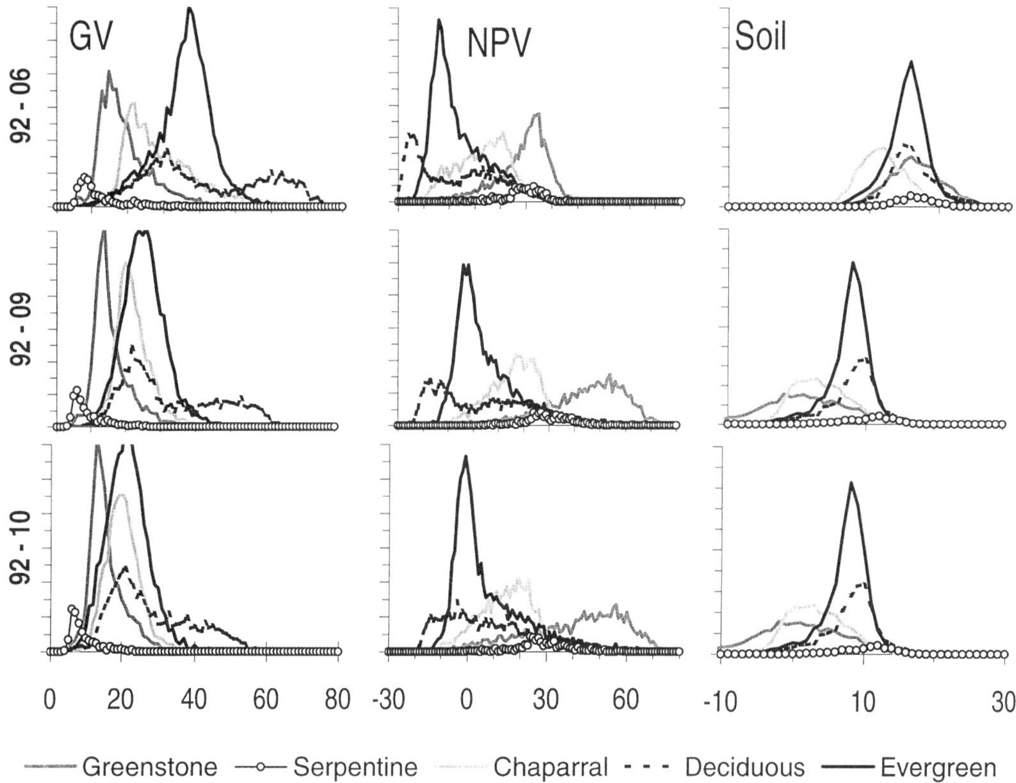

Figure 10.7. Changes in frequency distribution of spectral fractions for NPV, GV, and soil for each vegetation type on the three dates. Although GV decreases over the season and NPV increases, the distribution patterns for each vegetation type maintain a consistent relationship. This results in coherent spatial images that consistently differentiate vegetation types between dates for these communities. The amplitude depends on the number of pixels in each vegetation class. The horizontal axis represents the relative fractional abundance of each type.

season. Figure 10.7 provides a frequency distribution of GV, NPV, and soil fractions for each vegetation type for each date. The frequency distribution and amplitude are observed to vary with vegetation type. Clearly the distribution of GV fractions decreases over the season but the relative distribution of each vegetation type maintains roughly the same pattern throughout the summer. The NPV increases over these dates. Each of these SMA measures indicates a pattern consistent with expected phenological differences between these communities.

3.4 Temporal Changes in Surface Composition

Temporal changes in surface composition, as expressed by changes in endmember fractions, show a general pattern of increasing senescence and decreasing illumination (10.2c–e). A general seasonal decrease in the GV fraction is seen by the change in that fraction, displayed in the images by the green color. Most areas show a significant seasonal decrease in the GV fraction, with the most pronounced change in the forested wetlands. Areas that showed little to no change were restricted to sites having low GV, such as grasslands, or high GV sites, such as the irrigated golf course (not shown on image). In contrast, NPV increased between June and October. These changes are consistent with continued senescence of green foliage and accumulation of litter in autumn. Even the evergreen communities experience some loss of GV and increase in NPV. Temporal patterns in liquid water contents shown in Figure 10.5a–c are consistent with the decline observed in GV and increasing NPV (Figure 10.7) and expected leaf losses.

4.0 SUMMARY

Remote sensing applications for change detection using broad-band sensors are well established, yet comparatively little research has focused on applications of imaging spectrometry toward monitoring change. We examined spatial patterns of vegetation distribution at Stanford University's Jasper Ridge Biological Preserve using information derived from the AVIRIS and other GIS data and changes over a growing season. In this chapter we explored the use of the AVIRIS for monitoring seasonal and spatial changes in atmospheric water vapor, liquid water, and surface cover in the vicinity of the Jasper Ridge Biological Preserve, CA over late spring, summer, and autumn dates in 1992. Confidence in understanding remotely sensed estimates of regional canopy processes such as photosynthesis and evapotranspiration requires better experimental evidence at the scale of images. Such validation data could be improved using imaging spectrometers. Multitemporal analysis using imaging spectrometry places strict requirements on instrumental calibration and atmospheric correction. Apparent surface reflectance was retrieved and water vapor and liquid water mapped using a radiative-transfer based model that accounts for spatially variable atmospheres. Furthermore, interpretation of surface changes and temporal processes requires analysis tools that provide a standard frame of reference as well as a physical basis for interpretation. Using SMA, we found high correlations between spatial distributions of vegetation types mapped from other methods and AVIRIS-based assessments of vegetation. SMA was used to model fractional mixtures of GV, NPV, soil, and shade.

We found a near-linear, negative relationship between topography and column water vapor that varied significantly from the late spring to mid-autumn. Potential net solar radiation was strongly related to topographic position, and communities were stratified based on differences in the potential energy budgets. Retrieved apparent reflectance made it possible to study temporal changes in the surface independent of a temporally and spatially variable atmosphere. SMA and equivalent liquid water thickness were used to quantify surface composition and temporal changes. Although highly correlated, liquid water showed the greatest variability between vegetation types. Temporally, spectral fractions and liquid water showed a pattern of seasonally decreasing green vegetation, increasing senescence, and decreased illumination. Phenological patterns measured for vegetation in the area (Gamon et al., 1995) support the interpretation of seasonal LAI declines.

Imaging spectrometry, through improved atmospheric and surface characterization has been shown to be a powerful tool for change detection studies, providing stable temporal standardization and new physical parameters such as column water vapor and equivalent liquid water thickness for describing atmospheric and surface properties through time. Estimating the shade fraction and recognizing that it is a measure of canopy structure is unique to imaging spectrometry. When combined with SMA and the use of reference endmembers, imaging spectrometry offers a new approach for quantifying change where neither the surface nor the atmosphere can be assumed to be constant through space or time.

ACKNOWLEDGMENTS

Thanks to Nona Chiariello and the Jasper Ridge Biological Preserve for access and logistical support, GIS database, and meteorological data for Jasper Ridge. Lee Johnson and the Ecosystems Dynamics Branch of the NASA Ames Research Center provided use of the NIRS spectroradiometer. We wish to recognize the support of NASA from a SIR-C Grant JPL 958744 and the EOS program under Grants NAS5-31359 and NAS5-31714, and to the Sequoia 2000 program grant from the Digital Equipment Corporation that provided DEC 5000 and Alpha 3000 workstations and peripherals that were used to perform this research for SLU. Funding was also provided in part by NASA Grant NAGW-3195 and the University of California, which supplied start-up funds for computing facilities for DAR.

REFERENCES

Adams, J.B., M.O. Smith, and A.R. Gillespie. Imaging spectroscopy: interpretation based on spectral mixture analysis. In: *Remote Geochemical Analysis*, C.M. Pieters and P. Englert, Eds., Cambridge University Press, 1993, pp. 145–166.

Adams, J.B., D.E. Sabol, V. Kapos, R. Almeida Filho, D.A. Roberts, M.O. Smith, and A.R Gillespie. Classification of multispectral images based on fractions of endmembers: application to land-cover change in the Brazilian Amazon. *Remote Sens. Environ.*, 52, 137–154, 1995.

Allen, W.A. and A.J. Richardson. Interaction of light with a plant canopy. *J. Opt. Soc. Amer.*, 58(8), 1023–1028, 1968.

Borel, C.G. and S.A.W. Gerstl. Nonlinear spectral mixing models for vegetative and soil surfaces. *Remote Sens. Environ.*, 47, 403–416, 1994.

Curran, P.J. Remote sensing of foliar chemistry. *Remote Sens. Environ.*, 30, 271–278, 1989.

DeFries, R., C. Field, A. Fung, C. Justice, S. Los, P. Matson, M. Matthews, H. Mooney, C. Potter, K. Prentice, P. Sellers, J. Townshend, C. Tucker, S. Ustin, and P. Vitousek. Mapping the land surface for global atmosphere-biopshere models: toward continuous distributions of vegetation's functional properties. *J. Geophys. Research-Atmospheres*, 100(ND10) 20867–20882, 1995.

DeLong, R.K., T.E. Romesser, J. Marmo, and M.A. Folkman. Airborne and satellite imaging spectrometer development at TRW. *Proceedings SPIE Imaging Spectrometry*, Orlando, FL, April 17–18, 1995, pp. 287–294.

Elvidge, C.D. and F.P. Portigal. Change detection in vegetation using 1989 AVIRIS data. *Proceedings SPIE Imaging Spectroscopy of the Terrestrial Environment*, editor G. Vane, Orlando, FL, April 16–17, 1990, pp. 178–189.

Gallo, K.P., C.S.T. Daughtry, and M.E. Bauer. Spectral estimation of absorbed photosynthetically active radiation in corn canopies. *Remote Sens. Environ.*, 17, 221–232, 1985.

Gamon, J.A., C.B. Field, D.A. Roberts, S.L. Ustin, and R. Valentini. Functional patterns in an annual grassland during an AVIRIS overflight. *Remote Sens. Environ.*, 44, 239–253, 1993.

Gamon, J. A., C.B. Field, M.L. Goulden, K.L. Griffen, A.E. Hartley, G. Joel, J. Penuelas, and R. Valentini. Relationships between NDVI, canopy structure, and photosynthesis in three California vegetation types. *Ecol. Appl.*, 5, 28–41, 1995.

Gao, B.C. and A.F.H. Goetz. Column atmospheric water vapor and vegetation liquid water retrievals from airborne imaging spectrometer data. *J. Geophys. Res.*, 95, 3549–3564, 1990.

Gao, B.C. and A.F.H. Goetz. Retrieval of equivalent liquid water thickness and information related to biochemical components of vegetation canopies from AVIRIS data. *Remote Sens. Environ.*, 52, 155–162, 1995.

Gash, M., P. Hoepffner, M. Kabat, Y.H. Kerr, B. Monteny, S. Prince, F. Said, P. Sellers, and J.S. Wallace. HAPEX-Sahel: A large-scale study of land-atmosphere interactions in the semi-arid tropics. *Annales Geophysicae*, 12, 53–64, 1994.

Goetz, A.F.H., G. Vane, J.E. Solomon, and B.N. Rock. Imaging spectrometry for earth remote sensing. *Science*, 228, 1147–1153, 1985.

Goward, S.N., C.J. Tucker, and D.G. Dye. North American vegetation patterns observed with the Nimbus-7 Advanced Very High Resolution Radiometer. *Vegetatio*, 64, 3–14, 1985.

Graetz, R.D. and M.R. Gentle. The relationship between reflectance in the Landsat wavebands and the composition of an Australian semi-arid shrub rangeland. *Photogrammetric Engineering and Remote Sensing*, 48(11), 1721–1730, 1982.

Green, R.O., J.E. Conel, J.S. Margolis, C.J. Bruegge, and G.L. Hoover. An inversion algorithm for retrieval of atmospheric and leaf water absorption from AVIRIS radiance with compensation for atmospheric scattering. *Proceedings 3rd Annual JPL Airborne Geoscience Workshop*, Robert O. Green, Ed., Jet Propulsion Laboratory, Pasadena, CA, May 20–21, JPL 92-14, 1991, pp. 51–61.

Grossman, Y.L., S.L. Ustin, E. Sanderson, S. Jacquemoud, G. Schmuck, and J. Verdebout. Critique of stepwise multiple linear regression for the extraction of leaf biochemistry information from leaf reflectance data. *Remote Sens. Environ.*, 56, 182–193, 1996.

Hall, F.G., D.B. Botkin, D.E. Strebel, K.D. Woods, and S.J. Goetz. Large-scale patterns of forest succession as determined by remote sensing. *Ecology*, 72, 628–640, 1991.

Hall, F.H., K.F. Huemmrich, S.J. Goetz, P.J. Sellers, and J.E. Nickerson. Satellite remote sensing of surface energy balance: Success, failures, and unresolved issues in FIFE. *J. Geophys. Res.*, 97, 19, 061-19,089, 1992.

Hart, Q.J., S.L. Ustin, L. Duan, and G. Scheer. Estimating dry grass residues using landscape integration analysis. *Proceedings 4th Annual JPL Airborne Geoscience Workshop*, R.O. Green, Ed., Washington, DC, Oct. 25–27, Report No. 93-26, 1993, pp. 89–92.

Hart, Q.J., S.L. Ustin, G. Scheer, and L. Duan. Estimating dry grass biomass residues using AVIRIS image analysis. *Proceedings Int. Geoscience and Remote Sens. Symp. IGARSS '94*, August 8–12, California Institute of Technology, Pasadena, CA, 1994.

Holben, B.N. Characteristics of maximum-value composites images from temporal AVHRR data. *Int. J. Remote Sens.*, 7, 1417–1434, 1986.

Huete, A.R. Separation of soil-plant spectral mixtures by factor analysis. *Remote Sens. Environ.*, 19, 237–251, 1986.

Intergovernmental Panel on Climate Change (IPCC). Technical summary, the climate system. In: *An Overview, Climate Change 1995, The Science of Climate Change*, J.T. Houghton, L.G. Meira Filho, B.A. Callander, N. Harris, A. Kattenberg, and K. Maskell, Eds., Cambridge University Press, 1996, pp. 9–50.

Justice, C.O., J.R.G. Townshend, B.N. Holben, and C.J. Tucker. Analysis of the phenology of global vegetation using meteorological satellite data. *Int. J. Remote Sens.*, 6(8), 1271–1318, 1985.

Kneizys, F.X., E.P. Shettle, and L.W. Abreu. User's Guide to Lowtran7. *Air Force Geophysics Laboratory*, Report No. AFGL-TR-88-0177, Bedford, MA, 1988.

Kruse, F.A., A.B. Lefkoff, and J.B. Dietz. Expert system-based mineral mapping in northern Death Valley, California/Nevada using the Airborne Visible/Infrared Imaging Spectrometer (AVIRIS). *Remote Sens. Environ.*, 44(2/3), 309–336, 1993.

Lucas, R.M., M. Honzak, G.M. Foody, P.J. Curran, and C. Corves. Characterizing tropical secondary forests using multi-temporal Landsat sensor imagery. *Int. J. Remote Sens.*, 14, 3016–3067, 1993.

Malingreau, J.P., C.J. Tucker, and N. Laporte. AVHRR for monitoring global tropical deforestation. *Int. J. Remote Sens.*, 10, 855–867, 1989.

Myneni, R.B., S. Maggion, J. Iaquinta, J.L. Privette, N. Gobron, B. Pinty, D.S. Kimes, M.M. Verstraete, and D.L. Williams. Optical remote sensing of vegetation: modeling, caveats, and algorithms, *Remote Sens. Environ.*, 51, 169–188, 1995.

Myneni, R.B. and D.L. Williams. On the relationship between FAPAR and NDVI. *Remote Sens. Environ.*, 49, 200–211, 1994.

Page, B.M. and L.L. Tabor. Chaotic structure and decollement in Cenozoic rocks near Stanford University of California. *Geol. Soc. Am. Bull.*, 78, 1–12, 1967.

Palacios-Orueta, A. and S.L. Ustin. Multivariate statistical classification of soil spectra. *Rem. Sens. Environ.*, 57, 108–118, 1996.

Pech, R.P., R.D. Graetz, and A.W. Davis. Reflectance modeling and the derivation of vegetation indices for an Australian semi-arid shrubland. *Int. J. Remote Sens.*, 7(3), 389–403, 1986.

Ray, T.W. and B.C. Murray. Nonlinear spectral mixing in desert vegetation. *Remote Sens. Environ.*, 55, 59–64, 1996.

Roberts, D.A., M.O. Smith, D.E. Sabol, J.B. Adams, and S.L. Ustin. Mapping the spectral variability in photosynthetic and non-photosynthetic vegetation, soils and shade using AVIRIS. *Proceedings 3rd Annual Airborne Geoscience Workshop*, Robert O. Green, Ed., NASA Jep Propulsion Laboratory, Pasadena, CA, Pub. No. 92-14(1), 38–40, 1992.

Roberts, D.A., M.O. Smith, and J.B. Adams. Green vegetation, non-photosynthetic vegetation, and soils in AVIRIS data. *Remote Sens. Environ.*, 44, 255–269, 1993.

Roberts, D.A., R.O. Green, and J.B. Adams. Temporal and spatial patterns in vegetation and atmospheric properties from AVIRIS. *Remote Sens. Environ.*, 62, 223–240, 1997.

Roberts, D.A., G. Batista, J. Pererta, E. Waller, and B. Nolan. Change identification using multispectral spectral mixture analysis: applications in eastern Amazonia. In: *Remote Sensing Change Detection: Environmental Monitoring Applications and Methods*, R.S. Lunetta and C.D. Elvidge, Eds., Ann Arbor Press, Chelsea, MI, 1998.

Running, S.W. and E.R. Hunt. Generalization of a forest ecosystem process model for other biomes, BIOME-BGC, and an application for global-scale models. In: *Scaling Physiological Processes: Leaf to Globe*, J.R. Ehleringer and C.B. Field, Eds., Academic Press, New York, 1993, pp. 141–158.

Sellers, P.J. Canopy reflectance, photosynthesis and transpiration. *Int J. Remote Sens.*, 6(8), 1335–1372, 1985.

Sellers, P.J. Canopy reflectance, photosynthesis, and transpiration. II. The role of biophysics in the linearity of their interdependence. *Remote Sens. Environ.*, 21, 143–183, 1987.

Sellers, P.J., W.J. Shuttleworth, J.L. Dorman, A. Dalcher, and J.M. Roberts. Calibrating the simple biosphere model for Amazonian tropical forest using field and remote sensing data, Part 1: Average calibration and field data. *J. Appl. Met.*, 28, 727–759, 1989.

Sellers, P.J., J.A. Berry, G.J. Collatz, C.B. Field, and F.G. Hall. Canopy reflectance, photosynthesis, and transpiration, 3, Areal analysis using improved leaf models and a new canopy integration scheme. *Remote Sens. Environ.*, 42(3), 187–216, 1992.

Shimabukuro, Y.E., B.N. Holben, and C.J. Tucker. Fraction images derived from NOAA AVHRR data for studying the deforestation in the Brazilian Amazon. *Int. J. Remote Sens.*, 15, 517–520, 1994.

Skole, D. and C.J. Tucker. Tropical deforestation and habitat fragmentation in the Amazon: satellite data from 1978 to 1988. *Science*, 260, 1905–1910, 1993.

Smith, M.O., S.L. Ustin, J.B. Adams, and A.R. Gillespie. Vegetation in deserts I. A regional measure of abundance from multispectral images. *Remote Sens. Environ.*, 31, 1–26, 1990a.

Smith, M.O., S.L. Ustin, J.B. Adams, and A.R. Gillespie. Vegetation in deserts II. Environmental influences on regional abundance. *Remote Sens. Environ.*, 29, 27–52, 1990b.

Staenz, K.A. *Canadian Activities in Terrestrial Imaging Spectrometry.* http://www.eolists.ca/docmets/IS-Team-Canada, 1995.

Thomas, J.H. *Flora of the Santa Cruz Mountains of California, a Manual of the Vascular Plants.* Stanford University Press, Stanford, CA, 1961.

Tucker, C.J., I.Y. Fung, C.D. Keeling, and R.H. Gammon. Relationship between atmospheric CO_2 variations and a satellite-derived vegetation index. *Nature*, 319, 195–199, 1986a.

Tucker, C.J., C.O. Justice, and S.D. Prince. Monitoring the grasslands of the Sahel 1984–1985. *Int. J. Remote Sens.*, 7(11), 1571–1581, 1986b.

U.S. Global Change Research Program (USGCRP). *Our Changing Planet, A Supplement to the President's Fiscal Year 1996 Budget.* U.S. Government Printing Office, Washington, DC, 1996.

Ustin, S.L., M.O. Smith, D.A. Roberts, J.A. Gamon, and C.B. Field. Using AVIRIS images to measure temporal trends in abundance of photosynthetic and nonphotosynthetic canopy components. *Proceed-*

ings 3th Annual JPL Airborne Geoscience Workshop, Robert O. Green, Ed., NASA/JPL, Pasadena, CA, June 2–4, Report No. 92-14(1), pp. 5–7, 1992.

Ustin, S.L., L. Duan, and Q.J. Hart. Seasonal changes observed in AVIRIS images of Jasper Ridge, California. *Proceedings Int. Geoscience and Remote Sens. Symp. IGARSS '94,* August 8–12, California Institute of Technology, Pasadena, CA, 1994.

Ustin, S.L., M.O. Smith, and J.B. Adams. Remote Sensing of Ecological Processes: A strategy for Developing Ecological Models Using Spectral Mixture Analysis. In*: Scaling Physiological Processes: Leaf to Globe*, J. Ehlringer and C.B. Field, Eds., Academic Press, New York, 1993, pp. 339–357.

Ustin, S.L., Q.J. Hart, L. Duan, and G. Scheer. Vegetation mapping on hardwood rangelands in California. *Int. J. Remote Sens.*, 17, 3015–3036, 1996.

Vane, G., R.O. Green, T.G. Chrien, H.T. Enmark, E.G. Hansen, and W.M. Porter. The airborne visible/infrared imaging spectrometer (AVIRIS). *Remote Sens. Environ.*, 44, 127–143, 1993.

CHAPTER 11

Vegetation Change Detection Using High Spectral Resolution Vegetation Indices

Zhikang Chen, Christopher D. Elvidge, and David P. Groeneveld

1.0 INTRODUCTION

Arid and semiarid lands cover approximately 30 to 40 percent of the Earth's land surface. Because of the vast area covered, these lands play a major role in the energy balance and hydrologic, carbon, and nutrient cycles. The activity of vegetation in these lands typically undergoes wide seasonal and interannual fluctuations, largely regulated by the availability of water, and may be readily impacted by both climatic shifts and human activities such as grazing, wood gathering, and urbanization. Thus, monitoring the vegetation vigor of these lands is of interest for both scientific and resource management applications. In past research, satellite remote sensing has been successfully used to detect interannual variability such as the apparent expansion and contraction of the Saharan desert (Tucker et al., 1991). However, quantitative assessment of changes in green vegetation in these areas remains a challenge due to the relatively low spectral contribution by vegetation in pixels dominated by exposed rock, soil, and litter.

Vegetation indices are mathematical transformations designed to assess the spectral contribution of vegetation to multispectral observations. The most widely used green vegetation indices are formed with data from discrete red and near infrared (NIR) bands. These vegetation indices operate by contrasting intense chlorophyll pigment absorptions in the red against the high reflectivity of plant materials in the NIR (Tucker, 1979). The value of these vegetation indices lies in their potential use to estimate vegetation variables such as percent green cover, Leaf Area Index (LAI) or Absorbed Photosynthetically Active Radiation (APAR), which in turn can be used to analyze vegetation processes such as Net Primary Productivity (NPP) and evapotranspiration.

The vegetation indices developed in the 1970s are based on discrete red and NIR bands, and can be generally divided into two basic categories: ratios and orthogonal indices. The ratio-based indices include the Ratio Vegetation Index (RVI) and the Normalized Difference Vegetation Index (NDVI). Orthogonal indices include the Perpendicular Vegetation Index (PVI) and the Difference Vegetation Index (DVI). More recently a hybrid set of vegetation indices have emerged, such as the Soil Adjusted Vegetation Index (SAVI). The formulas and notable references for these vegetation indices are provided in Table 11.1.

Table 11.1. Red and NIR Vegetation Index Formulas.[a]

Abbreviation	Name	Vegetation Index	Reference
NDVI	Normalized difference vegetation index	$NDVI = \frac{NIR - RED}{NIR + RED}$	Rouse et al. (1973)
RVI	Ratio vegetation index	$RVI = \frac{NIR}{RED}$	Jordan (1969)
SAVI	Soil adjusted vegetation index	$SAVI = \frac{(NIR - RED)}{(NIR + RED + L)}(1+L)$	Huete (1988)
TSAVI	Transformed soil adjusted vegetation index	$TSAVI = \frac{a(NIR - aRED - b)}{RED + aNIR - ab}$	Baret et al. (1989)
$SAVI_2$	Soil adjusted ratio vegetation index	$SAVI_2 = \frac{NIR}{(RED + b/a)}$	Major et al. (1990)
PVI	Perpendicular vegetation index	$PVI = \frac{NIR - aRED - b}{\sqrt{1+a^2}}$	Richardson and Wiegand (1977)
DVI	Difference vegetation index	$DVI = NIR - RED$	Tucker (1979)
IDL_DGVI	First order derivative based green vegetation index derived using local baseline	$1DL\_DGVI = \sum_{\lambda_1}^{\lambda_n} \lvert \rho'(\lambda_i) - \rho'(\lambda_1) \rvert \Delta\lambda_i$	Elvidge and Chen (1995)
1DZ_DGVI	First order derivative based green vegetation index derived using zero baseline	$1DZ\_DGVI = \sum_{\lambda_1}^{\lambda_n} \lvert \rho'(\lambda_i) \rvert \Delta\lambda_i$	Elvidge and Chen (1995)
2DZ_DGVI	Second order derivative based green vegetation index derived using zero baseline	$2DZ\_DGVI = \sum_{\lambda_1}^{\lambda_i} \lvert \rho''(\lambda_i) \rvert \Delta\lambda_i$	Elvidge and Chen (1995)

[a] NOTE: The a (gain) and b (offset) terms are derived from the NIR vs RED rock-soil baseline. The L term (soil adjustment factor) in SAVI ranges from 0 to 1 and is typically set to 0.5.

DGVI legend: i=band number; λ_i=center wavelength at the ith band; λ_1=626 nm; λ_n=795 nm; ρ=reflectance; ρ'=first derivative reflectance; ρ''=second derivative reflectance; $\Delta\lambda_i$=bandwidth of the ith band.

It has been observed that variations in the spectral properties of background rock and soil materials can have adverse effects on vegetation indices, especially at low levels of vegetation cover. Three main types of rock-soil effects on vegetation indices have been identified:

1.1 Albedo Effect on Ratios

The brightness of the background materials can have a pronounced impact on the vegetation index values derived using ratio-based formulas (Huete et al., 1985; Elvidge and Lyon, 1985). On a red versus NIR diagram, lines connecting points having equal ratio values radiate from the origin. A spectral change near the origin will yield a much larger change in ratio value than the same spectral change occurring some distance away from the origin. As a result, the red and NIR reflectance of green vegetation over a bright background will yield lower vegetation index values than the same vegetation over a dark background. The albedo effect on ratio based indices can be demonstrated by calculating RVI or NDVI values for simple linear mixtures of pure vegetation and rock-soil spectra (Elvidge and Lyon, 1984). A graphical presentation of this phenomenon is available in Richardson and Weigand (1977).

1.2 Red-NIR Slope Effect

Variations in the slope from red to NIR reflectance in background materials (rock, soil, litter) can produce variations in vegetation index values (Elvidge and Lyon, 1985). Most background materials have slightly higher reflectance in NIR wavelengths than in the red region. This produces a positive slope from red to NIR. However, there are considerable variations in this red to NIR slope for different background materials. This effect can be demonstrated with the variations in vegetation index values observed for the spectra of bare rocks and soils, where no green vegetation is present.

1.3 Nonlinear Mixing Albedo Effect

NIR radiation transmitted through plant canopies can be either absorbed or reflected by background materials. For plant canopies over bright backgrounds there can be a measurable enhancement of the upwelling NIR radiation due to the reflectance of the transmitted radiation by the background (Huete et al., 1985; Roberts et al. 1993), creating nonlinear mixing. The magnitude of the nonlinear mixing is directly related to the brightness of the background and the wavelength dependent transmission of light through leaves. Where backgrounds are very dark, the level of nonlinear mixing will be negligible due to the nearly complete absorption of leaf transmitted radiation. Nonlinear mixing is not a major factor in the red region due to the opaque nature of green leaves at red wavelengths. This third effect acts in an opposite sense to the albedo effect on ratios, raising vegetation index values for plant canopies over bright backgrounds (Huete et al., 1985).

Several studies have indicated that high spectral resolution measurements of the chlorophyll red-edge region (700 to 795 nm) can be used to detect trace quantities (~5 percent cover) of green vegetation (Elvidge and Mouat, 1989; Elvidge et al., 1993). The red-edge feature is the long wavelength absorption wing of the chlorophyll pigment absorptions centered in the red wavelength region (Collins, 1978). This feature is absent in materials that lack chlorophyll such as soil and most litter. The chlorophyll red-edge exhibits the greatest change in reflectance per change in wavelength of any green leaf spectral feature in the visible and NIR. Given these facts, it is not surprising that many of the published chlorophyll red-edge studies have employed derivative analyses to measure wavelength position and red-edge amplitude. Derivative analysis of high-resolution spectra has been used widely in analytical chemistry to reduce the influence of extraneous background signals on measurements of the location and intensity of spectral features (Butler and Hopkins, 1970).

Early citations for derivative analysis of the chlorophyll red-edge focused on detecting wavelength shifts in the red-edge position associated with phenological changes and vegetation stress (e.g., Horler et al., 1983; Rock et al., 1988). More recently, derivative based measures of the chlorophyll red-edge amplitude have been proposed as solutions to background (rock, soil, litter) effects known to occur with broad-band vegetation indices (Hall et al., 1990; Demetriades-Shah et al., 1990). Both of these 1990 studies hypothesized that derivative based measures of the chlorophyll red-edge amplitude would be superior to traditional broad-band vegetation indices for the estimation of variables such as LAI and APAR. However, neither study included the direct comparison of the broad-band indices and the derivative based narrow-band indices.

More recently Elvidge and Chen (1995) showed that a high spectral-resolution derivative-based green vegetation index (DGVI) developed from continuous reflectance across the chlorophyll red-edge region and measuring the amplitude of the red-edge feature (Figure 11.1) is very effective at minimizing background (rock, soil, litter) impacts on the estimation of LAI and percent green cover. The derivative green vegetation index has been successfully applied to AVIRIS data of a Monterey pine plantation in California to map spatial variability in green cover levels in an area with a relatively uniform soil background (Chen, 1995). In this chapter we present a multitemporal comparison of DGVI images, demonstrating the successful detection of seasonal changes in green vegetation in an area with complicated background materials and vegetation species mixtures.

2.0 STUDY AREA

The study area is located southeast of Mono Lake, California. This region is semiarid and supports variable vegetation cover and composition consisting of shrubs (e.g., bitterbrush, sagebrush, and rabbitbrush) on alluvial slopes, salt grass on the lake margin, and Jeffrey pine (*Pinus jeffreyi*) forests on uplands. Bitterbrush and sagebrush are two dominant shrub species in the study area. The shrub vegetation at Mono Lake exhibits distinct annual cycles of LAI in response to the cold winters and dry summers. The temporal pattern of LAI values is generally Gaussian in shape, with the highest LAI values occurring shortly after the summer solstice in June: increasing from March to May and decreasing from August to October.

A total of 10 homogeneous single-species sampling stands were selected in the study area. One stand (site 1) was located in a gray-colored pumice-surfaced area with no vegetation cover. Nine others were covered by varying densities of bitterbrush (*Purshia tridentata*), which was selected as a reference species for monitoring DGVI's performance. The nine bitterbrush sample sites lacked significant herbaceous understory.

3.0 DATA ACQUISITION AND ANALYSIS

Two NASA Airborne Visible-Infrared Imaging Spectrometer (AVIRIS) data sets of Mono Lake area were collected separately on August 20 and October 7, 1992. These data sets were radiometrically corrected by the Jet Propulsion Laboratory. During 1992, AVIRIS data contained a spectral range of 390 nm–2499 nm with bandwidths of ~10 nm for the shortwave bands (390–1900 nm) and ~14 nm for the bands in the spectral range of 1900 to 2499 nm.

The two AVIRIS scenes covered the same series of preselected ground reflectance calibration targets. Using gains and offsets derived from the calibration targets, both images were calibrated to ground reflectance in the full AVIRIS spectral range by empirical line method.

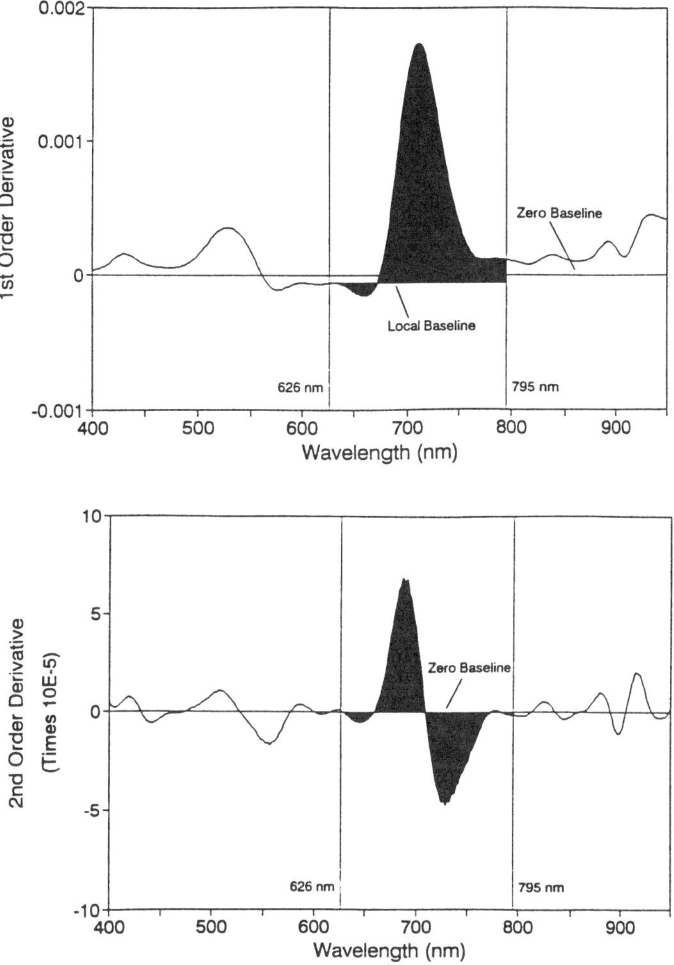

Figure 11.1. First and second order derivative for a pinyon pine canopy reflectance spectrum (LAI = 0.2858 with black gravel background). Derivative-based green vegetation indices were calculated by integrating derivative values between 626 and 795 nm.

Using a group of ground control points which were uniformly spread in both AVIRIS scenes, the two AVIRIS data sets acquired at different seasons were registered to each other by linear transform warping. AVIRIS reflectance spectra were extracted for the pixels for each site, averaged, and then smoothed to reduce noise. The first and second order derivative reflectance spectra were then calculated using a spectral distance between every other band as the derivative interval. Using the equation listed in Table 11.1, both 1DL_DGVI (1st order DGVI derived using local baseline) and 2DZ_DGVI (2nd order DGVI derived using zero baseline) were generated.

LAI for each of the 10 sites for the two image acquisition dates was estimated by multiplying LAI values measured in the field for individual shrubs by the percent canopy cover derived from low altitude aerial photography. In analyzing the AVIRIS DGVI values for the 10 sample sites it was apparent that a strong linear relationship existed between the 2DZ_DGVI and LAI

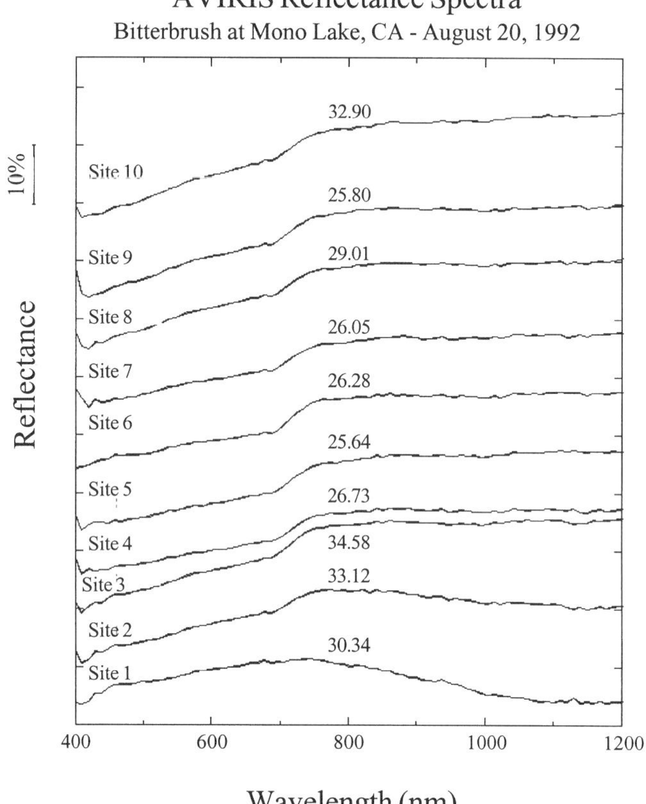

Figure 11.2. Raw mean reflectance spectra derived from August 20, 1992 AVIRIS data for the sites 1 through 10. The curves have been offset vertically to avoid overlap. The reflectance at 800 nm is provided for each spectrum.

values of the 10 sampling stands (r^2>0.93 for both AVIRIS data sets). However, 1DL_DGVI and LAI had very poor linearity due to contribution from negative derivatives of pumice material in barren shrub interspaces to the calculation of the 1DL_DGVI. This interference resulted in abnormally high 1DL_DGVI values for the gray-colored pumice gravel.

4.0 RESULTS

The retrieved AVIRIS reflectance spectra of the 10 sample sites are displayed in Figures 11.2 and 11.3. After weather became colder in October, the red-edge magnitudes decreased. Chlorophyll red-edge features were greatly enhanced and readily visible in derivative reflectance. The red-edge appeared bell-shaped in the first-order derivative spectra with maximal values located between 706–716 nm. Site 6 data exhibited the largest magnitude of the red-edge (strongest vegetation signal) in the derivative reflectance derived from both AVIRIS data sets. Site 1 was gravel soil lacking green vegetation features. As expected, site 1 appeared flat between 620–800 nm. Red-edge features of all other nine vegetated sites were less pronounced in October in comparison to August of 1992 due to loss of green leaves.

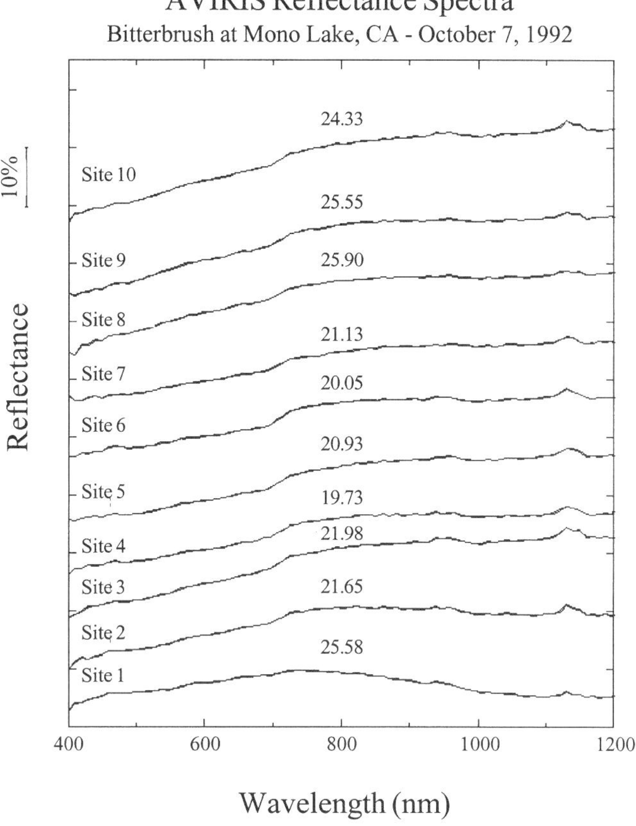

Figure 11.3. Raw mean reflectance spectra derived from October 7, 1992 AVIRIS data for the sites 1 through 10. The curves have been offset vertically to avoid overlap. The reflectance at 800 nm is provided for each spectrum.

Because of higher prediction accuracy of 2DZ_DGVI for estimating bitterbrush cover, this DGVI was applied to AVIRIS data of the entire study area. The 2DZ_DGVI maps derived from August's and October's AVIRIS data sets are displayed in Figure 11.4a and b. Beachfront areas with salt grass, adjacent to Mono Lake, yielded the highest 2DZ_DGVI values (bright). A mountainous region in the southwest corner and a burn scar on a flat pumice plain yielded the lowest 2DZ_DGVI values (dark). Green vegetation cover varied the most in the areas covered by Jeffrey pine which lies in north and east of the mountainous region. The wide range in color classes indicates different green cover signals influenced by variable density of pine trees and exposed soils, gravel, shrub, and grass cover. All other areas shown in the image were dominated by bitterbrush and sagebrush, demonstrating lower 2DZ_DGVI values. Similarity of the green vegetation distribution patterns also existed in the October 2DZ_DGVI map. But overall green vegetation cover levels of the study area decreased in October of 1992.

The vegetation index change image is displayed in Figure 11.4c (October DGVI minus September DGVI). In general, herbaceous species (e.g., salt grass) responded more sensitively to seasonal changes and had larger drop (darker) in green cover density from August to October. Cover of shrubs, including bitterbrush and sagebrush, changed less through seasonal loss of leaves (brighter). The lowest change in cover density between the two dates occurred in

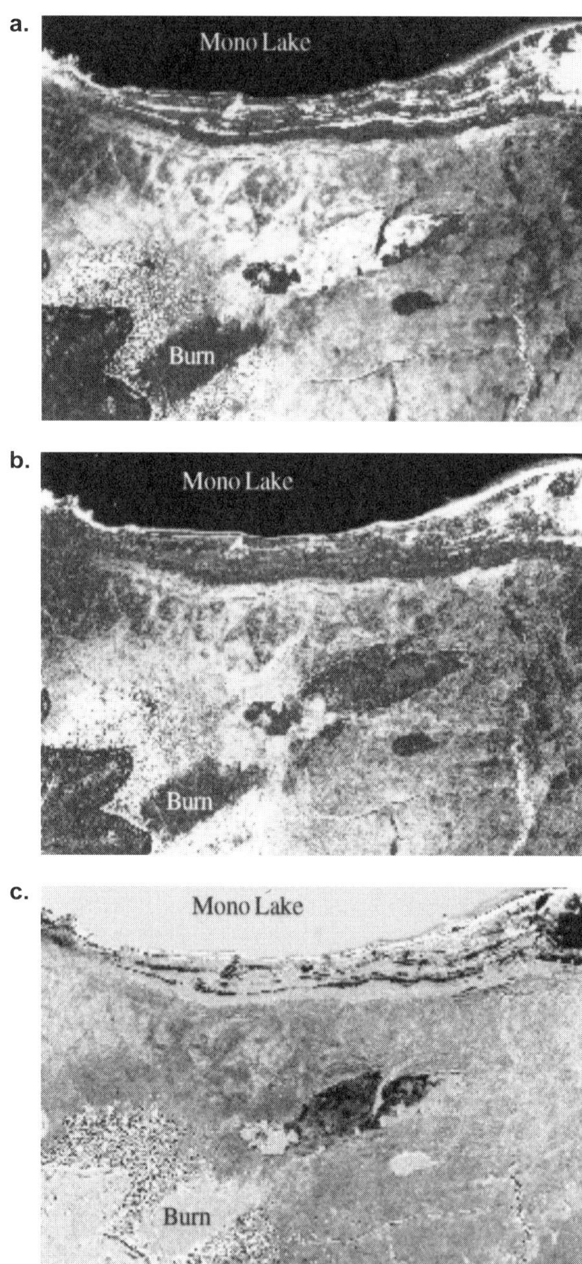

Figure 11.4. Images of the 2DZ_DGVI values from August 20 (a) and October 7, 1992 (b) along with difference image (c) indicating areas where green vegetation was either lost or gained.

regions covered by Jeffrey pines. There were noticeable patches within the Jeffrey pine forest region where strong declines in 2DZ_DGVI values were observed. Based on the aerial photography and field inspection of the study area, these are interpreted as grass and shrub dominated interspaces between trees. Barren areas showed no change in green cover between the two dates of imagery.

5.0 SUMMARY

This research was performed at Mono Lake of California, a study area containing complex background materials and mixtures of different vegetation species. Two AVIRIS data sets were acquired separately in August and October of 1992 over the study area. Derivative-based green vegetation index (DGVI) values were derived from retrieved AVIRIS reflectance spectra. The 2DZ_DGVI (2nd order DGVI derived in reference to zero baseline) showed a strong linear relationship (r^2>0.93 for August and October) with green cover of 10 bitterbrush (*Purshia tridentata*) sample sites having low leaf area index (LAI) values. The 2DZ_DGVI was applied to the two AVIRIS scenes acquired for the two periods to successfully quantify and compare green vegetation cover. Overall, herbaceous species showed the largest change in green vegetation cover from August to October. Shrubs, including bitterbrush and sagebrush, changed less during the same period. The lowest seasonal change in green cover occurred in areas covered by Jeffrey pine. The DGVI concept has potential for practical and accurate monitoring of ecosystems in arid and semiarid lands.

The authors acknowledge that there may be alternatives to the derivative methods for extracting the amplitude of the chlorophyll red-edge from high spectral resolution data. However, available evidence indicates that there is little benefit to having high spectral resolution data if vegetation indices are to be formed using narrow-band versions of broad-band formulas such as the NDVI, RVI, PVI, and SAVI.

Our results indicate that the monitoring of green vegetation in arid and semiarid lands will be substantially improved using imaging spectrometer data with a continuous series of narrow bands across the chlorophyll red-edge region. Such observations are already possible using airborne sensors such as the NASA/AVIRIS. Global application of this approach to monitoring arid and semiarid lands will have to await the deployment of imaging spectrometers in earth orbit.

ACKNOWLEDGMENTS

The authors would like to thank Mr. Y. Zhang for his help with field measurements. This research was done at the Desert Research Institute in Reno, Nevada. The research described here was partially supported by the U.S. Environmental Protection Agency through Cooperative Agreement CR-816826. The mention of trade names or commercial products does not constitute endorsement or recommendation for use.

REFERENCES

Baret, F., G. Guyot, and D.J. Major. TSAVI: A vegetation index which minimizes soil brightness effects on LAI and APAR estimation. *Proceedings of the 12th Canadian Symposium on Remote Sensing: IGARRS '90,* Vancouver, Canada, July 10–14, 3, 1355–1358, 1989.

Butler, W.L. and D.W. Hopkins. Higher derivatives analysis of complex absorption spectra. *Photochem. Photobiol.*, 12, 439–450, 1970.

Chen, Z. *Analysis of High Spectral-Resolution Remote Sensing Data for Quantifying Low Cover Levels of Green Vegetation.* University of Nevada, Reno. Ph.D. dissertation, University Microfilms International, Ann Arbor, MI, 1995.

Collins, W. Remote sensing of crop type and maturity. *Photogrammetric Eng. Remote Sensing*, 44, 43–55, 1978.

Demetriades-Shah, T.H., M.D. Steven, and J.A. Clark. High resolution derivative spectra in remote sensing, *Remote Sens. Environ.*, 33, 55–64, 1990.

Elvidge, C.D. and R.J.P. Lyon. Influence of background albedo on vegetation indices employing red and near-infrared. *Proceedings of the Seminar on Remote Sensing for Geologic Mapping*, IUGS-UNESCO Programme on Geological Applications of Remote Sensing, February 2–4, Orleans, France, 1984, p. 11.

Elvidge, C.D. and R.J.P. Lyon. Influence of rock-soil spectral variation on the assessment of green biomass. *Remote Sens. Environ.*, 17, 265–269, 1985.

Elvidge, C.D. and D.A. Mouat. Analysis of green vegetation detection limits in 1988 AVIRIS data. *Proceedings of the International Symposium of Remote Sensing of Environment, 7th Thematic Conference—Remote Sensing for Exploration Geology*, Calgary, Canada, 1989.

Elvidge, C.D., Z. Chen, and D.P. Groeneveld. Detection of trace quantities of green vegetation in 1990 AVIRIS data. *Remote Sens. Environ.*, 44, 271–279, 1993.

Elvidge, C.D. and Z. Chen. Comparison of broad-band and narrow-band red and near-infrared vegetation indices. *Remote Sens. Environ.* 54, 38–48, 1995.

Hall, F.G., K.F. Huemmrich, and S.N. Goward. Use of narrow-band spectra to estimate the fraction of absorbed photosynthetically active radiation. *Remote Sens. Environ.*, 32, 47–54, 1990.

Horler, D.N.H., M. Dockray, and J. Barber. The red-edge of plant leaf reflectance. *Int. J. Remote Sensing*, 4, 273–288, 1983.

Huete, A.R. A soil-adjusted vegetation index (SAVI). *Remote Sens. Environ.*, 25, 295–309, 1988.

Huete, A.R., R.D. Jackson, and D.F. Post. Spectral response of a plant canopy with different soil backgrounds. *Remote Sens. Environ.*, 17, 37–53, 1985.

Jordan, C.F. Derivation of leaf area index from quality of light on the forest floor. *Ecology,* 50, 663–666, 1969.

Major, D.J., F. Baret, and G. Guyot. A ratio vegetation index adjusted for soil brightness. *Int. J. Remote Sensing*, 11, 727–740, 1990.

Richardson, A.J. and C.L. Wiegand. Distinguishing vegetation from soil background information. *Photogrammetric Eng. Remote Sensing*, 43, 1541–1552, 1977.

Roberts, D.S., M.O. Smith, and J.B. Adams. Green vegetation, nonphotosynthetic vegetation and soils in AVIRIS data. *Remote Sens. Environ.*, 44, 255–269, 1993.

Rock, B.N., T. Hoshizaki, and J.R. Miller. Comparison of in situ and airborne spectral measurements of the blue shift associated with forest decline. *Remote Sens. Environ.*, 24, 109–127, 1988.

Rouse, J.W., R.H. Haas, J.A. Schell, and D.W. Deering. Monitoring vegetation systems in the great plains with ERTS. *Third ERTS Symposium*, NASA SP-351, 1, 309–317, 1974.

Tucker, C.J. Red and photographic infrared linear combinations for monitoring vegetation. *Remote Sens. Environ.*, 8, 127–150, 1979.

Tucker, C.J., H.E. Dregen, and W.W. Newcomb. Expansion and contraction of the Sahara desert from 1980 to 1990. *Science*, 253, 299–301, 1991.

CHAPTER 12

Monitoring Trends in Wetland Vegetation Using a Landsat MSS Time Series

Christopher D. Elvidge, Tomoaki Miura, Wally T. Jansen,
David P. Groeneveld, and Jayita Ray

1.0 INTRODUCTION

Land cover changes can be readily detected using satellite imagery based on changes in spectral reflectance or the introduction of spatial features such as roads or field outlines. This capability was first demonstrated with photographs taken from manned space missions in the early 1960s. Land cover changes can be analyzed either visually (Skole and Tucker, 1993) or through digital analysis (Yuan et al., 1998). In extreme cases, land cover changes can be unambiguously identified using a single date of imagery. In general, two dates of imagery are sufficient to document a set of land cover changes which have occurred between the two image acquisition dates.

The remote sensing of vegetation trends in the absence of land cover change is more challenging than standard land cover change analyses. Vegetation growth typically exhibits some type of annual cycle, with a period with low (or no) growth and a period of active growth and decline (Figure 12.1). This growth cycle is controlled by growth limiting factors, such as water availability (Figure 12.2), day length (Figure 12.3), or temperature (Figure 12.4). Variations in these primary growth-affecting factors result in growth responses of vegetation that vary from year to year. For instance, in a year where water is particularly plentiful in an arid or semiarid environment, growth limitations are reduced and more vegetation growth occurs. Imprinted on top of these primary growth-limiting factors are the impacts from external stressors, such as herbivory disease, or pollution.

Theoretically, trends in vegetation vigor can be discerned from minor changes in the annual growth cycle. However, a sufficient number of growth cycles must be included in the analysis to establish the normal pattern and to then discern deviations from normal. Can such an analysis be accomplished using a single image each year? Even if the imagery was acquired at or near the same date each year, the analysis may be ambiguous without confirmation of the trend through multiple observations within a growing season. An analysis of vegetation trends would be substantially improved if a sufficient number of images were included in the analysis to track the vegetation growth pattern through the growing season, year after year.

The analysis of vegetation trends in the absence of land cover change is an extremely important capability for evaluating the cumulative impacts of human activities on terrestrial ecosys-

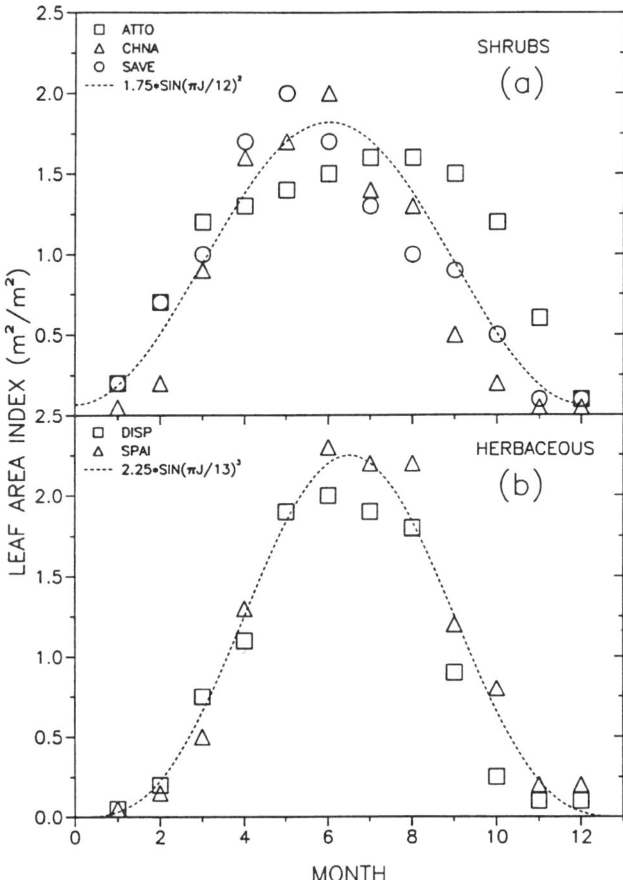

Figure 12.1. Measured and best-fit LAI curves for shrubs (a) and herbaceous species of Owens Valley, California (from Or and Groeneveld, 1994).

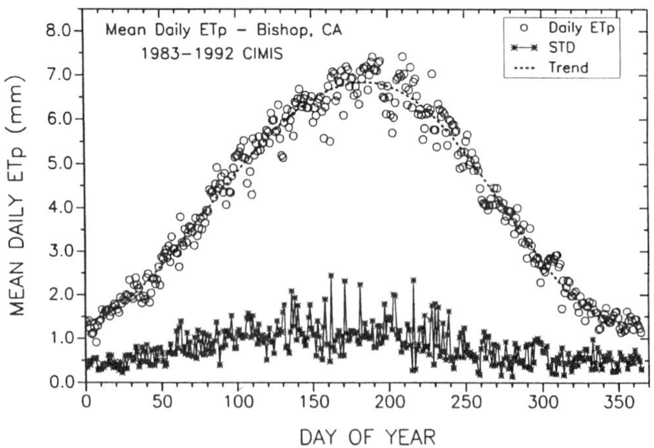

Figure 12.2. Daily evapotranspiration for a complete year of observations made at Bishop, California (from Or and Groeneveld, 1994).

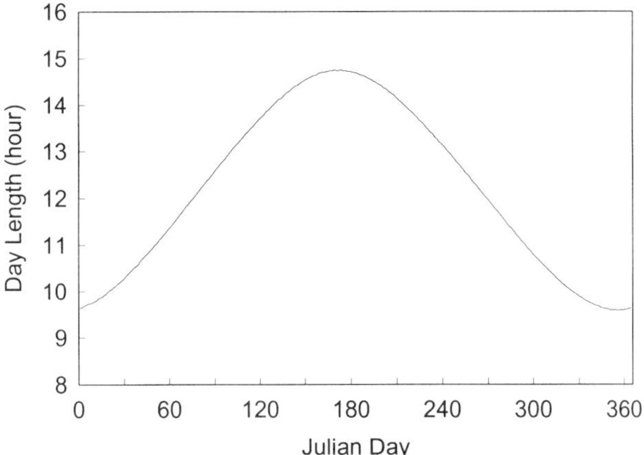

Figure 12.3. Day length curve for Owens Lake.

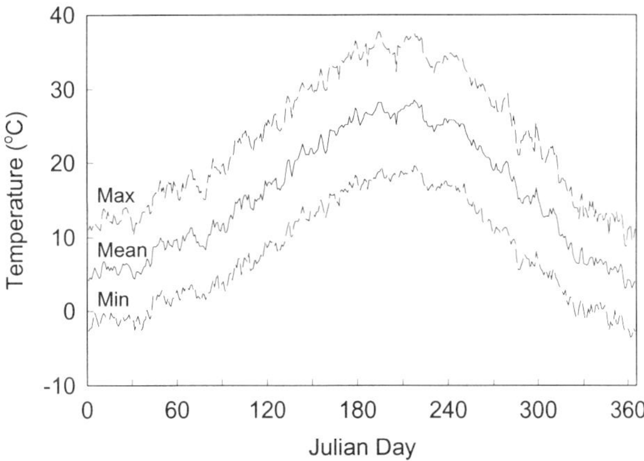

Figure 12.4. Mean, minimum, and maximum annual air temperature data for Owens Lake.

tems. However, there are very few examples of this, in part due to the dearth of high quality image time series. Display and analysis of an image time series as a coherent data set requires that the imagery be coregistered and calibrated to ground reflectance or radiometrically normalized. The U.S. Geological Survey (USGS), EROS Data Center (EDC), and National Aeronautics and Space Administration (NASA) have developed a global coarse resolution (1 km) time series of NOAA Advanced Very High Resolution Radiometer (AVHRR) observations optimized for the study of the Earth's land surface (Eidenshink and Faundeen, 1994). AVHRR data can be readily assembled into image time series because data are acquired globally on a daily basis. By compositing the cloud-free portions of successive passes, it is possible to assemble cloud-free observations of the land surface on weekly or biweekly increments. This high frequency of observations enables the tracking of seasonal vegetation growth cycles. The evenly spaced temporal increments facilitates display of the data and analysis of trends or patterns which occur over multiple years.

The 1-km AVHRR time series have been used successfully to develop continental scale maps of major vegetation units, based on the distinctive temporal cycles of regional scale vegetation units (Loveland et al., 1991). If extended over a greater number of years, the AVHRR time series would undoubtedly reveal useful information on regional trends in vegetation growth in response to factors such as climate variability and human impacts on terrestrial ecosystems. A basic limitation of the AVHRR time series is the spatial resolution, which is too coarse to resolve many important vegetation units, such as most wetlands. Higher spatial resolution data are available from sensors like Landsat or SPOT. As with AVHRR data, Landsat data time series can be assembled by coregistering images and calibrated to ground reflectance. Unlike AVHRR, it is difficult to assemble a cloud-free Landsat time series with short evenly spaced temporal increments, since the orbital repeat cycle is ~16 days instead of daily. Thus, in each year the vegetation growth cycle is sampled in a fragmentary manner. In some cases, there may be lengthy gaps in the observational record due to loss or limitations in satellite data acquisitions. All of this complicates the analysis, making it difficult to discern trends or gradual changes.

In this chapter we present a case study in the analysis of trends in wetland vegetation vigor using a time series of Landsat Multispectral Scanner (MSS) data spanning a 14-year time period. By manipulating the temporal arrangement of the data we have been able to resolve aggregate annual vegetation growth cycles and have identified upswings and downturns in wetland vigor related to regional precipitation patterns.

2.0 SITE DESCRIPTION

The Owens Lake basin in California, is located in the southern end of Owens Valley, between the Sierra Nevada to the west and the Inyo and Coso ranges to the east (see color Plate 18). It lies within the strong rain shadow of the Sierra Nevada, which rises to an elevation in excess of 4,000 meters above sea level. The arid Northern Mojave/Great Basin climate is characterized by hot summers and cold winters with a precipitation regime dominated by winter storms, concentrated between October and April, with substantial interannual variability. Precipitation amounts are low on the valley floor and increase with elevation. The average annual precipitation on the crests of the Sierra Nevada and the Inyo range are about 50 cm and 25 cm, respectively, while the annual precipitation on the valley floor rarely exceeds 15 cm (Saint-Amand et al., 1986).

The basements of the basin consist of the Sierran block on the west side, which is composed primarily of a prequaternary granitic to dioritic complex, and a mixture of undifferentiated metamorphic, sedimentary, and granitic rocks on the east side (Bateman and Merrian, 1954; Hollett et al., 1991).

The basin consists of aquifers that were derived from the sediments on the Owens River delta and from the surrounding alluvial fans (Lopes, 1988). The north central part of the basin consists of well-sorted, fine- to medium-sand aquifers while the marginal areas are composed of those derived from the alluvial fans. At the edge of the lakebed these porous alluvial sediments are interleaved with the less permeable clay layers which characterize the lakebed (Lopes, 1988).

Owens Lake is the terminus of a hydrologically closed basin (Lopes, 1988). Its inflow components are: precipitation, runoff from the surrounding mountains, surface flow from Owens River, and subsurface flow from the upper Owens Valley and Centennial Flat while its outflow components are basically only evaporation from the lake bed and evapotranspiration from the vegetated areas (Lopes, 1988). Although nearly full at the turn of the century (1901), Owens Lake is now largely dry due to diversion of the Owens River and smaller tributaries to the City of Los Angeles.

Springs and seeps are found along the margins of the lakebed. They are generally located between the 1,085 m and the 1,097 m elevation contours where the alluvial fans from the mountain fronts grade into the lakebed with an abrupt change in slope and sediment permeability (Feeney Hall, 1996). Wetland/riparian vegetation at Owens Lake occurs in association with these springs and seeps and the Owens River delta. Bair et al. (1995) described these wetland communities using two broad floristic categories defined by the California Natural Diversity Data Base of natural communities (Holland, 1986). "Alkali Meadows and Seeps" are associations of low growing perennial grasses and herbs with relatively few species, including *Distichlis spicata* var. *stricta* (saltgrass), *Anemopsis californica* (anemopsis), *Juncus balticus* (wirerush), and *Nitrophila occidentalis* (nitrophila). "Transmontane Alkali Marshes" are typically found clustered around a source of flowing water. The characteristic species in this community is *Scirpus americanus* (American bulrush). A detailed description regarding these communities can be found in Bair et al. (1995).

Saltgrass is, by far, the dominant species in the wetlands and seeps which occur along the lakebed margins. Depending on temperature, this species initiates leafing from the middle of March to April, depending on temperature, with seed drop occurring from late-May to mid-July. Even after seed drop, saltgrass may remain green through October or later, depending on temperature and water availability.

3.0 SATELLITE IMAGE DATA

The image time series used in this study consisted of 64 Landsat MSS scenes, covering a period from May 12, 1979 to October 10, 1992 (Table 12.1). In most of the represented years, four to six images were included. Thirteen images were acquired during 1992, the last year MSS data of the U.S. were acquired. No Landsat MSS data of the area were available for 1989, 1990, or 1991. The images were selected based on their low levels of cloud cover, especially over the Owens Lake study area. The 64 scenes represent approximately half of the total available low cloud cover Landsat MSS scenes held by the EDC for the 1972–92 history of MSS data acquisitions. As can be seen in Figure 12.5, there is a relatively even spread of data coverage across all seasons when the full suite of images are considered. In contrast, significant gaps exist within the records for individual years.

4.0 SATELLITE DATA PROCESSING

Preprocessing done by EDC included conversion to an oblique Mercator projection (USGS, 1979). The assembly of the image time series suitable for analysis required the following two transformations:

- The geometric alignment of the images. All the images were geometrically aligned to one of the images which had been fitted to the Universal Transverse Mercator 11 (UTM 11) grid.
- The radiometric alignments of the images. Radiometric changes not related to those actually occurring on surfaces were removed by the means of calibration to surface reflectance.

4.1 Image-to-Image Coregistration

In order to monitor or analyze surface changes using an image time series data set, it is necessary to geometrically align all the images to either the coordinate system of one of the images in the temporal data set (image-to-image coregistration), or a geodetic coordinate sys-

Table 12.1. List of 64 Landsat MSS Images Used in the Study (the data acquisition dates, satellite numbers, and solar elevation/azimuth angles are given).

ID No	Date of Image	Landsat No	Solar Illumination Geometry	
			Elevation	Azimuth
64	92/10/10	5	39	143
63	92/09/24	5	44	137
62	92/09/08	5	48	131
61	92/08/23	5	52	123
60	92/08/15	4	51	116
59	92/07/22	5	57	111
58	92/07/14	4	55	105
57	92/06/28	4	57	103
56	92/06/20	5	60	107
55	92/06/04	5	60	109
54	92/05/27	4	57	107
53	92/05/11	4	55	112
52	92/04/25	4	51	117
51	88/10/15	5	39	147
50	88/06/09	5	62	110
49	88/04/06	5	50	130
48	88/03/05	5	38	138
47	87/11/14	5	30	152
46	87/09/27	5	44	140
45	87/06/07	5	61	109
44	87/04/20	5	53	124
43	86/10/10	5	39	142
42	86/08/23	5	52	123
41	86/07/06	5	59	107
40	86/05/03	5	57	120
39	86/02/28	5	36	138
38	85/12/26	5	24	150
37	85/11/08	5	32	151
36	85/07/03	5	61	109
35	85/05/16	5	60	117
34	85/03/29	5	47	133
33	85/02/25	5	36	140
32	84/06/30	5	61	108
31	84/05/29	5	61	112
30	84/05/13	5	59	117
29	84/04/03	4	48	130
28	84/01/14	4	25	147
27	83/12/29	4	24	150
26	83/10/10	4	41	145
25	83/07/22	4	59	112
24	83/06/20	4	62	108
23	83/05/19	4	60	116
22	83/05/03	4	57	122
21	83/04/17	4	53	127
20	83/01/11	4	25	148
19	82/12/10	4	25	152
18	82/10/05	3	42	142
17	82/06/01	3	61	110
16	82/03/21	3	43	133

Table 12.1. List of 64 Landsat MSS Images Used in the Study (The data acquisition dates, satellite numbers, and solar elevation/azimuth angles are given). (Continued)

ID No	Date of Image	Landsat No	Solar Illumination Geometry	
			Elevation	Azimuth
15	82/01/17	2	25	145
14	81/10/19	2	36	144
13	81/06/15	2	60	106
12	81/05/10	2	57	116
11	81/04/22	2	53	122
10	81/04/04	2	48	128
9	81/02/27	2	35	136
8	80/10/24	2	35	147
7	80/06/02	3	60	109
6	80/05/06	3	56	116
5	79/12/14	3	24	149
4	79/10/30	2	33	147
3	79/05/30	3	60	110
2	79/05/21	2	58	110
1	79/05/12	3	58	116

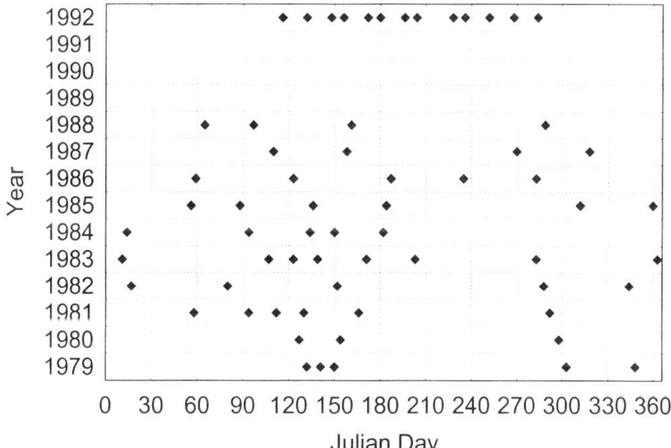

Figure 12.5. Julian date versus year for the Landsat MSS images used.

tem (georeferencing). In this manner, image data from each date can be compared for any location within the coverage area. In this study, the image-to-image coregistration was employed to accomplish the geometric alignments. As the reference image we utilized the Landsat MSS image of Owens Lake from July 3, 1985, which had been georeferenced to the UTM 11 grid by the USGS as part of the U.S. Environmental Protection Agency and USGS North American Landscape Characterization data set assembly (Lunetta et al., 1993). By coregistering to this reference image, all of the coregistered images were georeferenced to the UTM 11 grid.

In practice, a geometric alignment or transformation can be divided into three steps: (1) selection of a set of tie-points (ground control points) that relate the coordinate system of an image to a reference image or coordinate system, (2) a coordinate transformation, and (3) resampling (Moik, 1980). To reduce the computational magnitude of the coregistration effort, the Owens Lake area was extracted from each of the Landsat scenes, effectively reducing the coverage area and data volume by 75 percent. Output images were 1100 by 1100 pixels, with 60 meter pixels.

Image-to-image control points were selected from small features perennially observable on both the subject and reference images. Man-made features such as road intersections, runways, or other objects are ideal for use as control points; however, only a small number of such features were consistently recognizable in the sparsely populated study area at the MSS pixel resolution. Only two man-made features were used as image-to-image control points: the edge of an airport and a road intersection. The bulk of the 23 image-to-image control points were small geographical features such as small playas and small hills. Great attention was paid to the accurate selection of the individual ground control points. In most images only a subset of the 23 could be identified and used for performing the coregistration.

For the coordinate transformation, we utilized a first-order polynomial combined with an affine transformation. Affine transformations include four different geometric operations: rotation, displacement, scaling, and skewing (Moik, 1980). Factors which lead to this selection include: (1) All the images had been fitted to an Oblique Mercator projection during the preprocessing performed by the EDC (USGS, 1979). Through experimentation we found that some degree of rotation was required in each case to accomplish the coregistration. (2) The size of the reduced images were small enough that a first-order polynomial was found to be sufficient for the coregistration. And (3) use of higher level polynomial transformations resulted in extremely distorted (and inaccurate) output images due to the fact that the ground control points were not regularly spaced. For resampling, a nearest-neighbor interpolation was used to avoid loss of radiometric accuracy (Moik, 1980).

The results of the image-to-image coregistration were examined using three different methods: (1) root mean square (RMS) error, (2) image overlays, and (3) image flickering. The average RMS error had to be lower than one pixel for a coregistered image to be considered acceptable for use in the study. When this condition was met, the 1100-by-1100 images were overlain onto the reference image and misregistrations or displacements were visually investigated. Then the 1100-by-1100 images and the reference image were flickered on the same image display to visually assess misregistration. Because RMS error only evaluates the fit of the image-to-image control points, the image overlay and flickering were used to evaluate the overall accuracy of the coregistration. In cases where the coregistration results were not satisfactory, additional image-to-image control points were sought and the coregistration was performed on the original data and the results were examined again using the methods described above. This process was repeated until a satisfactory coregistration was achieved.

Average RMS errors were less than 0.60 pixel for all images. Misregistration or displacements of the resultant images relative to the reference were, in all but one case, less than 1.0 pixel (visually examined). In the one case which did not meet the above specifications, an image with a 2.0 pixel misregistration was generated.

4.2 Calibration to Surface Reflectance

The second preprocessing step needed to render the images in the image time series comparable is to radiometrically align the images. The purpose of the radiometric alignment is to remove, to the extent possible, radiometric differences between the scenes caused by sensor conditions, scene-to-scene variations in illumination, and atmospheric effects (scattering and absorption).

In order to remove those radiometric differences, the images were calibrated to surface reflectance using the empirical line calibration method (e.g., Roberts et al., 1985; Elvidge, 1988; Conel, 1990). This method utilizes image data from two or more spectrally distinct homogeneous and geographically flat calibration targets to develop linear relationships relating digital numbers to reflectance;

$$R_{t,b} = G_b * \mathrm{DN}_{t,b} + O_b \tag{1}$$

where $R_{t,b}$ is the laboratory or field reflectance in band b for the calibration target t, $\mathrm{DN}_{t,b}$ (digital number) is the relative uncalibrated encoded radiance in band b of an image pixel for the calibration target t, the intercept term (offset) O_b accounts for the atmospheric scattering and the instrumental dark current in band b, and the slope term (gain) G_b includes the atmospheric transmission and the sensor gain in band b. Although the calibration coefficients O_b and G_b were developed for a set of calibration targets, they were applicable to the image data by assuming that the atmosphere at the time of the data acquisition was homogeneous through the entire scene. In this manner, the images can be compared to reflectance measured in the laboratory or in the field in addition to the other images in the image time series data set.

Ideally, radiometric calibration targets would be Lambertian reflectors, with constant reflectance at all illumination and viewing angles. For Lambertian targets, only one reflectance spectrum is required for use of the target in calibrating images acquired at different illumination/viewing geometries. In practice, few surfaces are truly Lambertian.

The area covered by the 1100-by-1100 images contained six prospective calibration targets. Two targets, the Red Cinder (RED) and the Centennial Flat (CENT), are not flat, are a few hundred feet above the study site, and are composed of volcanic cinders and arkose sediments, respectively. The Olancha Dune (SAND) covers a large area south of the lakebed. This dunefield contains both flat sheets of bare sand and sets of hummocky dunes. The North Beach (NOBE) and the West Beach (WEBE) are nearly flat granitic gravel surface at the same elevation as the study site. The Haiwee reservoir (LAKE) was used as a dark target (zero reflectance) in the near-infrared wavelength bands (MSS 3 and 4). The number of image pixels extracted for each calibration target are listed in Table 12.2.

An experiment was designed to examine the bidirection reflectance properties of the calibration target materials. Samples were taken from these sites (except from the Haiwee reservoir) and placed onto trays. Nadir-oriented reflectance measurements were made from these trays across the range of solar elevation angles encountered in the Landsat time series (Table 12.1: 24 to 62 degrees) with an Analytical Spectral Devices - Personal Spectrometer 2 (PS-2). The PS-2 acquires spectra in the 400 to 1000 nm wavelength region with 4 nm bandwidths and 1.4 nm sampling intervals.

At the beginning of each set of the reflectance measurements, the PS-2 was calibrated to reflectance by measuring dark currents and a white calibration panel. In addition, the solar elevation angle was recorded. These time records were later used to develop relationships between solar elevation angles and the time of the day. For each measurement, five individual spectra were acquired. After completing the spectra for one target, trays were manually switched to place the next in position. Each set of measurements was completed in about five minutes. This procedure was repeated every 15 minutes, under clear sky conditions, while the solar elevation increased from 20 to 65 degrees.

The reflectance spectra were convolved to the four MSS bandpasses using MSS spectral response curves. These reflectance values and corresponding solar elevation angles were plotted for each of the five prospective calibration targets (Figure 12.6). RED was the only target

Table 12.2. The Names of Calibration Targets, their Compositions, and a Number of Pixels Extracted from Each.

Calibration Target	Composition	Number of Pixels
Red Cinder (RED)	Red Basalt	16
Centennial Flat (CENT)	Arkose Sediment	12
Olancha Dune (SAND)	Sand	Not Used
North Beach (NOBE)	Granitic Gravel	25
West Beach (WEBE)	Granitic Gravel	20
Haiwee Reservoir (LAKE)	Water	25

Table 12.3. Number of Pixels Extracted for the Wetland Areas.

Name of Wetland Area	Number of Pixels
Northwest Corner	92
Cottonwood Springs	121
Owens River Delta	121
Millsite	109
Tubman	30

that could be assumed to be a Lambertian surface (Figure 12.6a). CENT was nearly a Lambertian surface (Figure 12.6b). The reflectance of the other targets decreased as solar elevation angles increased (Figures 12.6c, d, and e).

In order to compensate for the anisotropy (non-Lambertian properties) of the calibration targets, linear regression analysis was applied to derive equations for predicting the reflectance of each target at varying solar elevation angles. For each Landsat scene, the solar illumination angle was used to estimate the ground reflectance of each target in the four MSS spectral bands. Then the empirical line calibration methods were applied to derive equations for estimating surface reflectances within the Landsat scenes. For MSS 1 and 2, four calibration targets, RED, CENT, NOBE, and WEBE were used to develop calibration coefficients. For MSS 3 and 4, five calibration targets, LAKE in addition to the four targets above, were used. LAKE was assumed to have zero reflectance through all the images. SAND was not used for developing calibration coefficients because of the target's anisitropic behavior and the difficulty encountered in identifying flat portions of the dune area in the images.

Examples of the calibration results for the image of July 6, 1986 are shown in Figure 12.7. This image was taken at the solar elevation of 59 degrees. Because the illumination was high at that time, digital numbers cover nearly the full dynamic ranges for all the four bands. The maximum retrievable reflectances, which correspond to the maximum DN (127), are 0.51 for MSS 1, 0.43 for MSS 2, 0.46 for MSS 3, and 0.60 for MSS 4. Any surface materials having reflectance more than these values were saturated in this image. An example of the calibration results obtained for an image acquired with a low solar elevation angle (25 degrees) is shown in Figure 12.8.

The empirically derived calibration coefficients were applied to each of the coregistered subscenes, converting the data to surface reflectance. This procedure assumes that the atmosphere over the scene area was homogeneous.

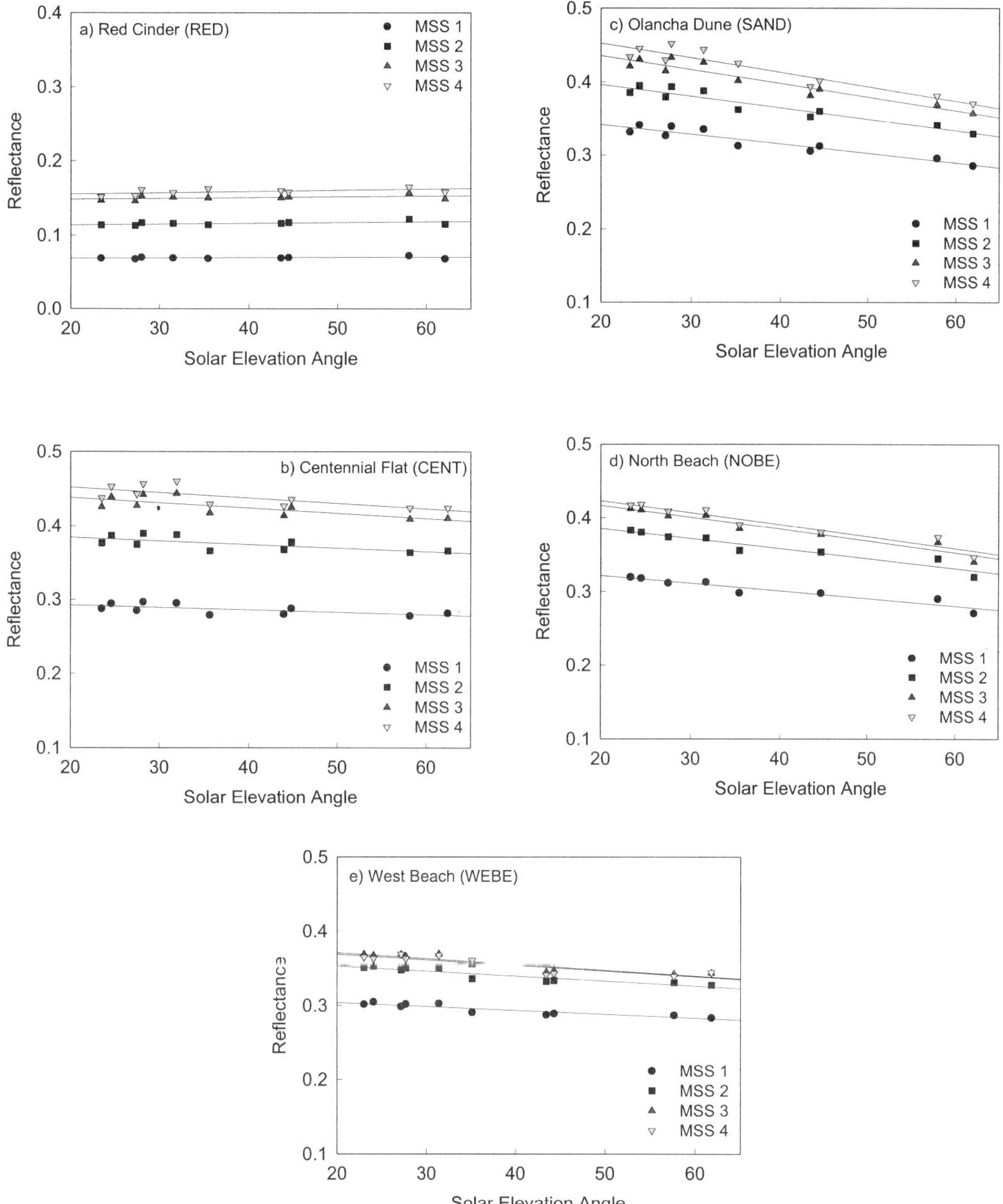

Figure 12.6. Relationship between calibration target reflectance and solar elevation angle: (a) Red Cinders, (b) Centennial Flat, (c) Olancha Dunes, (d) North Beach, (e) West Beach.

4.3 Perpendicular Vegetation Index

In order to measure the intensity of the green vegetation spectral contribution to the image data, the images were converted to Perpendicular Vegetation Index (PVI) values (Richardson and Wiegand, 1977; Elvidge and Lyon, 1985);

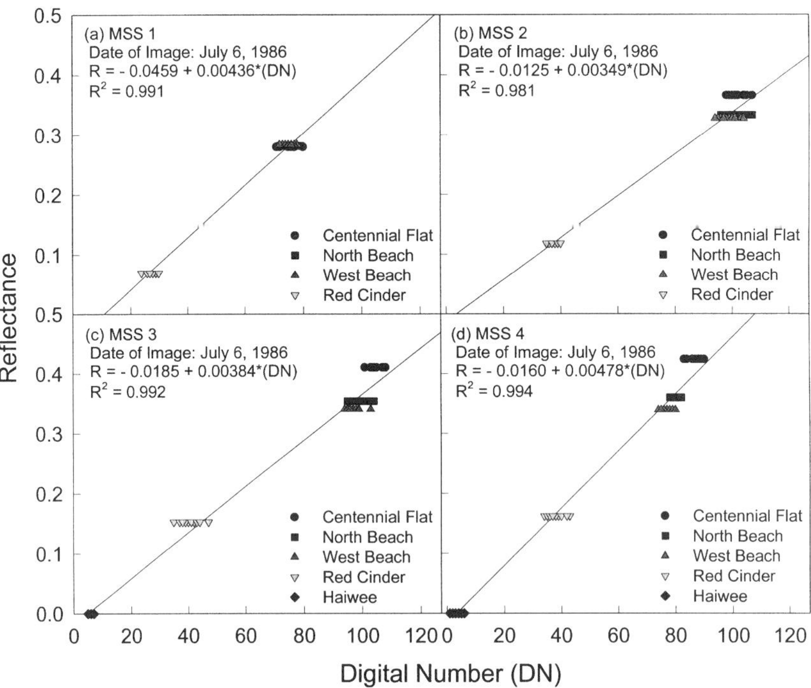

Figure 12.7. Calibration results for the image from July 6, 1986, acquired under a 59° solar elevation angle.

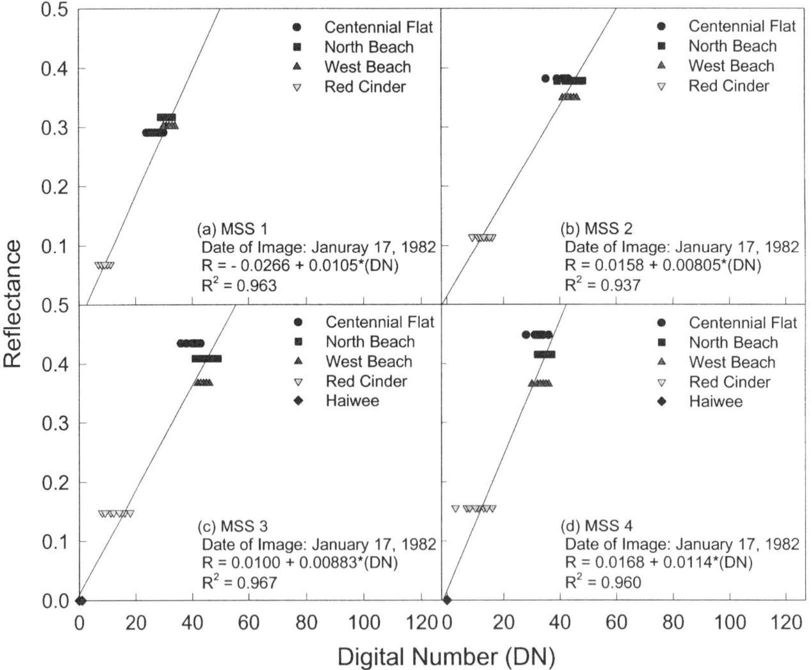

Figure 12.8. Calibration results for the image from January 17, 1982, acquired under a 25° solar elevation angle.

$$\text{PVI} = a * R_{NIR} - b * R_{red} - \textit{PVI Residual} \tag{2}$$

where R_{NIR} and R_{red} are reflectance in near-infrared (NIR) (MSS 4) and red (MSS 2) bands, respectively. The PVI coefficients, a and b, are for the orthogonal transformation to measure the distance from the rock-soil baseline and derived from the slope of the rock-soil baseline. The additive term, *PVI Residual*, compensates for the perpendicular vegetation index (PVI) value of the baseline associated with the nonzero intercept of the rock-soil baseline (Elvidge and Lyon, 1985).

Image data from three calibration targets (CENT, SAND, and NOBE) and a barren basalt area located between the Inyo and Coso ranges were used to develop the rock-soil baselines for each image. The other calibration targets, RED, WEBE, and LAKE were not used because their spectral shapes were found to be different from the typical soils of the lakebed. Pixel values of the targets were extracted from an image and the linear regression was applied to a set of red (MSS 2) and NIR (MSS 4) values to derive the coefficients of the rock-soil baselines. This procedure was performed for the individual images. An example of the derived rock-soil baseline is shown in Figure 12.9, which is for the image from July 6, 1986.

5.0 SATELLITE DATA ANALYSIS

The processed time series imagery was used to study a set of five wetlands located on the margins of the lakebed. In addition, the area of open water present at Owens lake was derived for each of the 64 images using a multispectral classification. Below is a description of the vegetation analyses.

5.1 Chronological PVI Series for the Wetlands

PVI data from five wetland areas were extracted: (1) Northwest Corner, (2) Cottonwood Springs, (3) Millsite, (4) Tubman, and (5) Owens River Delta. Each of the wetland areas was visually identified on the July 6, 1986 image (Plate 18) and outlined with a polygon. PVI values for pixels inside the polygon were extracted from each image date. The PVI data from the extracted pixels were averaged to generate a chronological set of mean PVI values for each wetland (Figure 12.10). While each of the annual sets of PVI data exhibit low values in the winter and high values in during the middle of the year, it is difficult to discern in which years the wetlands were greener and in which years the wetlands underwent declines in vegetation growth. Due to the fragmentary nature of the vegetation growth cycles in the available set of image data, it is not possible to simply overlay the annual growth curves and draw conclusions regarding the relative vigor of the vegetation in one year versus another.

5.2 Modeling Mean Annual PVI Curves

By modeling the mean annual PVI growth curve for the individual wetlands, it may be possible to observe vegetation trends by analyzing the departures from the modeled growth curve. To prepare the PVI data for the modeling, we rearranged the order of the PVI data from chronological order (Figure 12.10) into Julian date order (Figure 12.11). The mean annual PVI growth curve was then fit using a day length curve (Figure 12.3), which follows a sine function. The fitting, which was accomplished using linear regression analysis, included adjustment of the Julian date of the peak and changes in the amplitude of the curve.

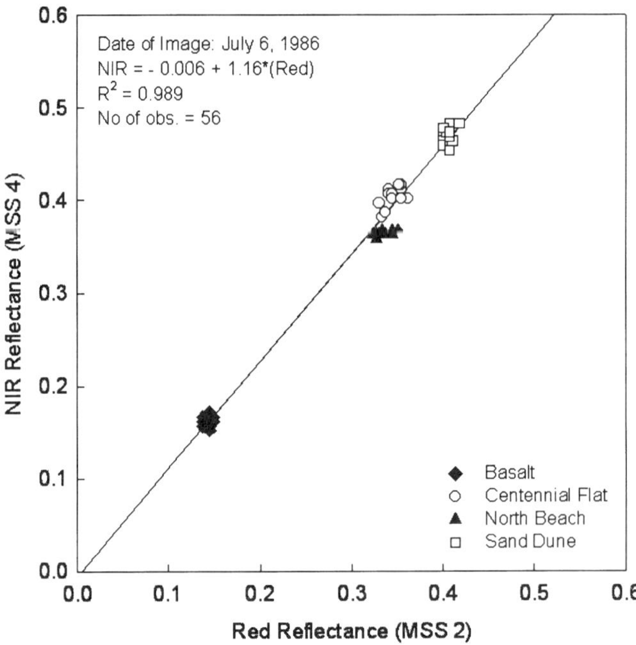

Figure 12.9. Red versus NIR rock-soil baseline for the image from July 6, 1986.

Note that the growth peaks for the wetlands varied by as much as three weeks. The growth peak for Cottonwood Springs occurred at Julian date 222, while the Owens River Delta and Northwest Corner wetlands had their growth peaks at Julian date 237. The latest growth peak was for the Tubman wetland, at Julian date 243, three weeks after the growth peak for Cottonwood Springs. In general, the position of the peak came earlier for the wetlands on the west side of the lakebed than those on the drier east side (Plate 18).

5.3 Analysis of Vegetation Trends

Once the mean annual growth curve of the wetlands had been derived, we subtracted the modeled PVI values from the growth curve from the actual PVI values for each date of imagery, generating a series of residual PVI values. When plotted in chronological order (Figure 12.12), the residual PVI data can be analyzed for trends. In reviewing the residual PVI data it can be observed that the most stable wetland, over time, is Cottonwood Springs. This site has the most stable and continuous water supply (Feeney Hall, 1996). All of the wetlands had above average trend in vegetation growth during 1981. A negative trend in vegetation growth cycles was observed for the Northwest Corner, Owens River Delta, Millsite, and Tubman wetlands during the second half of 1982 until the beginning of 1984. Following this decline, the Owens Delta and Northwest Corner wetlands had two years with positive trends in vegetation growth (1984–85). During 1985 the Tubman wetland exhibited a decline in vegetation growth. The Millsite wetland had a year with a positive trend in vegetation growth during 1986.

Local precipitation records (Figure 12.13a) indicate that 1979 was a dry year; 1980, 1982, 1983, 1986, and 1991 were wet years; 1981, 1984, 1985, and 1992 were typical years, while 1987 to 1990 was a period of extreme drought. We believe that the wetland response at the lake margin springs and seeps to these meteorological variations is frequently delayed by one or

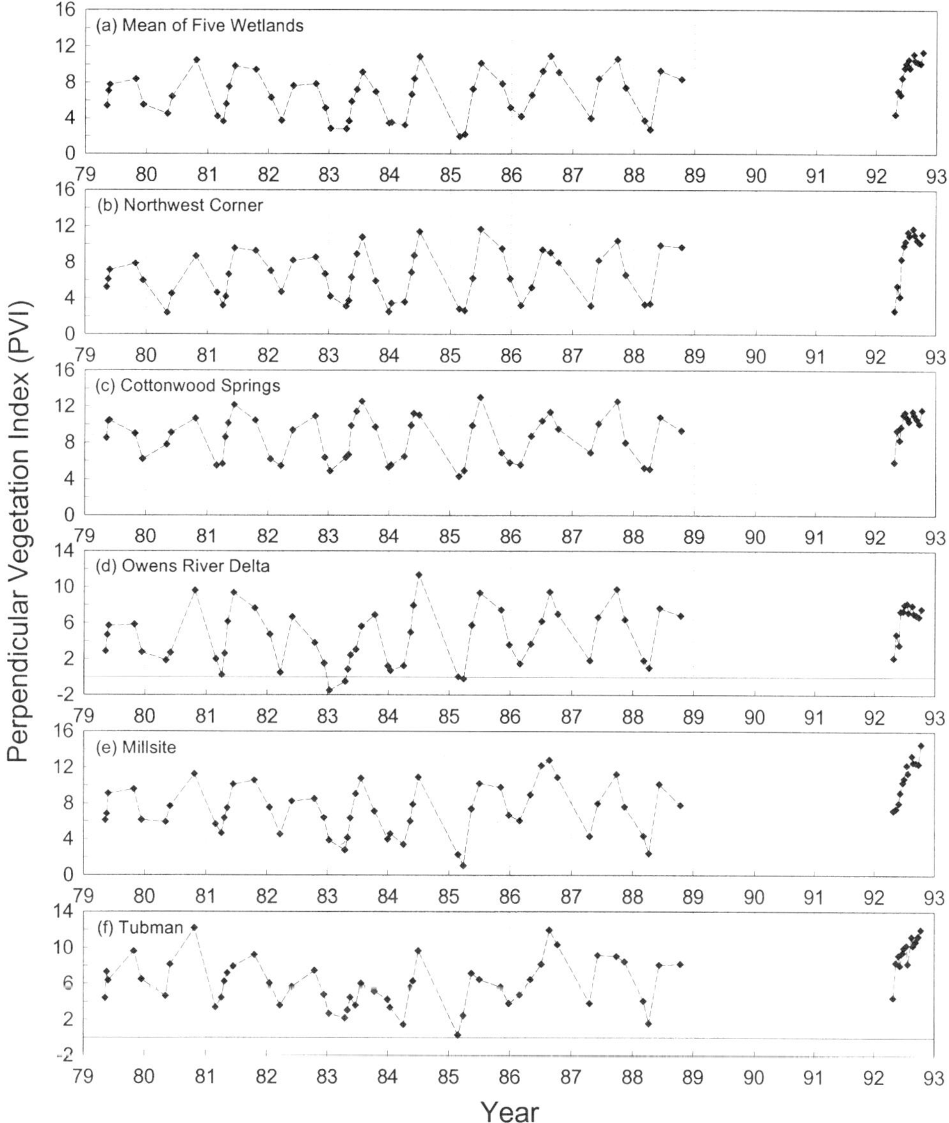

Figure 12.10. Chronological series of mean PVI values for each wetland.

more years while water moves from the higher elevations through the alluvial sediments and emerges at the ground surface. For instance, there was very little precipitation from March 1980 until the end of 1982, a period of more than two years. The downward trend in wetland growth in response to this relatively dry period appears to have been hydrologically delayed until 1982 and 1983. Heavy regional precipitation in 1982–83, associated with El Nino, did not appear to enhance the vegetation growth until the middle of 1984. In 1986 it appears that heavy precipitation on the valley floor at Owens Lake was sufficient to result in an increase in vegetation growth later the same growing season for the Millsite and Tubman sites. The intensity of the local rainfall in early 1986 is reflected in the rapid filling of the lower lakebed with water, observed in the May 3, 1986 Landsat scene (Figure 12.13c).

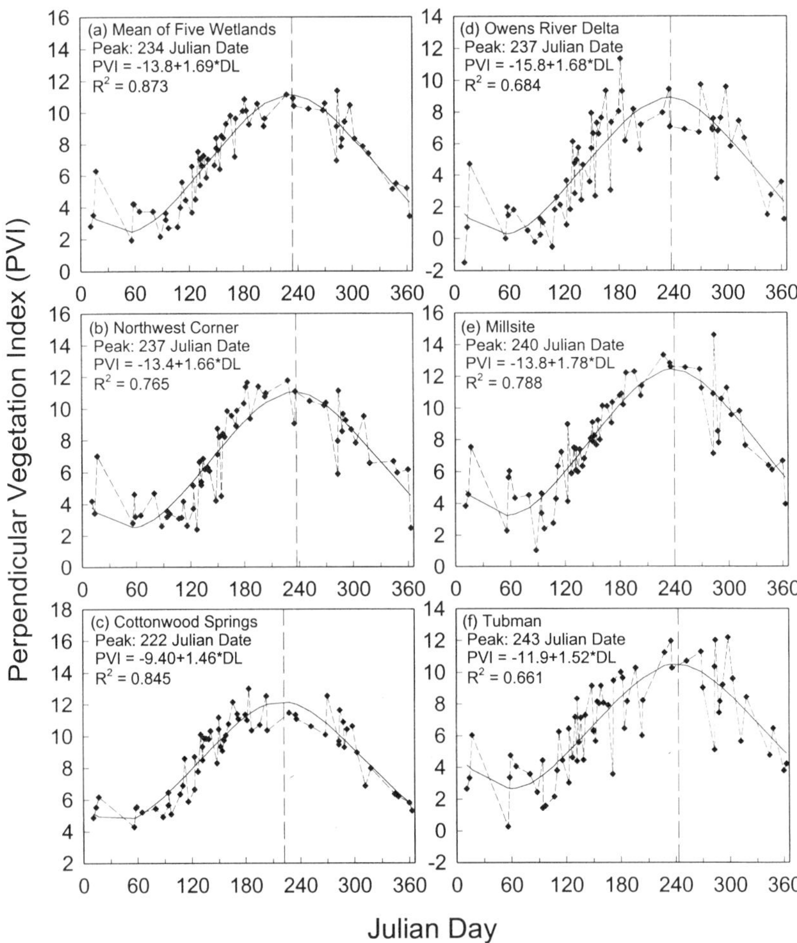

Figure 12.11. Julian date order mean PVI values for each wetland with modeled mean annual growth curves.

The wetlands of the Owens River Delta are dependent on water releases into the lower Owens River by the Los Angeles Department of Water and Power. The massive downward trend of the Owens River Delta vegetation from late 1982 until early 1984 appears to have been due to inundation of the delta by water releases made into the lower Owens River during the same time period (Figure 12.13b). Once past the delta, this water enters the brine pond at the bottom of the lakebed, at which point it is below the elevation of the other wetlands. The highly positive trend of the Owens River Delta wetland during 1984 is a growth response to the infusion of water from the releases made from late 1982 until early 1984.

6.0 SUMMARY

Vegetation growth over most of the world's land surfaces is characterized by seasonal growth cycles which are controlled by growth-limiting factors, such as water availability, day length, and temperature. While day length is stable, water availability is subject to large interannual variability, which can influence plant growth in either a positive or negative manner. Imprinted

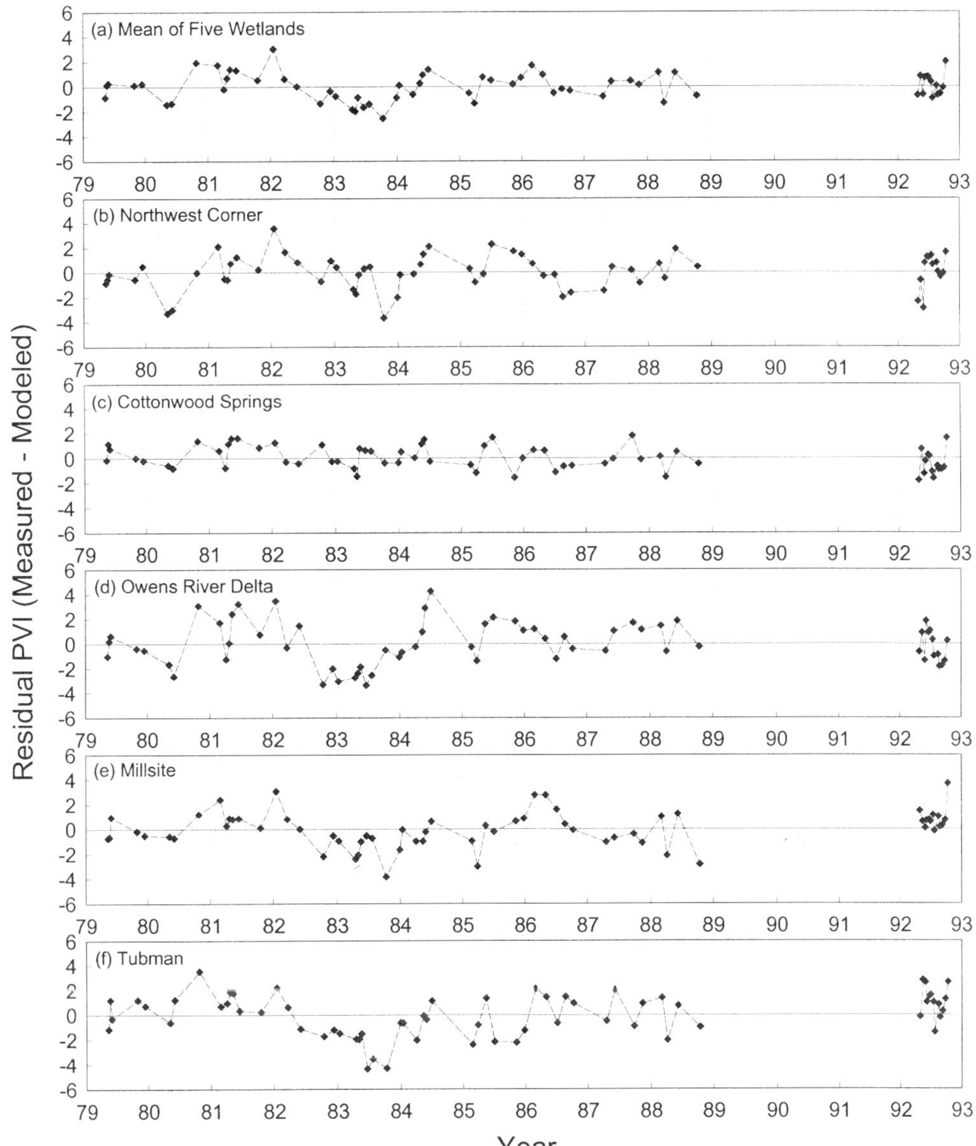

Figure 12.12. Chronological series of residual PVI values for the wetlands.

on top of the plant growth effects from variation in the primary growth-limiting factors are the impacts from external stressors, such as herbivory disease, or pollution.

Vegetation growth cycles can be observed using time series of satellite observations made in red and near-infrared wavelengths. For this application the satellite data must be aligned both geographically and radiometrically. In addition, the time series must include a sufficient number of growth cycles to establish the normal pattern to then discern deviations from normal. This can be most readily accomplished using coarse resolution satellite data from sensors which acquire data on a daily basis, such as the NOAA-AVHRR. By compositing cloud-free AVHRR observations it is possible to assemble vegetation index time series on weekly or biweekly

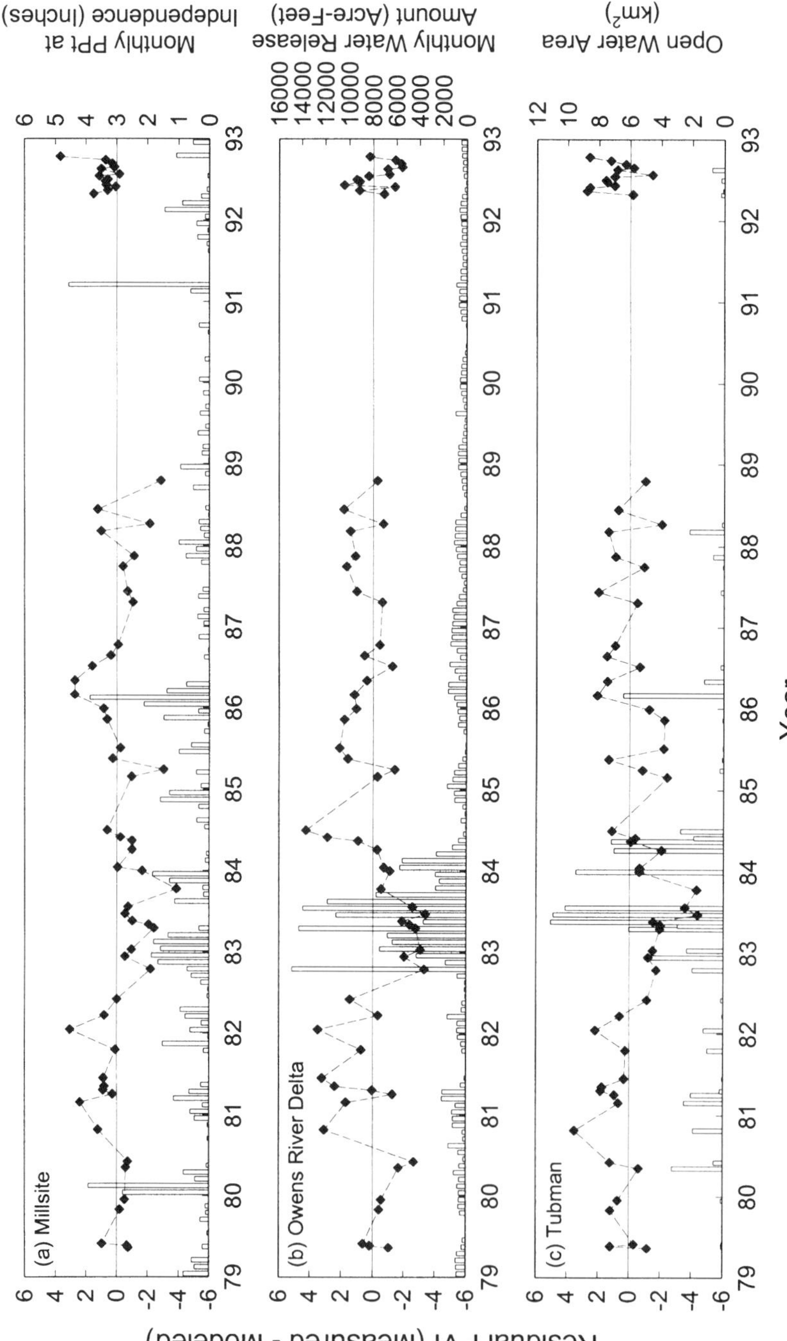

Figure 12.13. Chronological series of (a) monthly precipitation from Independence, California with residual PVI values from the Millsite wetland, (b) monthly water releases into the lower Owens River with residual PVI values from the Owens River Delta wetland, and (3) area of open water derived from the Landsat time series with residual PVI values from the Tubman wetland.

increments, giving sufficient temporal resolution to track vegetation growth through complete cycles for many years.

The situation is different for high spatial resolution imagery, such as that generated by systems such as Landsat. The advantage of high spatial resolution imagery is that it is possible to observe vegetation units, such as wetlands, that are below the spatial resolution limits of AVHRR and other coarse spatial resolution systems. However, archives for Landsat and other high resolution systems are too sparse to provide cloud-free imagery on weekly or biweekly increments. Generally it is only possible to have a fragmentary sampling of the vegetation growth cycle.

We have demonstrated that annual trends in vegetation vigor can be observed from Landsat data time series, even though the vegetation growth cycles for individual years are sampled in a fragmentary manner. By rotating vegetation index time series data from chronological order to Julian date order it is possible to model the mean annual vegetation index growth curve. Residual differences between the observed and modeled vegetation index values are then rotated back to chronological order for interpretation. For the wetlands of Owens Lake we were able to observe both downward and upward trends in vegetation growth and relate these to hydrologic events.

The analysis of vegetation trends in the absence of land cover change is an extremely important capability for documenting the cumulative impacts of human activities on terrestrial ecosystems. Our results indicate that this type of long-term vegetation analysis can be accomplished with high spatial resolution satellite imagery, despite fragmentary sampling of the annual growth cycles.

ACKNOWLEDGMENTS

The authors wish to acknowledge the State of California, Great Basin Air Pollution Control District, Bishop, California, and the U.S. Environmental Protection Agency, Environmental Monitoring Systems Laboratory, Las Vegas, Nevada for partial support of this project.

REFERENCES

Bair, J.J., G.F. Cochran, and T.M. Mihcvc. *Evaluation of Wetland Vegetation Response to Ground Water Drawdown from Production Wells at Owens Lake, California.* Desert Research Institute Water Resources Center, Publication Number 41146, 1995.

Bateman, P.C. and C.W. Merriam. *Geologic map of the Owens Valley region, California.* Map sheet 11, in: Geology of Southern California (R.H. Janhs, Ed.), California Division of Mines Bulletin 170, 1954.

Conel, J.E. Determination of surface reflectance and estimates of atmospheric optical depth and single scattering albedo from Landsat Thematic Mapper data. *Int. J. Remote Sensing*, 11(5) 783–828, 1990.

Eidenshink, J.C. and J.L. Faundeen. The 1-km AVHRR global land data set: first stages in implementation. *Int. J. Remote Sensing*, 15, 3443–3462, 1994.

Elvidge, C.D. Vegetation reflectance features in AVIRIS data. In: *Proceedings of the Sixth Thematic Conference on Remote Sensing of Exploration Geology*, ERIM, Houston, TX, May 16–19, pp. 169–182, 1988.

Elvidge, C.D. and R.J.P. Lyon. Influence of rock-soil spectral variation on the assessment of green biomass. *Remote Sens. Environ.*, 17:265–279, 1985.

Feeney Hall, M.P. Investigation of spring and seep discharge at Owens Lake, California. Master of Science thesis, University of Nevada, Reno, 1966.

Holland, R.F. *Preliminary Description of the Terrestrial Natural Communities of California.* California Department of Fish and Game, Sacramento, CA, 1986.

Hollett, K.J., W.R. Danskin, W.F. McCaffrey, and C.L. Walti. Geology and Water Resources of Owens Valley, California, *U.S. Geological Survey-Supply Paper*, Number 2370, 1991.

Lopes, T.J. *Hydrology and Water Budget of Owens Lake, California*. DRI Publication Number 41107, Report prepared for Los Angeles Department of Water and Power, Los Angeles, CA by Desert Research Institute, Reno, NV, 1988.

Loveland, T.R., J.W. Merchant, D.O. Ohlen, and J.F. Brown. Development of a land-cover characteristics database for the conterminous U.S. *Photogrammetric Engineering and Remote Sensing*, 57(11) 1453–1463, 1991.

Lunetta, R.S., J.G. Lyon, J.A. Sturdevant, J. L. Dwyer, C.D. Elvidge, L.K. Fenstermaker, D. Yuan, S.R. Hoffer, and R. Weerackoon. North American Landscape Characterization (NALC) Research Plan. USEPA Report Number EPA/600/R-93/135, 1993.

Moik, J.G. *Digital Processing of Remotely Sensed Images*. NASA Special Paper 431, U.S. Government Printing Office, Washington, DC, 330, 1980.

Roberts, D.A., Y. Yamaguchi, and R.J.P. Lyon. Calibration of Airborne Imaging Spectrometer data to percent reflectance using field spectral measurements. In: *Proceedings of the 19th International Symposium on Remote Sensing of the Environment*, ERIM, Ann Arbor, MI, October 21–25, 1985, pp. 679–688.

Saint-Amand, P., L.A. Matthews, C. Gains, and R. Reinking. Dust storms from Owens and Mono Valleys, California. *China Lake Naval Weapons Center*, Technical Publication Number 6731, 1986.

Skole, D.L. and C.J. Tucker. Tropical deforestation and habitat fragmentation in the Amazon: satellite data from 1978 to 1988. *Science*, 260, 1905–1910, 1993.

United States Geological Survey (USGS). *Landsat Data Users Handbook*. Branch of Distribution, U.S. Geological Survey, Arlington, VA, 1979.

Yuan, D., C.D. Elvidge, and R.L. Lunetta. Survey of Multispectral Methods For Land Cover Change Analysis. In: *Remote Sensing Change Detection: Environmental Monitoring Methods and Applications,* R.S. Lunetta and C.D. Elvidge, Eds., Ann Arbor Press, Chelsea, MI, 1998.

CHAPTER 13

Radar Remote Sensing of Wetlands

Elijah W. Ramsey III

1.0 INTRODUCTION

Since the launch of the Seasat spacecraft with an L band HH polarization synthetic aperture radar (SAR), researchers and resource managers have been examining, fitting, and developing physical models to explain active microwave (radar) returns from the earth's landscapes (Way and Smith, 1991). Following the success of Seasat, three spaceborne imaging radar (SIR-A, B, and C) missions were conducted. Three operational satellite radar systems are now functioning and more are planned for launch in the next few years. Throughout the development of this satellite network, airborne and ground-based radar systems of various types were, and still are, obtaining data for algorithm verification, calibration, and monitoring (e.g., Ulaby et al., 1981, 1982; Ulaby and Dobson, 1989).

This model development and verification, and the availability of calibrated satellite data should lead to more widespread use of radar by environmental researchers and, ultimately, resource managers and policy makers. In other words, general knowledge of the interaction mechanisms and the nature of radar data collection can now allow use of the expanding volume of radar data by a much wider research- and application-oriented community. In that vein, this chapter is an attempt to generally define critical knowledge about radar pertinent to remote sensing of wetlands. It is at a level of detail used by the author as an application researcher, and the theory and evidence compiled by others are used to reach a solution posed within the central theme of habitat and resource management.

2.0 WETLANDS

Wetlands comprise a diverse array of biophysical systems that exhibit extreme variations in areal extent, temporal duration, and spatial complexity. Within this diversity of scale, wetland vulnerability and intensity of use can result in regionally or locally complex spatial and temporal patterns of alteration.

Wetlands are extremely productive ecosystems. In order of total biomass production, only agriculture areas exceed marshes and mangroves (Ramsey and Jensen, 1996). They are linked to much of the productivity of the ocean and provide critical nursery areas for multitudes of coastal to oceanic life. Recent concern over global warming has led to identifying these permanently to intermittently flooded landscapes, which are producers of a primary greenhouse gas, methane (Stofan et al., 1995; Waring et al., 1995).

Wetlands are caught between forces driving expanded agricultural, oil and gas, and waterfront (e.g., ocean, lakes, rivers) developments. As transitional environments, wetlands are balanced between open water and upland landscapes; the direction in which these transitional areas progress depends on changes in the soil production (in place or imported); flooding dynamics; or, in coastal areas, relative sea level. Disrupting the balance of these forces changes the nature of the wetland. Whether these changes are of benefit to or in conflict with the interests of the human population may be debatable. Clearly, however, development of methods that provide a more accurate picture of the extent and dynamics of wetlands will provide a firmer basis for discussion between all interested groups. These groups include those that are directly linked to extraction of natural resources, those that benefit from availability of these goods and the recreational benefits linked to wetlands, and those associated with government and nongovernment agencies and organizations concerned with the sustained use of wetlands as a natural resource.

During the last half of this century—especially since the passage of the National Environmental Policy Act in 1970 and more recently the 1990 executive order by President Bush with the goal of no net loss of wetlands—there has been an intensifying awareness of environmental degradation induced by human influences. Federal jurisdiction, administration, protection, mitigation, and management of wetlands rely on determining hydrology, soil character, and vegetation type. Examples of the agencies and their responsibilities include the U.S. Army Corps of Engineers in wetland evaluation and delineation, the U.S. Environmental Protection Agency in wetland regulation and Section 404 of the Clean Water Act and water quality management, the U.S. Department of Agriculture administration of the 1985 Food Security Act (farmed lands), and the U.S. Fish and Wildlife Service administration of the National Wetland Inventory Program. These agencies rely on information about the hydrology, soil character, and vegetation of an area to define and monitor wetlands (Lyon, 1993).

Apart from government regulation, however, development of better monitoring methods is needed to increase our knowledge of the physical and biological characteristics of each wetland resource, and to gain, from this knowledge, a better understanding of wetland dynamics and their controlling processes. Discussions based on accurate knowledge and increased awareness of wetland issues can then begin to develop management strategies (to protect, restore, and/or mitigate) that account for the function and value of all wetland resources in the face of natural (e.g., global climate change) and socioeconomic forces (e.g., protection from sea-level rise, development), while continuing to satisfy critical resource needs of the human population.

3.0 REMOTE SENSING OF WETLANDS

Remote sensing is the only approach that can provide timely, economical monitoring of wetland status over large areas, and it is capable of generating environmental data on various spatial and temporal scales by integrating a wide variety of sensors and sensor platforms (Wickland, 1991). Optical sensors such as the Landsat Thematic Mapper (TM), the Airborne Visible Infrared Imaging Spectrometer (AVIRIS) of the National Aeronautics and Space Administration (NASA), and the Advanced Very High Resolution Radiometer (AVHRR) of the National Oceanic and Atmospheric Administration (NOAA) have shown promise in identifying wetlands and monitoring acute stress (fire, herbivory) and storm impacts (e.g., Ramsey et al., 1994, 1997; Ramsey and Jensen, 1995; Ramsey and Laine, 1997). Ground-based optical and hydrological measurements have been used to relate the Leaf Area Index (LAI, a biomass estimator) and canopy structure to wetland type and variation and to calibrate remotely sensed imagery (e.g., Ramsey et al., 1992, 1993). These studies showed that remotely sensed data can be used to identify and monitor wetland hydrology, type, and landscape changes (e.g., fire, herbivory, storm impact, drought, harvesting).

A number of wetland definitions cover jurisdiction criteria, agricultural development, and mapping convention (Lyon, 1993). Even though various definitions stem from different perspectives, all definitions include three general characteristics. Wetlands contain (1) a predominately hydric soil that is (2) saturated or covered with water (at some frequency dependent on definition) and supports (3) predominately vegetations adapted for life in these conditions (Lyon, 1993). If we limit our monitoring efforts to surface and near surface features (and exclude such hydrological variables as depth to groundwater, although passive microwave may prove useful for measuring this as well; Shutko, 1992; Shutko et al., 1997), we may be able to determine soil moisture content, existence of standing water or flooding, and/or type of vegetation present, thus remotely identifying wetlands as well as measuring important wetland variables.

Optical sensors, in many cases, can aid in determining each of these factors defining a wetland. Most applications, however, deal with classifying the type of vegetation present, the variations in soil moisture, or the presence of standing water dampening the optical return to the sensor. An additional factor for developing a viable monitoring program is the ability to consistently, and with predetermination, measure each of these wetland characteristics. Collections with optical systems are often constrained by cloud cover and atmospheric haze, especially in subtropical to tropical environments.

4.0 RADAR REMOTE SENSING

Radar improves our ability to collect environmental data. It can operate day and night and during almost any weather condition. Although in a strict sense, shorter radar wavelengths (<4 cm) can be attenuated by clouds in an intense rainfall (Ulaby et al., 1981; Figures 13.1a and b), the occurrence of this interference is minimal when compared with weather obscuring optical collections. This nearly unimpeded source of environmental data not only improves the ability to sustain a consistent monitoring program, but because of the range of radar wavelengths and selectable polarizations (Tables 13.1 and 13.2), it also offers a very different monitoring viewpoint than that of passive optical sensors. While remote sensing with optical and infrared equipment is mainly concerned with interaction based on molecular resonances, radar responds more to the geometric (structural) and bulk dielectric properties of the vegetation or soil (Ulaby et al., 1981). Objects of comparable size to the radar wavelengths (mainly 1 to 30 cm in satellite imaging of earth's resources) most strongly influence the radar scatter (Evans et al., 1988; Hess and Melack, 1994; Stofan et al., 1995). Aside from complexities of interpretation, in general, this variety of sensors results in greater control in monitoring different aspects of the vegetation features. It also allows a greater ability to profile the response from various depths within the vegetation canopy as compared to optical sensors (Dobson et al., 1995a).

Of all the world's landscapes, certain characteristics of wetlands make them unique for radar monitoring of status and trends. One important characteristic is the usually high moisture content of wetland soils. Another is the variability of flooding related to the inundation depth, frequency, and duration. The nearly flat terrain of wetlands is also an anomaly in remote sensing applications. Finally, wetlands are unique in their limited variety of vegetation versus other landscapes (Hess et al., 1990). This uniqueness is due to the specialized adaptation of wetland plants to various frequencies of flooding and dramatic changes in soil moisture conditions. In coastal wetlands, this specialization extends to the adaptation of plants to survive in various levels of salinity through exclusion or isolation.

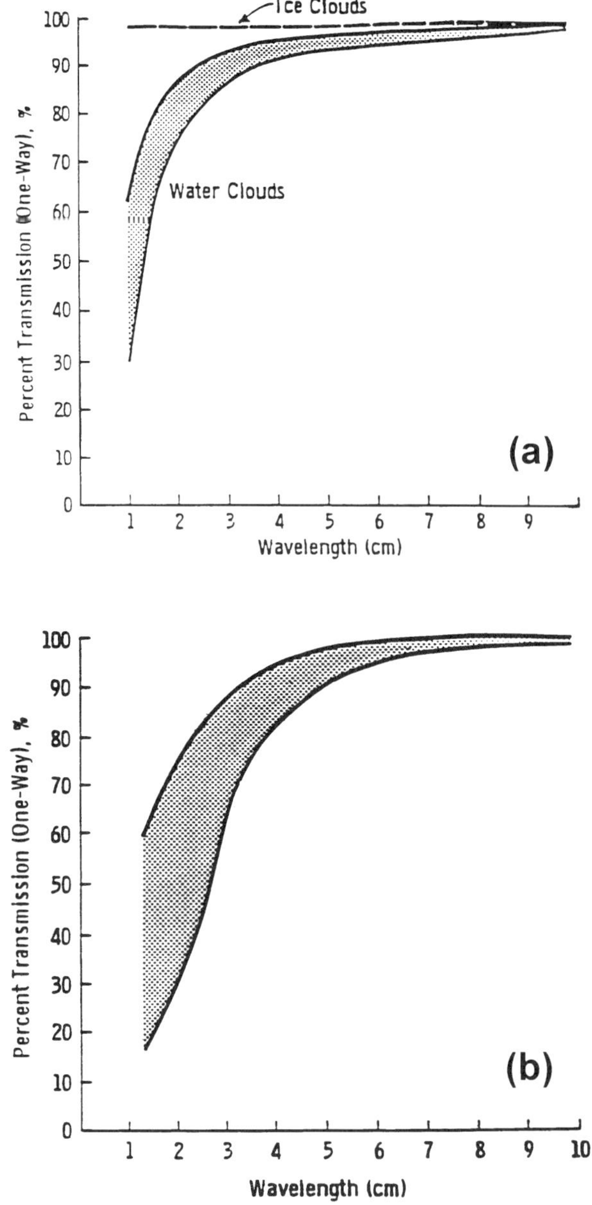

Figure 13.1. Transmittance of radar through (a) clouds and (b) rain (Ulaby et al., 1981, by permission of Artech House, Inc.). Notice the effect of ice clouds.

Parameters describing how radar functions in an environment can be grouped according to the properties associated with the vegetation and surface (soil in this case) and those associated with the radar configuration (Lin et al., 1994). The radar configuration includes the viewing geometry in relation to the terrain, the sensor wavelength, and wavelength polarization (Tables 13.1 and 13.2; Figure 13.2). Commonly, vegetation properties are described in terms of size, shape, orientation, and density, and soil surface properties as roughness and composition. In general, the magnitude of surface roughness (as compared to the wavelength) determines the

Table 13.1. Radar Wavelengths.

Band	Wavelength (cm)	Frequency (Ghz)
KA	0.8 to 1.1	27 to 40
K	1.1 to 1.7	18 to 27
KU	1.7 to 2.4	12 to 18
X	2.4 to 3.8	8 to 12.5
C	3.8 to 7.5	4 to 8
S	7.5 to 15	2 to 4
L	15 to 30	1 to 2
P	30 to 100	0.3 to 1

Table 13.2. Polarizations.

Like-Polarized	HH =	Transmit Horizontally / Receive Horizontally
	VV =	Transmit Vertically / Receive Vertically
Cross-Polarized	HV =	Transmit Horizontally / Receive Vertically
	VH =	Transmit Vertically / Receive Horizontally

amount of energy scattered in different directions or the type of scattering that ranges from isotropic to specular (Elachi, 1988; Hess and Melack, 1994; Dobson et al., 1995b). Increases in the dielectric permittivity tends to increase the strength of this scattering (Hess and Melack, 1994; Dobson et al., 1995b). In most natural materials the dielectric permittivity varies according to the moisture content of the material (Ormsby et al., 1985; Dobson et al., 1995b). Radar scattering depends, to a large degree, on the relationship between surface roughness and soil moisture content. For example, the rougher a soil surface—relative to the wavelength—the more scattering of the incident wave; increases in soil moisture will further increase this scattering (Figure 13.3). A consequence of increased soil moisture, however, is that the depth of soil penetration by the radar signal decreases (Figure 13.4a), reducing the potential for information from deeper in the soil. Changes in the radar system viewing geometry further complicate interpretation of the return signal (Figure 13.4b).

The interaction of these properties and the radar configuration is highly complex and is the subject of numerous books and tutorials (e.g., Ulaby et al., 1981, 1982; Elachi, 1987; Werle, 1988; Ulaby and Dobson, 1989; Curlander and McDonough, 1991). However, if examined from an applied-user perspective, the general abilities of imaging radar to evaluate the status and monitor the trends in wetlands can be discussed. As mentioned, one factor related to most wetlands is the near flatness of the terrain (Ramsey et al., 1997; Ramsey, 1995), simplifying the problems caused by variable relief, such as rolling hills, mountains, etc. The flatness of the terrain and assumed azimuthal symmetry of the vegetation (Freeman et al., 1995) may diminish the requirement of obtaining radar images with the same look (azimuth) direction for comparisons to be made. Another factor that may also tend to simplify the interpretation of the radar image of wetlands is the diminished ground surface roughness. Except for incising channels (e.g., tidal, river), ponds, and lakes, the wetland ground surface is normally fairly flat and at times may be considered smooth with respect to longer radar wavelengths (> L band). This may not, however, always be the case. As seen in Figure 13.3, even a small increase in surface roughness (1.1-cm to 4.1-cm root-mean-square variation) can produce a considerable change in the backscatter cross section. Wang and Imhoff (1993) suggested the roughened soil surface under mangrove wetlands used in their study could be considered diffuse reflectors at L band

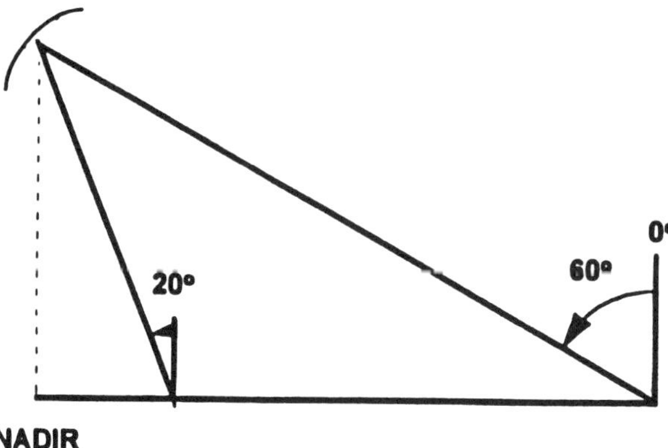

Figure 13.2. Incidence angles at 20° and 60°. The incidence angle is the angle formed by the radar beam and the normal to the ground at the beam's point of incidence. Normally, the incidence angle increases from about 20° (near range) to about 60° (far range).

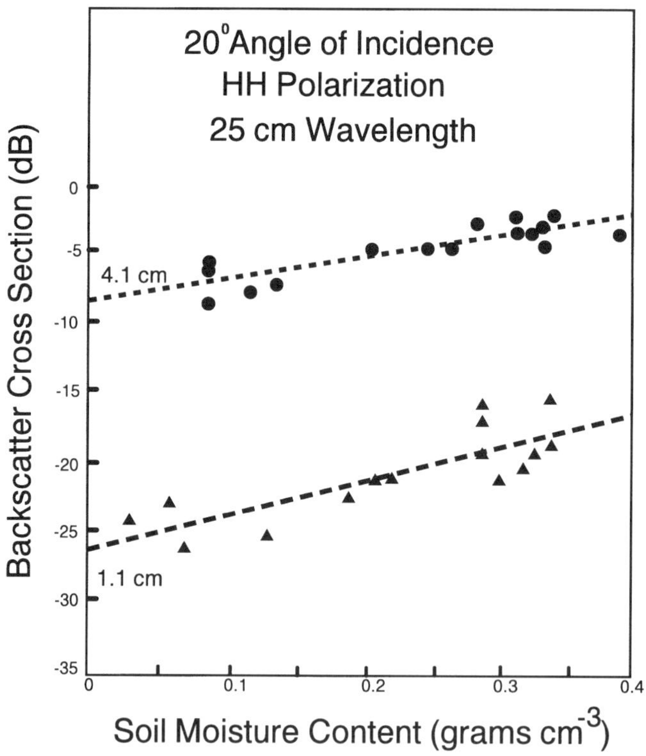

Figure 13.3. Radar backscatter as a function of surface soil roughness and moisture content (Ulaby and Dobson, 1987, by permission of Artech House, Inc.). Soil roughness is expressed as a root mean square (rms) of the standard deviation of the surface from the average surface. The surface correlation length (usually reported as an additional surface roughness measure, not reported here) is the length after which two points are statistically independent (the autocorrelation function decreases to 1/e; Elachi, 1988). Notice the change in backscatter from an rms surface roughness of 1.1 cm to 4.1 cm is about 18 dB, whereas the range in backscatter is around 8 dB to 9 dB from very dry (0 grams/cm^3) to very wet (0.40 grams/cm^3) soils at each soil roughness.

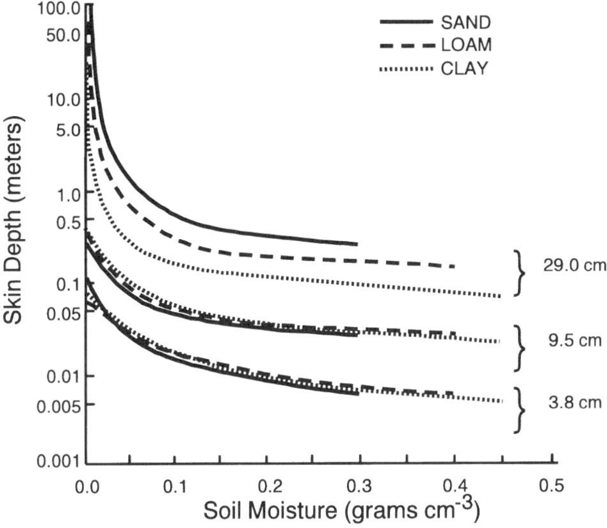

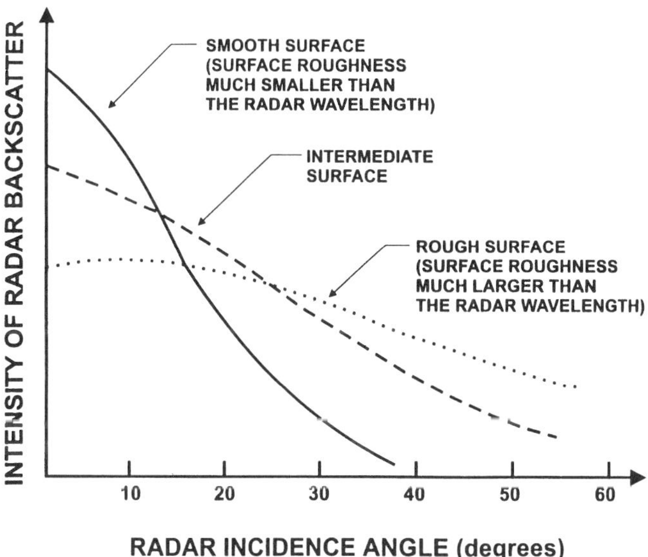

Figure 13.4. (a) Radar penetration as a function of volumetric water content, frequency, and soil type (Ulaby et al., 1981, by permission of Artech House, Inc.). (b) Radar return as a function of incidence angle and surface roughness. Notice the steeper to more level response of the radar backscatter as the surface roughness increases (adapted from notes from the Spectral Imagery Processing Course, Geodynamics, Washington, DC).

wavelengths. An added complexity would be the presence of a substantial understory, or aerial root systems (Hess et al., 1990; Wang and Imhoff, 1993).

A further simplification of interpreting radar returns in the remote sensing of wetlands is that the HV and VH cross-polarizations can be interchanged; the return magnitude can be considered the same in either polarization when sensing terrain features (Freeman et al., 1995). Other properties of the radar system and unique properties of wetlands as related to radar remote sensing will be discussed in the context of emphasizing a particular wetland feature within the constraints of competing and interrelated canopy (vegetation, soil, and water) interactions with the incident radar wave. When appropriate for interpreting the radar response, the radar return will be reported as the backscatter cross section. This cross section is defined as the log of the ratio of energy received by the sensor over the energy received by isotropic scatters. A positive number implies focusing energy back toward the sensor and a negative number implies focusing energy away from the sensor (Elachi, 1988).

4.1 Detecting Flooding Under Forest Canopies

There has been a long history of remotely detecting flooding under vegetation canopies. An excellent summary and detailed interpretation of radar detection of forest flooding is contained in Hess et al. (1990), in which the ability to detect flooding was related to an enhanced radar return compared to returns from nonflooded forests. The accepted mechanism for the enhancement is related to the double bounce interaction (MacDonald et al., 1980; Engheta and Elachi, 1982; Richards et al., 1987; Figure 13.5). The signal penetrates the canopy, reflects off the water surface, and is subsequently reflected back toward the sensor by a second reflection off the tree trunk (or the converse path). The return signal enhancement, of course, only works if there is a high penetration of the canopy. As a rule, the longer wavelengths have higher penetration; thus, L bands and bands associated with longer wavelengths most clearly show the enhancement of the return in flooded forested areas (e.g., Ormsby et al., 1985; Imhoff et al., 1986). These studies and others have reported enhanced L band radar returns in semipermanently to permanently flooded wetland forest, including cypress, cypress-tupelo, willow; in seasonally flooded bottomland hardwood forests; and in flooded coniferous forests (Lyon and McCarthy, 1981; Wedler and Kessler, 1981; Waite et al., 1981; Ford et al., 1983, 1986; Place, 1984; Hoffer et al., 1985a; Evans et al., 1986; Harris and Digby-Argus, 1986; Wu and Sader, 1987; Hess et al., 1990). Along with more recent studies (Hess and Melack, 1994; Hess et al., 1995), these studies support the preferred use of L band HH data to detect flooding beneath forest canopies.

Some studies have reported that L band HH is nearly insensitive to changes in the incidence angle when detecting forest flooding (Ormsby et al., 1985; Harris and Digby-Argus, 1986). Hess et al. (1990) cite reports of no enhancement of the radar return from a closed-canopy mangrove forest and flooded cypress stands with dense undergrowth. In a study using SIR-B (L band HH) SAR collected at three incidence angles (28°, 45°, and 58°), flooding of forests dominated by cypress was related to high returns at all incidence angles, while returns from a cypress-tupelo/hardwood mixed forest were higher at 28° than either 45° or 58° incidence angles (Hoffer et al., 1985b). Inspection of the multi-incident radar images, however, show bright returns at 28°, decreasing at 45°, and no enhancement of the return at 58° from the presumably flooded swamp areas (visually inspected in Ford et al., 1986). In a later description of the same data set, Lozano-Garcia and Hoffer (1993) report the same comparison between the cypress swamp and a mixed forest swamp area. Ford and Casey (1988) observed lower SIR-B (L band HH) SAR returns at higher incidence angles from a presumably flooded mangrove swamp.

At lower incidence angles and during leaf off conditions (from seasonal senescence in deciduous trees, herbivory, storm impact, death), C band, X band, and K band radars also may penetrate to the water surface and be enhanced by the double bounce mechanism (e.g., Waite and MacDonald, 1971; Pultz et al., 1991; Ustin et al., 1991). Clearly the canopy density,

Figure 13.5. Double-bounce returns from an upland pine forest and flooded cypress swamp (Hess et al., 1990, by permission of Taylor and Francis). This is not the only, or possibly not even the dominate, mechanism of the incident wave interaction with the canopy, but in cases where the radar incident wave penetrates through the canopy and flooding exists, the double-bounce mechanism will usually enhance the overall return signal. The enhancement being a complex function of radar system and vegetation parameters.

structural differences between tree types, and moisture content will modify the relative enhancement. Even though longer wavelengths are preferred for flood detection under forest canopies, shorter wavelength radars with low incidence angles may also provide flood detection under certain conditions. Systems that include longer wavelength radars (e.g., Jet Propulsion Laboratory's AirSAR with P band [67 cm]) should increase the detection ability (Hess et al., 1990).

4.2 Mangrove Forests

Mangroves, a forested coastal wetland type, occupy nearly 75 percent of the world's coastlines between 25°N and 25°S latitudes (Heald and Odum, 1975). In terms of net productivity, they are deemed the world's most productive ecosystems, followed by sea grasses, marsh grasses, other coastal ecosystems (Odum et al., 1982), and then by all other ecosystems combined (Mitsch and Gosselink, 1986). Many studies using optical sensors have attempted to map the extent and classify the type of mangrove wetlands (e.g., Bina et al., 1978; Untawale et al., 1980; Butera, 1983; Jensen et al., 1991). Determining mangrove classification is necessary because different mangrove species generate different levels of biomass production. The extent and location of the different species types seem to be primarily a function of freshwater input and frequency of flushing. In a series of optical studies using direct observation and model evaluation of new world mangrove species, it was shown that, even though different visibly, separation of the three major mangrove species by leaf optical properties was not possible with Landsat TM and SPOT type sensors (Ramsey and Jensen, 1995; 1996). However, structural aspects related to Leaf Area Index were correlated to a simple vegetation transformation of the canopy reflectance. Reported successful differentiation of mangroves—at least new world species—was probably due more to differences in canopy structure than differences in the leaf optical properties. Combined with the prevalent clouds in this subtropical to tropical environment, these problems provide an opportunity for radar remote sensing of mangrove species.

Modeled and collected SIR-B SAR (L band HH) data taken at various incidence angles were used to study flood detection under mangrove canopies (Wang and Imhoff, 1993). Results of this study found that canopy volume scattering dominated at nonflood conditions, but that the double bounce mechanism nearly equaled the volume return at smaller incidence angles. The addition of the double bounce return to the volume return enhanced the overall return of the flooded mangroves. Specialized adaptation of the mangrove root systems (prop roots in red mangrove, pneumatophores in black mangrove) substantially attenuated the return signal (Wang and Imhoff, 1993). Attenuation due to the mangrove root systems should, however, be minimized or canceled during flooding. Another factor that may cause ambiguities in flood detection is spatial variability in flooding and surface water pooling unrelated to the species type distribution (Hess and Melack, 1994).

4.3 Detecting Flooding Under Grass Canopies

In contrast to forested wetlands, L band returns from marshes (including peat bogs) are usually associated with a diminished radar return (e.g., Wedler and Kessler, 1981; Mouginis-Mark et al., 1984; Harris and Digby-Argus, 1986; Evans et al., 1988). However, moderate to bright radar L band returns have been noted (Ford et al., 1983). In cases linked to flooding, the marsh grasses were thought to calm the water surface, accentuating specular reflection (Lyon and McCarthy, 1981; Ormsby et al., 1985). L band, not impeded by the marsh vegetation, was thought to be specularly reflected away from the sensor, while X band, attenuated in the marsh canopy, never reached the water surface (Ormsby et al., 1985; Harris and Digby-Argus, 1986). There have been cases in which bright C band returns from flooded marshes have been reported (Hess and Melack, 1994).

In a marsh dominated by *Juncus roemerianus* (black needlerush), X band returns were extremely high from low marsh and moderate from higher marsh and forested areas (Ormsby et al., 1985). In this study, Ormsby et al. suggested that the change in return strength was related to the relative strengths of the volume and double-bounce interactions. In the taller marsh, the return was almost solely due to volume scattering in the canopy, while in the shorter marsh, the double-bounce return added to the volume return, increasing the overall return. Wedler and Kessler (1981) found X band HH returns were enhanced, but not X band HV returns, from certain stands of grasses and reeds. They attributed the enhanced returns to strong scattering and signal reinforcement at the vegetation water interface. An enhanced response from shorter vegetation was also observed for C band returns at low incidence angles, but the shorter vegetation was not identified as marsh or scrub vegetation (Ormsby et al., 1985). Lewis and Peterson (1991) state X band radars can detect flooding under brush and scrub vegetation and K band radars can detect flooding under grass vegetation.

At shorter wavelengths, the mechanisms for radar interactions with vertically oriented marsh grasses may mimic those of longer wavelengths ($\geq$ L band) with wetland forests (Ferrazzoli et al., 1997). Relevant to wetlands, a study of radar returns from various rice canopies may have simulated radar sensing of a fairly vertical marsh canopy during flood conditions. Decreasing ERS-1 SAR returns with an increasing leaf area index (plant number per area) were observed from flooded rice fields (Durden et al., 1995; Figure 13.6). To explain the inverse relationship, a model was constructed that predicted that a double-bounce (direct-reflect) mechanism initially dominated the returns at lower LAIs (as related to density) but decreased with increases in LAI. It was suggested that the diminished double-bounce returns resulted from increased canopy

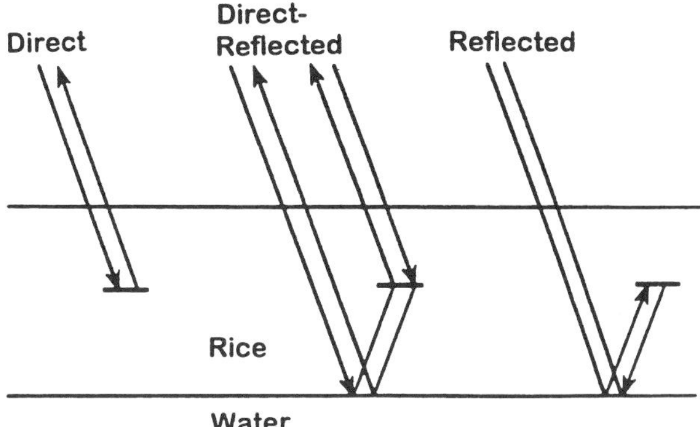

Figure 13.6. A proposed model for scatter mechanisms from a flooded rice field (Durden et al., 1995, © IEEE). The direct-reflected (double-bounce) return was thought to dominate at lower canopy biomass (LAI) but decrease with LAI increase until the lower direct (volume) returns became dominant.

attenuation. This decrease in double-bounce returns continued until direct, or volume returns, fairly low and constant from about 1.0 to 6.0 LAI, became the highest return mechanism after 4.3 LAI. Returns from the reflected interaction were diminished because of the multiple interactions with the canopy before return.

A study of ERS-1 SAR returns from a black needlerush marsh in coastal Florida showed marsh flooding diminished the C band VV returns with respect to nonflooded marsh (Ramsey, 1995; Figure 13.7). This marsh was characterized by thin, nearly vertical stalks of different densities, which were most likely dependent on marsh maturity and height above sea level. In addition to the field verification relating flooding to diminished ERS-1 SAR returns in the black needlerush marsh, a graphical depiction of model results predicted that increases in biomass were associated with asymptotic decreases in SAR return under various soil wetness conditions (Dobson et al., 1996). Decreases were attributed to increasing attenuation at higher biomass levels and loss of soil interactions. In a flooded black needlerush marsh, however, the returns were lower than nonflooded marsh (exceptionally so at low marsh biomass), and returns asymptotically increased with biomass instead of decreasing as in the nonflood case. Field studies suggest that there is a decrease in ERS-1 SAR returns with increases in black needlerush biomass (Ramsey et al., unpublished data), and as stated above, show a high contrast between returns from nonflooded and flooded marsh. These observations seem consistent with Dobson et al.'s (1996) results, but field results could also be partly in accordance with Durden et al.'s (1995) results. Part of the dilemma is because Durden et al.'s (1995) study did not compare flooded and nonflooded rice fields. In Durden et al.'s (1995) model changes in the returns linked to the double bounce mechanism do show decreases in SAR returns, but the decrease of return with increasing biomass under flood conditions seems contrary to Dobson et al.'s (1996) model results. Reports of a study examining ERS-1 returns from southern Florida landscapes indicated that increases in new green biomass at flooded black needlerush sites did increase the radar return (Kasischke and Bourgeau-Chavez, 1997).

4.4 Complexities in Flood Detection

The same surface flooding that enhances the identification of forested wetlands may, in certain cases, simplify the radar return by negating the ground surface interaction (Hess et al.,

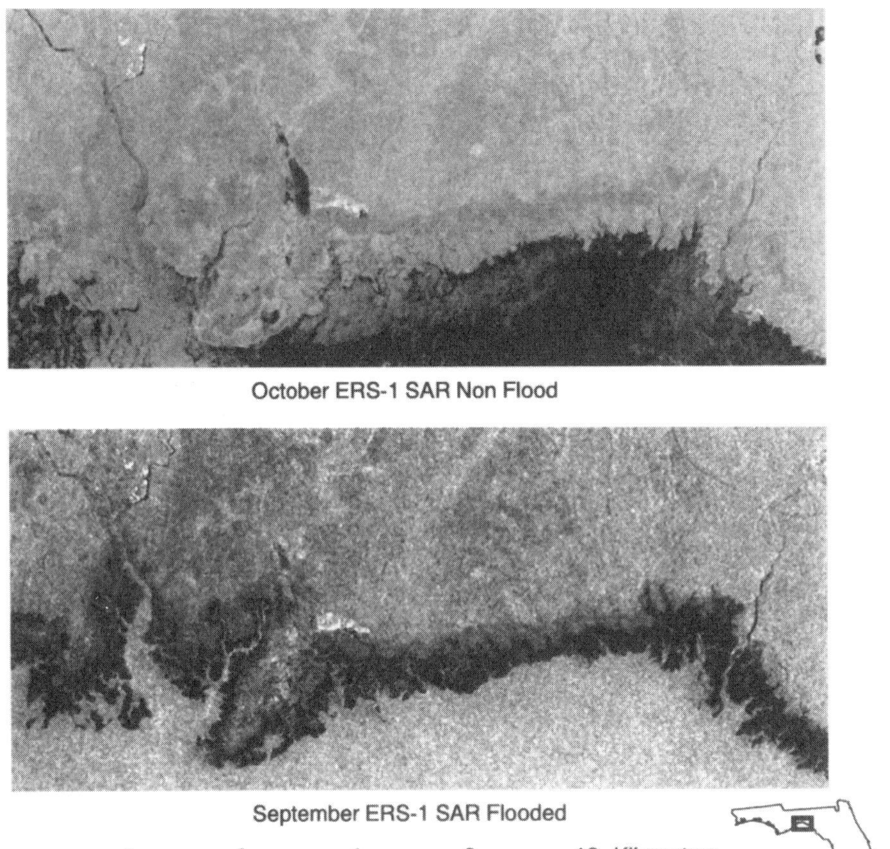

Figure 13.7. ERS-1 returns from a flooded and nonflooded black needlerush marsh within the St. Marks National Wildlife Refuge in the Big Bend area of coastal Florida (Ramsey, 1995, by permission of Taylor and Francis). The top image was collected in October 1993 (nonflooded) and the bottom image in September 1993 (flooded). The lowered returns are from the flooded marsh.

1990; Kasischke and Bourgeau-Chavez, 1997). Permanently flooded cypress-tupelo forests certainly lend themselves to this mechanism of identification. On the other hand, intermittent flooding can confuse the identification or differentiation of wetland types (Hess and Melack, 1994; Ramsey, 1995). If a flooding event is not spatially uniform, the presence or absence of flooding under the canopy may result in variability of class signatures based on the radar return. One cause of the spatial variability of radar return might be the timing of the radar collection and the temporal variability in the flood extent. This timing problem is highly probable in coastal regions influenced by diurnal or semidiurnal flooding, but can also be problematic in inland areas experiencing changing flood conditions (Imhoff et al., 1986; Hess et al., 1990). Spatial variability could also be a direct or indirect result of the increased rate of flooding preceding the radar collection. Relatively higher rates of flooding in a coastal wetland may result in spatially intermittent pooling of water (Ramsey, 1995). In that study, marsh areas with and without pooling of water were mixed within the sensor spatial resolution, resulting in variable and diminished returns as compared to nonflooded marsh. This type of pooling is most likely observed in tidally influenced wetlands, but may also be observed in inland wetlands. One other result of flooding that may not be as obvious, especially in forested wetlands, is the potential interaction of the decrease in biomass with the radar incident wavelengths due to changing flood heights (Hess et al., 1995). This interaction is especially pertinent in wetlands where elevation changes slowly within the flooded area.

4.5 Soil Moisture and Open Water

Radar is sensitive to soil moisture under short vegetation canopies (Dobson et al., 1995b). In cases of a moderate vegetation overstory, however, the detection of soil moisture is not straightforward, but rather a function of the relative strength of the canopy and the incident wave interaction (Dobson et al., 1995a, 1995b). Many studies have examined the ability of radar to estimate soil moisture under various types and configurations of vegetation canopies; results are varied (a good summary of models is in Dawson et al., 1997). Soil moisture retrieval algorithms have been generated for C band radar at a 10° incident angle for bare and vegetated soil surfaces (Elachi, 1988). More generally, empirical algorithms based on L band radar have proved effective in estimating soil moisture content of nonvegetated soils (Dobson et al., 1995b). In sensing mature forests, ERS-1 C band radar is insensitive to soil moisture changes, but in grass canopies (≤ 40 cm tall) the return from the soil may dominate the total return (Dobson et al., 1992). Chauhan et al. (1992) found that in grass canopies of 30-cm average height, L band (HH) returns were dependent on the soil surface properties, but they suggested that in taller grass canopies the contribution to the return from the larger stems could be significant. Dobson and Ulaby (1986) suggested that surface roughness or canopy cover must be accounted for when attempting retrieval of soil moisture from single date observation. In agricultural crops, low correlation was found between measured soil moisture and C band, with possibly slight improvement from L to P bands (Baronti et al., 1993; Pardipuram et al., 1993; van den Broek and Groot, 1993). Problems in the detection of soil moisture were thought to be related to properties of surface roughness and vegetation. To accurately estimate soil moisture from radar, the influence of these properties on the radar signal must be estimated (Figures 13.3 and 13.4).

Saatchi and Moghaddam (1994) modeled radar remote sensing of soil moisture content in grasslands both with and without thatch layers, which are prominent in many marsh wetlands. Model results were reported to be in reasonable agreement with observations, although the agreement was lower when thatch was present, and a higher bias resulted when the thatch was moist rather than dry. In another study examining the detection of soil moisture using C, L, and P bands with full polarization (VV, HH, VH, HV) under grass canopies, L band VV and HH polarizations were found least sensitive to the grass canopy and soil roughness and, thus, the best predictors of soil moisture changes (Lin et al., 1994). Similarly, in forest leaf off conditions, L band VV and HH returns had a molted pattern mimicking changes in soil moisture; no pattern was discerned with L band VH returns (Evans et al., 1986). Using an iterative inversion model and polarimetric radar, Polatin et al. (1994) reported high detection accuracy of soil moisture when compared to simulated data.

The high dielectric permittivity of water limits radar's ability to penetrate it, and calm water acts as a specular reflector, voiding the radar return. Returns from radar operating with wavelengths between 1.0 to 40 cm are scattered from short gravity waves and small capillary waves (Elachi, 1988). A rougher surface scatters more of the incident radar wave and increases the potential of return. Scattering from roughened, open water can limit the ability to differentiate land and water areas. This limitation was shown in a study mapping the impact of Hurricane

Andrew to a coastal marsh in Louisiana (Ramsey et al., 1994). On before-and-after ERS-1 SAR (C band VV) images, open waters roughened by winds were highly confused with marsh (Figure 13.8). Use of multidate images, and possibly a horizontally polarized sensor such as Radarsat, might lessen the confusion (Pultz et al., 1991).

4.6 Vegetation Orientation and Radar Polarization

If the diameter of the target material is much less than the wavelength, interaction of the radar incident wavelength increases as the orientation of the target material is aligned with the polarization of the radar wavelength (Ulaby and Wilson, 1985). A vertically oriented target (i.e., reference is perpendicular to ground) interacts with a vertically polarized incident wave but has little to no interaction with a horizontal incident wave (Elachi, 1988, Figure 13.9). This occurs due to the alignment of the dipoles with the incident wave (Dobson et al., 1995b), which intensifies the interaction and potential return. In addition, if a target was oriented vertically, returns from cross-polarizations (HV or VH) would be very low (Ferrazzoli et al., 1997).

Increases in number density (analogous to leaf area index) can increase or decrease the potential radar return. Early experiments examining the attenuation—as a function of absorption and scattering—of radar through various types of material (e.g., wheat, soybean, and corn canopies) showed attenuation was a function of wavelength, polarization, and incident angle (Ulaby and Wilson, 1985; Ulaby and El-Rayes, 1987). In radar sensing of forests (tall canopies) with L and P bands, HH, and in some cases VV, polarized waves will penetrate further into the canopy than HV polarized waves (Wu and Sader, 1987; Evans et al., 1988; Rignot et al., 1995). HV (or VH), and occasionally VV, polarizations will interact mostly with the subcanopy volume (larger and lower primary branches), decreasing the effectiveness of these polarizations for soil moisture mapping or flood detection (Wu and Sader, 1987; Evans et al., 1988; Hess and Melack, 1994; Rignot et al., 1995). For canopies with smaller stalks (i.e., grasses), these same effects may be shifted to shorter wavelengths.

Contrast between wetland and upland forests has been found consistently greater at L band HH and VV polarizations than cross-polarizations (either HV or VH), while discrimination between wetland forests and marshes may be better at L band VH polarization than HH and VV polarizations (Hoffer et al., 1985a, 1990). C band cross-polarized (HV) returns increase from bare- to short-grass-covered soils (Ferrazzoli et al., 1997).

4.7 Incidence Angle, Orientation, and Polarization

The incidence angle has been the subject of many reports and does warrant increased attention when analyzing radar data. From studies of attenuation of the radar signal at different incidence angles through vegetation canopies, it has been shown that attenuation tends to increase and the radar return tends to decrease as the incidence angle increases (Elachi, 1988; Figures 13.10 and 13.11). In ground scatterometer studies of wheat and soybean crops, losses in the vertical (V) and horizontal (H) polarizations were nearly equal for X, C, and L bands at a 24° incidence angle, but at a 56° incidence angle the losses associated with V polarization were always greater than losses related to H polarization. Leaves were of secondary importance to stalks and branches in reducing V polarization transmission through the soybean canopy, about equal at L and C bands H polarization, and dominant at X band H polarization (Ulaby and Wilson, 1985). This decrease in penetration related to polarization and incidence angle was attributed to increased alignment of the polarized incident wave with the mostly vertical stalks

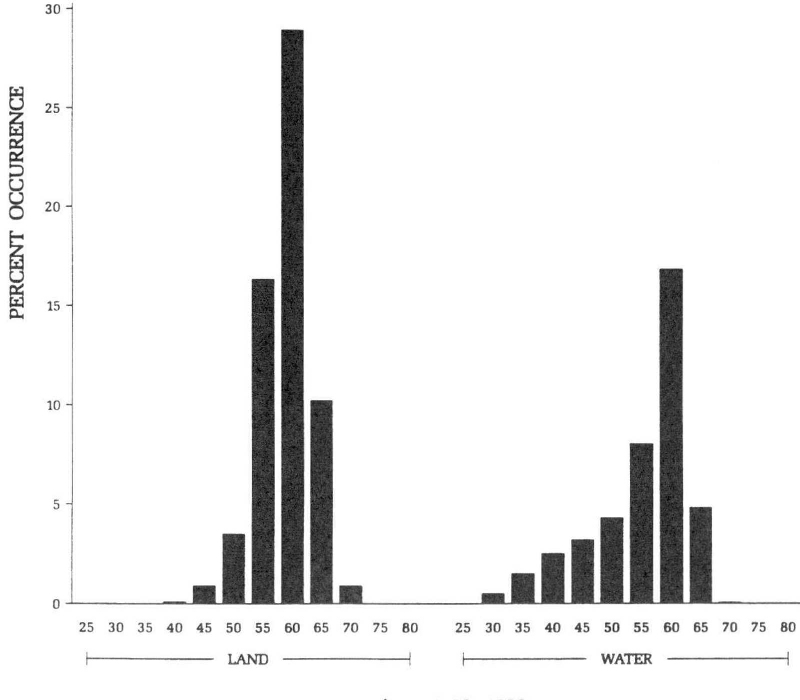

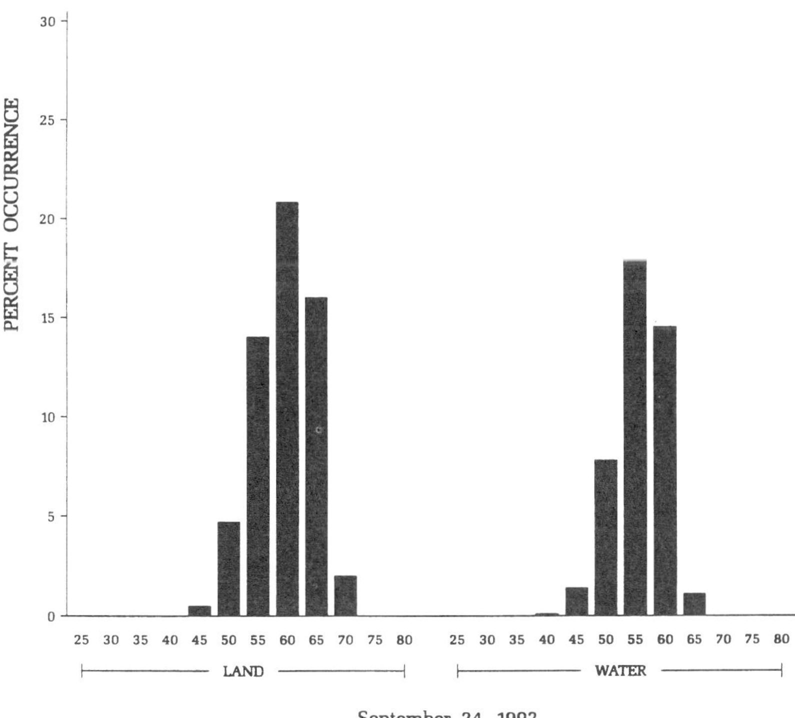

Figure 13.8. The distribution of returns from land and water from (top) before and (bottom) after the hurricane passage. The similarity of histograms describing ERS-1 SAR (C band VV) returns from the water and land (marsh) show the inability to distinguish these areas at the time of the collections (adapted from Ramsey et al., 1994).

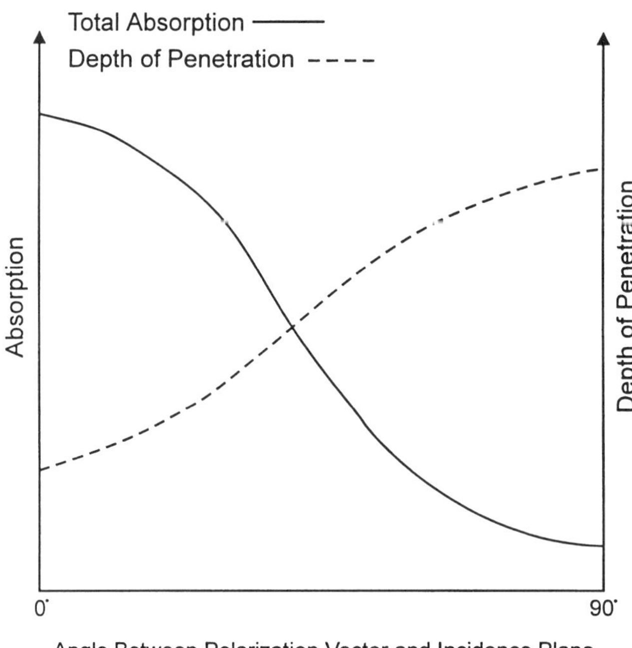

Figure 13.9. Total absorption (continuous line) and mean depth of penetration (dashed line) on a layer of vertically oriented scatters. The direction of the polarization vector is relative to the incident plane (adapted Elachi, 1987, by permission of John Wiley & Sons, Inc.).

(Ulaby and Wilson, 1985). In a separate ground scatterometer study of corn crops at a 60° incidence angle, X band attenuation was similar at V and H polarizations, being dominated by the leaves (Ulaby et al., 1987). L band loss was much higher at V than H polarization, while losses related to V and H polarizations at C band were about equal; however, the variance of means differed, being higher at V polarization than at H polarization. Again, for L and C bands, this relationship between polarization and attenuation was attributed to strong coupling between the canopy and incident wave orientation. This suggests that for radar interactions with canopies of comparable size (greater than wavelengths used), differences between X, C, and L bands at both V and H polarizations are reduced at lower incidence angles and increased at higher incidence angles (Figure 13.12).

Satellites, such as the ERS-1, ERS-2, and Radarsat, with C band VV and HH polarization SARs operating at a low incidence angle of 23° seem to follow this premise to improve canopy penetration. Additionally, X band, and increasingly at C band, signals will not penetrate through most canopies (especially at higher incidence angles) and will primarily interact with the canopy volume. V polarized waves, especially shorter wavelengths and higher incidence angles, will be more attenuated than H polarized waves in most natural canopies (due to the prevalence in most canopies of a more vertical orientation).

In some cases, however, the high angles of incidence tend to accentuate differences in relief, as shown by X band (HH polarization) data collected at very large incidence angles (70° to 80°). The combination of high incidence angle and short wavelength results in little penetration of the canopy (Lillesand and Keifer, 1979). The return is then dominantly from the uppermost portion of the canopy, accentuating the fine structure of the vegetation and possibly the land-

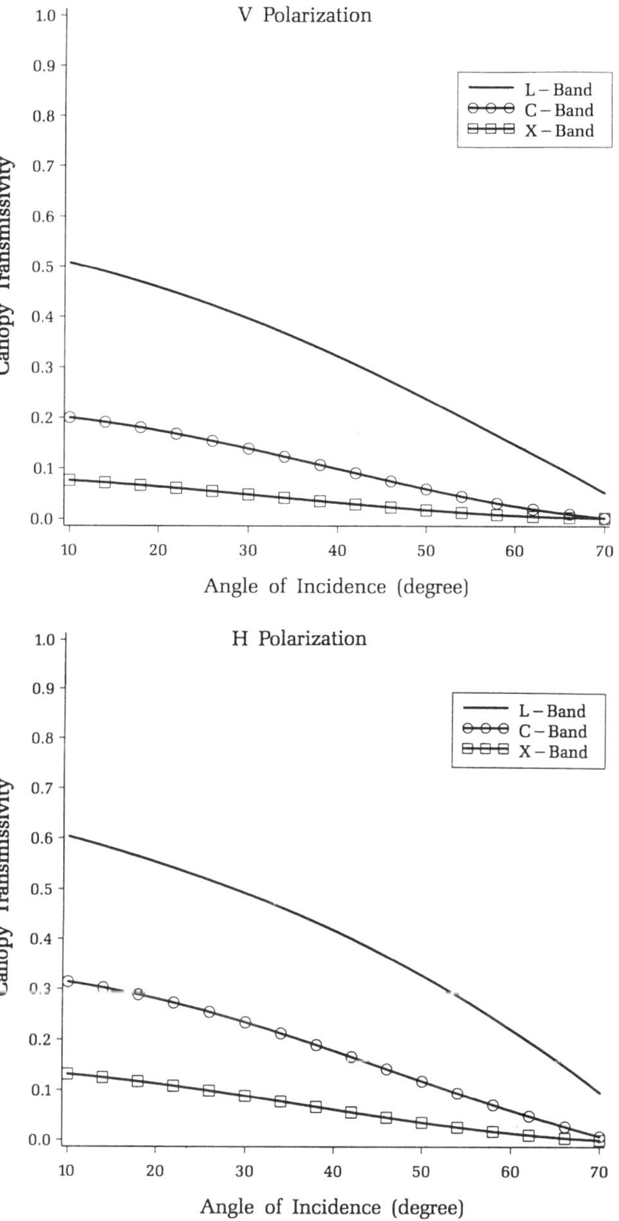

Figure 13.10. (top) The transmissivity of an aspen canopy versus incidence angle at V and (bottom) at H polarization (Ulaby et al., 1990, by permission of Taylor and Francis). Note that variations (sometimes high) exist for each curve that can mask these general trends.

water and marsh-forest interfaces (Mouginis-Mark et al., 1984). One advantage of this accentuation is the possibility of determining differences in regrowing vegetation, such as regrowth of burned marsh grasses. In a study of marsh burn recovery using X band HH polarization USGS Star-1 SAR, areas of burned black needlerush marsh were clearly distinguishable from nonburned marsh within an incidence angle range of 70° to 74°, but outside this range, burns were not clearly identified (Ramsey, unpublished data; Figure 13.13). The lack of detection

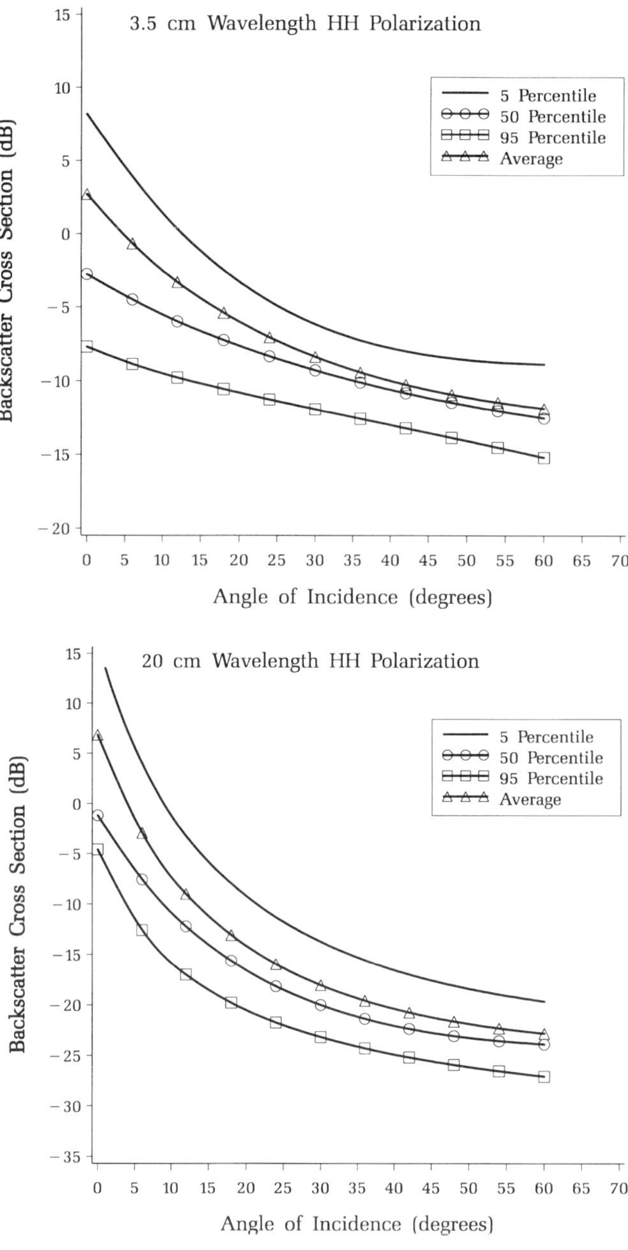

Figure 13.11. (top) The radar return from vegetation versus incidence angle for approximately 3.5-cm and (bottom) 20-cm wavelength (adapted from Elachi, 1988, © IEEE). Note that variations (sometimes high) exist for each curve that can mask these general trends.

beyond 74° may be due to the decrease in return with increase in incidence angle (Elachi, 1988; Lin et al., 1994). The accentuated radar returns were from burned sites that ranged between 0 to 400 days since burn and had canopy heights that mostly ranged from 50 cm to over 80 cm less than normal mature marsh. The time-since-burn was not a clear factor in the burn detection. The exact mechanism that accentuated the return of this radar from burned marsh is unclear;

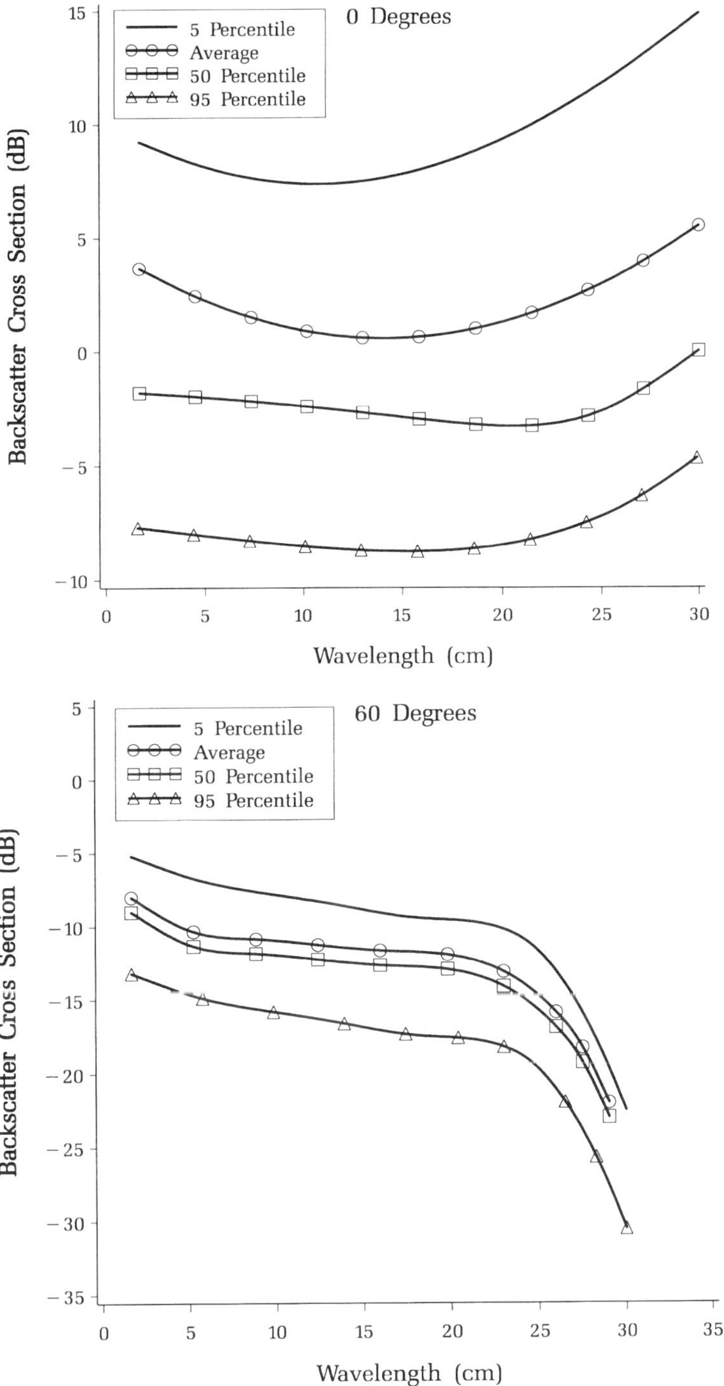

Figure 13.12. (top) The variation of radar returns from vegetation at 0° and (bottom) 60° incidence angles versus wavelength (Elachi, 1988, © IEEE). Notice the scatter is highest at 0° and lowest at 60°. Note: A dependency on wavelength does not become apparent until at about 20° incidence angle. Also note these curves are not specific and the actual response may differ with vegetation types.

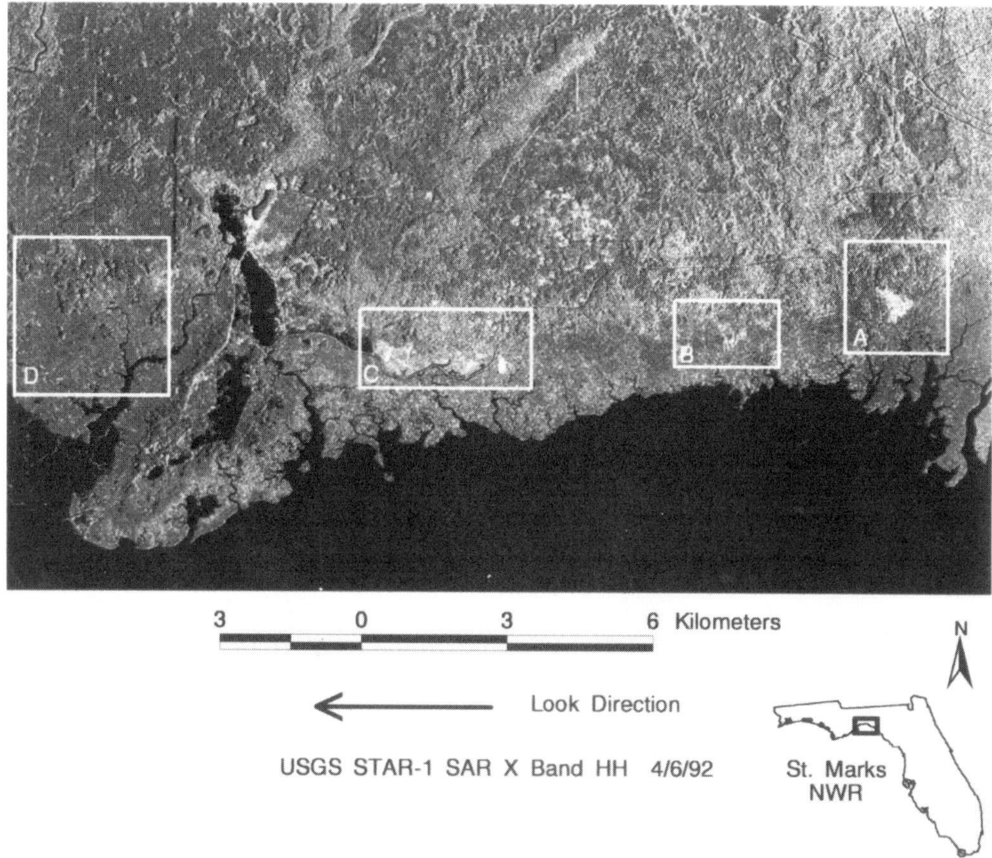

Figure 13.13. A USGS Star-1 X band HH polarization radar image (April 1992) of the St. Marks National Wildlife Refuge in the Big Bend area of coastal Florida (Ramsey, unpublished data). The boxes are centered on burn areas. Burn areas within boxes A (near range), B, and C show high contrast with the surrounding marsh, while a large burn within box D (far range) is not visible. Note the lower return from the water (black) areas.

however, this tool may provide valuable information about burn detection in this coastal marsh environment and similar environments.

4.8 Radar Wavelength

Of the system characteristics, the wavelength of the radar sensor is critical in influencing the interaction of the radar with the sensed material. This is not always the case, however, as the radar response to changes in wavelength may be only weakly tied to changes in certain environmental factors; for instance, changes in soil moisture. In general, holding all other factors constant, the longer the wavelength, the higher the canopy penetration of the radar (Elachi, 1988; Figure 13.14). In many instances this results in wavelengths such as L and P interacting with more of the canopy (depth of canopy penetration) than shorter wavelength bands such as C and X (Ferrazzoli et al., 1997). In forests, C and X bands are thought to interact primarily with the upper canopy, gaining penetration to the understory or ground surface through canopy gaps (Kasischke et al., 1994).

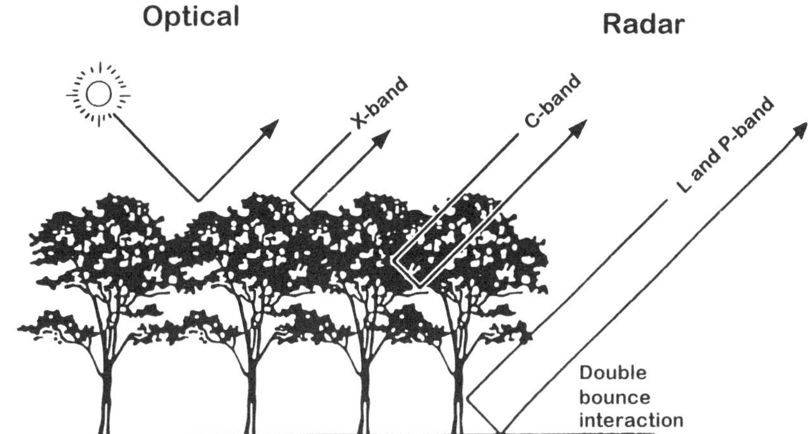

Figure 13.14. Optical reflectance and primary interactions of X, C, L, and P band radars with forest canopies (Waring et al., 1995, © American Institute of Biological Sciences). Notice the depiction is generalized and nonspecific and cannot be used solely to explain specific examples.

Wavelength also affects the saturation level of the radar. C band radars (e.g., ERS-1 and 2 SARs) have shown significant saturation at an aboveground biomass around 2 kg/m^2; L band radars (e.g., JERS-1) have a dynamic range of around 4 kg/m^2; while P band radars have a saturation limit of around 10 kg/m^2 (Imhoff, 1995a). L and especially P bands are thought to have a number of interaction pathways including any one of, or a combination of canopy, ground, and trunk. These interactions are commonly grouped into volume (direct), double bounce (direct-reflect), direct ground, and crown/ground (Dobson et al., 1992; van Zyl et al., 1990; Figure 13.15). This model seems to work well in most forests, but experimentation with C and L band radars has shown that differences in canopy structures can substantially alter the expected relative intensity of returns. Tree canopies with convoluted branching could attenuate the L band nearly as much as the C band (Paris and Ustin, 1990). In studies examining forest biomass, it was shown that stratification of the forest by type and/or age was necessary before accurate estimates of forest biomass could be made (Kasischke et al., 1994; Imhoff, 1995a, 1995b; Ferrazzoli et al., 1997). Simulating a range of incidence angles and C, L, and P band (all polarizations) returns, Imhoff (1995b) suggested forest canopy consolidation (diffuse—composed of many small components, consolidated—made up of fewer but larger components) could produce up to 18 dB differences in radar returns in forests of equal biomass. These differences due to structure could mask substantial differences in biomass, but may be useful in forest classification. The problem would be decoupling the changes in biomass from changes in forest structure, both parameters changing with forest type and age. In fact, stratification of the forest according to structural category has been reportedly used before estimating the forest type (Dobson et al., 1995a).

4.9 Classification Using Radar Systems

Classification of the radar return image can follow the commonly used procedures of optical classifications, i.e., clustering, component and discriminant analyses, maximum likelihood, linear combinations, and neural network analysis (e.g., Drieman, 1987; Foody et al., 1994; McCulloch and Yates, 1993; Lin et al., 1994; Ranson et al., 1997). Each of these image-based point classi-

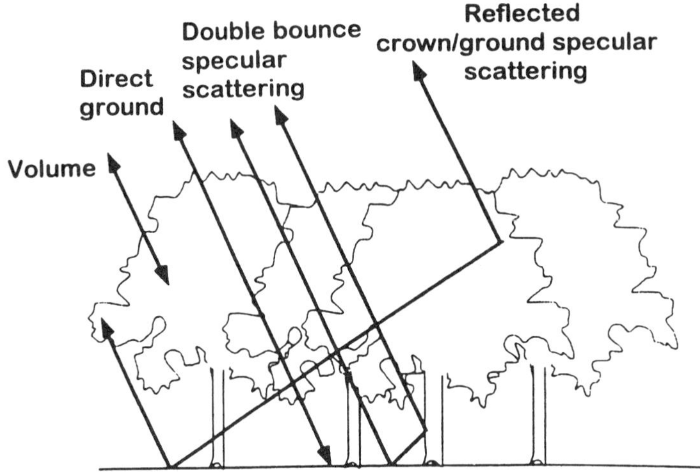

Figure 13.15. Dominate radar scattering interactions in a forest canopy as: crown (volume), direct ground, trunk/ground specular scattering (double-bounce) and crown/ground specular scattering (Dobson et al., 1992, © IEEE). The relative magnitude of the return from any one interaction mechanism can change as a function of the radar system and vegetation parameters.

fiers can use some form of knowledge-based approach either in the pre-classification (e.g., maximum likelihood analysis, Boolean logic, and supervised classification) or post-classification (e.g., grouping and identifying the spectral clusters or components). Even interferometry, a particularly unique radar function, has been used as a classification technique (Wegmuller and Werner, 1997).

Even though current research seems to be spurring the use of multifrequency and multipolarization radar systems, knowledge can be gained by using single frequency and polarimetric radar systems. In a study linking expected mechanisms of radar interaction to different landcovers, differences in two dates (spring-dry and fall-wet) of ERS-1 SAR images from an area in southern Florida encompassing mangrove forests and marsh were documented (Kasischke and Bourgeau-Chavez, 1997). Differences in the radar returns for the two dates were generally attributed to changes in vegetation biomass, soil moisture content, and flood conditions. Although slight in most cases, the increase in mean returns from forested sites (except for one high biomass site) was assumed to be due to soil moisture increases or increased flooding. Marsh sites, assumed flooded at both collection dates, were associated with slight changes in the mean return. An increase or decrease in live biomass was suggested as the reason for these changes (Kasischke and Bourgeau-Chavez, 1997).

The use of texture or context inputs into the various classifiers of optical imagery has been reported. However, this technique does not seem to have widespread acceptance for operational classification. In radar remote sensing, the addition of a texture measure to the classification is more commonplace. Texture in the radar image is a combination of the system and terrain features (Pierce et al., 1994). System texture (speckle) results from the statistical nature of the radar return and can be improved by increasing the number of looks (statistical averages of the radar returns; van Zyl et al., 1987; Elachi, 1988). The system texture can be estimated by averaging the radar return over undisturbed bodies of water. Removing this system component allows a texture measure—directly related to the terrain—to be estimated and used as part of the classification process (Pierce et al., 1994). Copolarized (e.g., HH/VV) and cross-polarized (e.g., VH/VV) ratios (and differences) may also be important inputs into the classification. Similar to optical ratios and differences, combining different polarizations may enhance certain features of the terrain (vegetation canopy) and diminish the influence of others. Wu and Sader (1987) found using polarization ratios increased the distinction between nonflooded cypress-tupelo and hardwood forests.

A more direct method of using the functionality of the radar system to classify terrain is developed on knowledge- (or rule-) based logic (Durden et al., 1989; van Zyl, 1989; Hess and Melack, 1994; Pierce et al., 1994, 1995; Polatin et al., 1994; Rignot et al., 1995; Dobson et al., 1996; Ferrazzoli et al., 1997). Single or combinations of SAR sensors (e.g., ERS-1, JERS-1, JPL AirSAR) can be trained on known landscape features. The structural characteristics of the training sites are then combined with expected strengths of the different wavelength and polarization interactions in order to generate a progressive classification hierarchy. This method has shown the potential for high classification accuracies when using a limited number of landcover types (tall, short, and surface landscape classes). Using only these three classes, classifications with ERS-1 and JERS-1 SARs separately yielded about a 75 percent and 98 percent accuracy, respectively (Dobson et al., 1995b). A further refinement of the classification separated the tall class into coniferous and deciduous forests (four classes total). The reported classification accuracies were about 76 percent for both SARs (Dobson et al., 1995b). Combining ERS-1 and JERS-1 data resulted in a reported composite accuracy of about 94 percent for the four-landcover classification (Dobson et al., 1995b).

Another type of rule-based classification uses the full polarimetric signature of one or a combination of radar frequencies (van Zyl et al., 1987, Evans et al., 1988). In this type of classification, the properties of the radar incident and scattered waves are accounted for by including the rotational sense, the ellipticity angle representing linear to circular polarizations, the orientation angle for the linear case representing horizontal and vertical polarizations, and phase information. The classification operates by first predicting the mechanism of interaction of the radar incident wave with various terrain features (open water, buildings, forests, etc.) and then predicting what polarimetric signature (based on inspection of the scattered wave properties) will result from this interaction. The terrain features are then grouped according to rules based on the expected set of return signatures. This approach has also shown promise in classifying general terrain features, but little or no information exists defining the accuracy of this classification method in different terrain types and complexities.

Another approach combines information from optical and radar sensors. This approach has been used in combinations of Landsat Thematic Mapper (TM) and SIR-B SAR (Rebillard and Evans, 1983; Lozano-Garcia and Hoffer, 1993; Haack and Slonecker, 1994). In these studies a reported increase in the classification accuracy was obtained by using the combined optical and radar image sources. An approach using a progressive classification methodology (Ramsey and Laine, 1997) and a simple combination of Landsat TM, high resolution color photographs, and an ERS-1 SAR image of a coastal Florida wetland and adjacent upland forest resulted in about 75 percent classification accuracy (Ramsey et al., in press; Table 13.3). In this detailed classification that included discrimination of marsh zones (estimated high to low marsh) within the same marsh type, singularly the TM data (the six reflective bands) explained 44 percent of the wetland variance, the photographic data about 31 percent, and the SAR about 16 percent (Figure 13.16). The SAR data were most useful in discriminating wetland from forest areas, but were generally poor in discriminating the different marsh zones. This was a very simple approach to integrating optical and SAR data; however, it provides information on the performance of this SAR system in detailed wetland classification.

Table 13.3. Accuracy Assessment of Progressive Classification.

Progressive Classification	Color Infrared Photography (Reference Data)													Total	Correctly Classified %
	1	2	3	4	5	6	7	8	9	10	11	12	13		
1. Hardwood	**41**	0	3	4	0	0	0	0	0	0	0	0	0	**48**	85.42
2. Pine	0	**42**	2	3	0	0	0	0	0	0	3	0	0	**50**	84.00
3. Scrub shrub/Pine Palmetto	0	0	**43**	6	0	0	1	0	0	0	0	0	0	**50**	86.00
4. Fresh marsh	0	2	8	**36**	3	0	1	0	0	0	0	0	0	**50**	72.00
5. High marsh	0	2	9	1	**75**	3	2	1	0	1	4	0	0	**98**	76.53
6. Medium high marsh	0	0	3	0	2	**109**	16	12	2	1	2	0	0	**147**	74.15
7. Medium marsh	0	1	0	0	0	3	**104**	15	11	2	10	0	0	**146**	71.23
8. Low marsh	0	0	1	0	0	0	4	**75**	13	1	3	0	0	**97**	77.32
9. Channel water and mud	0	0	1	0	1	6	10	10	**66**	0	4	0	0	**98**	67.35
10. Shallow coastal water	0	0	0	0	0	0	0	0	1	**49**	0	0	0	**50**	98.00
11. Water	0	0	0	7	6	1	0	0	2	2	**28**	0	0	**46**	60.87
12. Sand flats	1	0	9	3	11	4	1	1	9	9	9	**31**	0	**61**	50.82
13. Oyster bars/Sand bars	0	0	0	0	0	0	0	0	0	1	0	2	**47**	**50**	94.00
Total	**42**	**47**	**79**	**60**	**98**	**126**	**139**	**114**	**95**	**57**	**54**	**33**	**47**	**991**	**75.28**

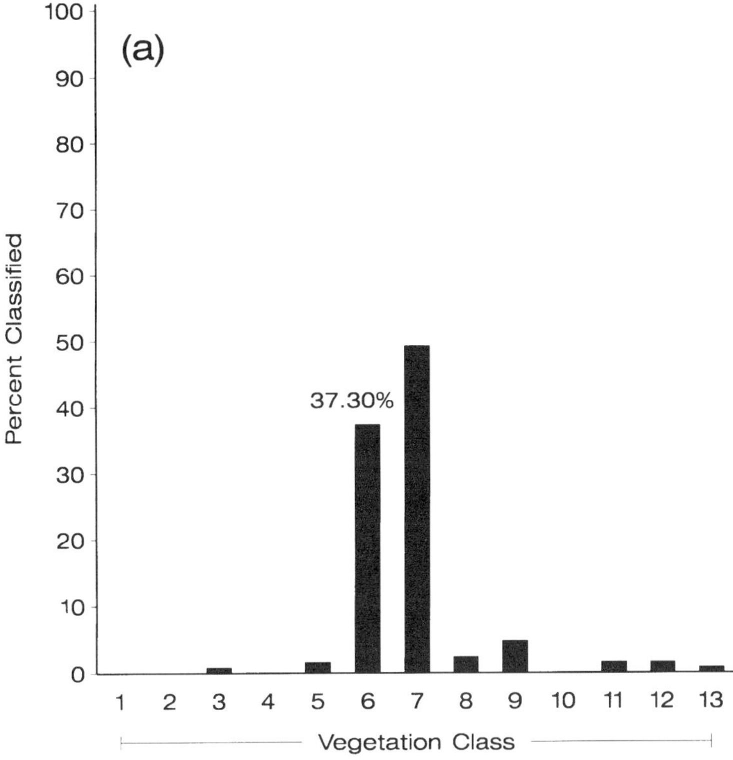

Figure 13.16a. Histograms showing how the percentages of correct classification increased by combining TM, CIR, and SAR as classifying variables. Correct classification of medium high marsh, class 6, (in %) by using (a) TM only, (b) TM and SAR (ERS-1), and (c) TM, CIR (green), and SAR (ERS-1) as classifying variables. Notice that the misclassified class (medium marsh, i.e., class 7) progressively moving toward the correct class (medium high marsh) from figure (a) to (c).

5.0 SUMMARY

A plethora of observations and numerous model developments have shown the potential of radar remote sensing for mapping the status and trends of wetland resources. Three critical parameters defining a wetland—hydrology, soil character, and vegetation type—can be mapped with radar remote sensing. Consistent, accurate mapping of these critical parameters is required for operational use; however, such accuracy is dependent on complex interactions between the various radar system parameters and the wetland canopy characteristics (vegetation, soil, water). The potential application user must decide whether the use of radar techniques is more appropriate than the use of optical or ground-based techniques, or if a combination of systems is required. Once the decision to use radar is made, the user must next choose the radar system that provides the most consistently accurate information, a necessity for operational applications. The job of choosing a radar system then simplifies to estimating how much of a given system's radar return signal (or lack of) is related to the vegetation canopy and how much is related to the ground beneath the canopy. Once this broad determination is made, further refinements to the chosen radar systems can be made. For example, higher incidence angles should produce more return from the canopy volume than lower incidence angles accenting lower canopy components.

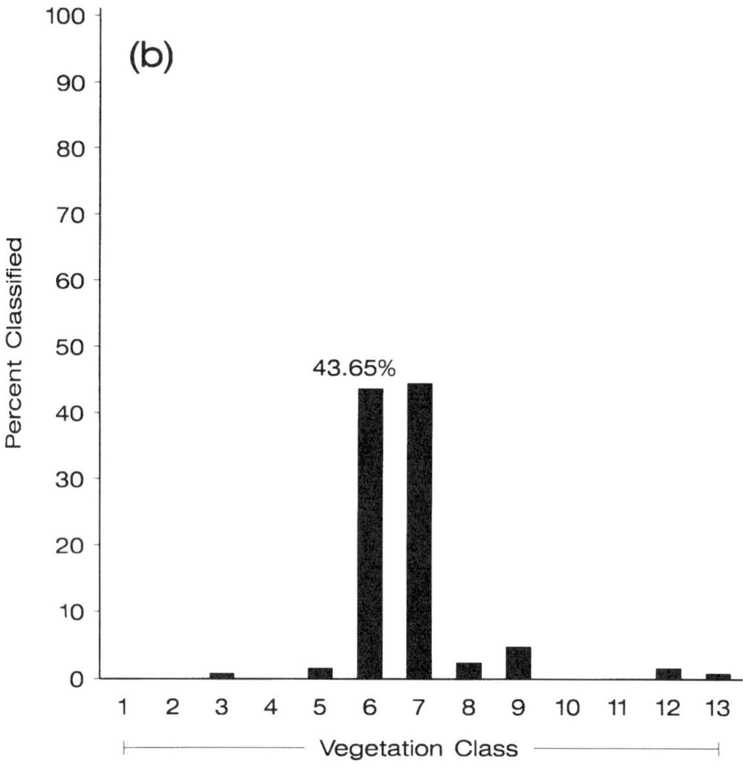

Figure 13.16b. Histograms showing how the percentages of correct classification increased by combining TM, CIR, and SAR as classifying variables. Correct classification of medium high marsh, class 6, (in %) by using (a) TM only, (b) TM and SAR (ERS-1), and (c) TM, CIR (green), and SAR (ERS-1) as classifying variables. Notice that the misclassified class (medium marsh, i.e., class 7) progressively moving toward the correct class (medium high marsh) from figure (a) to (c).

For flood detection, the radar return must be moderately insensitive to canopy variations and enhanced sufficiently by flooding beneath the canopy. In closed forest canopies, L band, and increasingly P band, seem to offer the highest potential for canopy penetration, and thus, flood detection under forest canopies. Restricted to operational satellite sensors (P band sensing from satellite altitudes has serious limitations), L band HH polarization radar systems (JERS-1) at low to moderate incidence angles seem to be the preferable. If not readily available, however, shorter wavelength (e.g., C band) systems (ERS-1, ERS-2, Radarsat) with steep incidence angles can also offer a flood mapping tool in certain canopy types and closures. Initially, in either case, simultaneous field and radar data collections should be performed during wet and dry, and leaf-on and leaf-off conditions to confirm and quantify the variability in the radar return in the area of interest. Spatially variable pooling, understory canopy, and specialized root systems should also be considered in the verification.

Compared to flood detection under forest canopies, there is limited information about radar's ability to detect flooding under various types of grass canopies. There are reports of diminished L and C band returns from flooded marshes, while others report bright C and X band returns from flooded marsh. In many cases, however, no simultaneous field data were collected to confirm the assumed flooding. In one case, simultaneous field measurements confirmed the attenuation of ERS-1 C band VV radar in a coastal marsh. Model predictions collaborated the

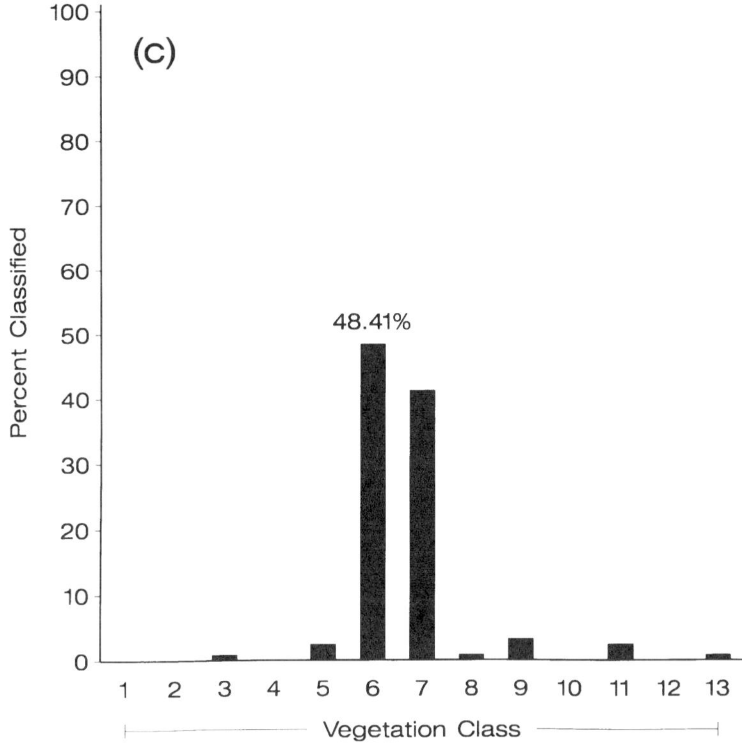

Figure 13.16c. Histograms showing how the percentages of correct classification increased by combining TM, CIR, and SAR as classifying variables. Correct classification of medium high marsh, class 6, (in %) by using (a) TM only, (b) TM and SAR (ERS-1), and (c) TM, CIR (green), and SAR (ERS-1) as classifying variables. Notice that the misclassified class (medium marsh, i.e., class 7) progressively moving toward the correct class (medium high marsh) from figure (a) to (c).

observations. This was only one type of marsh, however. More observations with extensive and simultaneous field measurements should be conducted before radar flood detection under marsh canopies can be carried out operationally. These should include observing how the radar return changes in response to changing marsh types and biomass and in variable conditions of flooding, spatially discontinuous pooling, and soil moisture content.

Soil moisture content mapping with radar has similar requirements as in flood detection under vegetation canopies; however, the sensitivity requirement is much higher. The radar system must not only detect changes in the soil moisture content, it must quantify the changes, at least to general categories, to be a useful tool for determining if an area is a functional wetland. This requirement is not as critical in areas where distinct wetland vegetation exists, but becomes increasingly important in marginal wetlands that cannot be determined by vegetation type. Furthermore, monitoring changing soil moisture conditions resulting from alteration of a region's hydrology requires more sophisticated techniques than in flood detection. Useful algorithms are available if the application is restricted to bare surfaces, and possibly low vegetation cover. Many studies report promising results when relating soil moisture content to the radar return; however, others cite problems with the vegetation overstory and soil roughness complicating soil moisture detection. It seems, perhaps, that longer wavelengths (L or P bands) should be used for soil moisture mapping under forest canopies, but that shorter wavelength (C band) radar systems with steeper incident angles may be useful in certain grass canopies. This is an active area of research. Currently, however, retrieval of soil moisture content data under changing vegetation canopies using radar systems does not seem operationally feasible.

As discussed throughout this chapter, the capabilities of radar systems and nature of the vegetation canopy (background included) combine to form a fairly complex set of interaction mechanisms. Generally speaking, the penetration of the canopy increases as either the wavelength increases or the incidence angle decreases; however, the polarization of the radar becomes increasingly important if the canopy has a preferred orientation. It seems that in most cases cross-polarized radar interacts more with the canopy volume (branches and stems) than like-polarized radars. Of the single-polarization, HH seems to have a higher potential for penetrating the canopy than VV. In applications to forest biomass, then, L or P band HH radars at moderately steep incidence angles would be preferred. Dependent on the nature of the marsh (height, density, live/dead) and radar incident angle, C or L band could be used for biomass estimation. The choice of polarization is dependent on the relative strengths of the returns.

The number of possible combinations of radar system parameters provides a tool rich for monitoring, discriminating between, and classifying wetland and adjacent vegetations. Unique information about the different canopy components, such as leaves, stems and branches, and stalks and trunks, may possibly be obtained by profiling the canopy (different depths of penetration) with increasingly longer wavelengths, with increasingly steeper incidence angles, or with different polarizations at a single or multiple wavelength(s). Similarly, the density and orientation changes of a wetland canopy may potentially be estimated by proper selection of system parameters or by using known variability of the canopy (e.g., changes in wetland type within a region or regrowth of a wetland through time). Knowledge of signal-target interactions—based on ground-based and aircraft-based observations and model developments—has led to a number of methods that use various forms and levels of rule-based classification techniques. Integration of these techniques with optical, thermal, and passive microwave remote sensing techniques could eventually provide a more consistent and detailed landscape characterization, and thus, improved tools for monitoring the status and trends of critical wetland environments. The challenge for wetland researchers and resource managers is to rationally apply these tools in a way that advances the capabilities of these remote sensing systems, and thereby provides greater detail about the wetland environment, thus increasing the potential for sustainable use in the face of growing detrimental pressures.

REFERENCES

Baronti, S., F. Del Frate, P. Ferrazzoli, and S. Paloscia. SAR polarimetric features of agricultural areas. In: *Proceedings of the Twenty-Fifth International Symposium, Remote Sensing and Global Environmental Change,* Environmental Research Institute of Michigan, Ann Arbor, MI, pp. 2639–2656, 1993.

Bina, R.T., R. Jara, B. de Jesus, Jr., and E. Lorenzo. Mangrove inventory of the Philippines using the Landsat multispectral data and image 100 system. In: *Proceedings of the Twelfth International Symposium on Remote Sensing of Environment*, Environmental Research Institute of Michigan, Ann Arbor, MI, pp. 2343–2359, 1978.

Butera, M.K. Remote sensing of wetlands. *IEEE Trans. Geoscience Remote Sensing*, GE-21(3), 383–392, 1983.

Chauhan, N., P. O'Neill, D. Le Vine, R. Lang, and N. Khadr. Space remote sensing. In: *International Geoscience and Remote Sensing Symposium,* IEEE Press, New York, 1992.

Curlander, J.C. and R.N. McDonough. *Synthetic Aperture Radar.* John Wiley & Sons, Inc., New York, 1991.

Dawson, M.S., A.K. Fung, and M.T. Manry. A robust statistical-based estimator for soil moisture retrieval from radar measurements. *IEEE Trans. Geoscience Remote Sensing,* 35, 57–67, 1997.

Dobson, M.C. and F.T. Ulaby. Preliminary evaluation of the SIR-B response to soil moisture, surface roughness, and crop canopy cover. *IEEE Trans. Geoscience Remote Sensing,* GE-24, 517–526, 1986.

Dobson, M.C., L.E. Pierce, and F.T. Ulaby. Knowledge-based land-cover classification using ERS-1/JERS-1 SAR composites. *IEEE Trans. Geoscience Remote Sensing,* 34(1), 83–99, 1996.

Dobson, M.C., F.T. Ulaby, L.E. Pierce, T.L. Sharik, K.M. Bergen, J. Kellndorfer, J.R. Kendra, E. Li, Y.C. Lin, A. Nashashibi, K.L. Sarabandi, and P. Siqueira. Estimation of forest biophysical characteristics in northern Michigan with SIR-C/X-SAR. *IEEE Trans. Geoscience Remote Sensing*, 33, 877–895, 1995a.

Dobson, M.C., F.T. Ulaby, and L.E. Pierce. Land-cover classification and estimation of terrain attributes using synthetic aperture radar. *Remote Sensing Environ.,* 51, 199–214, 1995b.

Dobson, M.C., L. Pierce, K. Sarabandi, F.T. Ulaby, and T. Sharik. Preliminary analysis of ERS-1 SAR for forest ecosystem studies. *IEEE Trans. Geoscience Remote Sensing*, 30, 203–211, 1992.

Drieman, J.A. Evaluation of SIR-B imagery for monitoring forest depletion and regeneration in western Alberta. *Can. J. Remote Sensing*, 13(1) 19–23, 1987.

Durden, S.L., J.J. van Zyl, and H.A. Zebker. Modeling and observation of the radar polarization signature of forested areas. *IEEE Trans. Geoscience Remote Sensing*, 27, 290–301, 1989.

Durden, S.L., L.A. Morrissey, and G.P. Livingston. Microwave backscatter and attenuation dependence on leaf area index for flooded rice fields. *IEEE Trans. Geoscience Remote Sensing*, 33, 807–810, 1995.

Elachi, C. *Introduction to the Physics and Techniques of Remote Sensing.* John Wiley & Sons, Inc., Toronto, Canada, 1987.

Elachi, C. *Spaceborne Radar Remote Sensing: Applications and Techniques.* IEEE Press, New York, 1988.

Engheta, N. and C. Elachi. Radar scattering from a diffuse vegetation layer over a smooth surface. *IEEE Trans. Geoscience Remote Sensing,* GE-20, 212–216, 1982.

Evans, D.L., T.G. Farr, J.P. Ford, T.W. Thompson, and C.L. Werner. Multipolarization radar images for geologic mapping and vegetation discrimination. *IEEE Trans. Geoscience Remote Sensing*, GE-24, 246–257, 1986.

Evans, D.L., T.G. Farr, J.J. van Zyl, and H.A. Zebker. Radar polarimetry: analysis tools and applications. *IEEE Trans. Geoscienc Remote Sensing*, 26, 774–789, 1988.

Ferrazzoli, P., S. Paloscia, P. Pampaloni, G. Schiavon, S. Sigismondi, and D. Solimini. Radar sensitivity to tree geometry and woody volume: a model analysis. *IEEE Trans. Geoscience Remote Sensing*, 35, 5–17, 1997.

Foody, G.M., M.B. McCulloch, and W.B. Yates. Crop classification from C-band polarimetric radar data. *Int. J. Remote Sensing*, 15, 2871–2885, 1994.

Ford, J.P., J.B. Cimino, and C. Elachi. *Space Shuttle Columbia Views the World with Imaging Radar: The SIR-A Experiment.* Jet Propulsion Laboratory (#82-95), Pasadena, CA, 1983.

Ford, J.P., J.B. Cimino, B. Holt, and M.R. Ruzek. *Shuttle Imaging Radar View the Earth from Challenger: The SIR-B Experiment.* Jet Propulsion Laboratory (#86-10), Pasadena, CA, 1986.

Ford, J.P. and D.J. Casey. Shuttle radar mapping with diverse incidence angles in the rainforest of Borneo. *Int. J. Remote Sensing*, 9(5), 927–943, 1988.

Freeman, A., M. Alves, B. Chapman, J. Cruz, S. Shaffer, J. Sun, E. Turner, and K. Sarabandi. CIR-C data quality and calibration results. *IEEE Trans. Geoscience Remote Sensing*, 33, 848–857, 1995.

Haack, B.N. and E.T. Slonecker. Merged spaceborne radar and thematic mapper digital data for locating villages in Sudan. *Photogrammetric Eng. Remote Sensing*, 60, 1253–1257, 1994.

Harris, J. and S. Digby-Argus. The detection of wetlands on radar imagery. In: *Proceedings of the Tenth Canadian Symposium on Remote Sensing*. Canadian Aeronautics and Space Institute, Ottawa, Ontario, Canada, pp. 529–543, 1986.

Heald, E.J. and W.E. Odum. The contribution of mangrove swamps to Florida fisheries. In: *Proceedings of the Gulf and Caribbean Fisheries Institute.* Institute of Marine Science, University of Miami, Coral Gables, FL, pp. 130–135, 1975.

Hess, L.L. and J.M. Melack. Mapping wetland hydrology and vegetation with synthetic aperture radar. *Int. J. Ecol. Environ. Sci.*, 20, 197–205, 1994.

Hess, L.L., J.M. Melack, S. Filoso, and Y. Wang. Delineation of inundated area and vegetation along the amazon floodplain with the SIR-C synthetic aperture radar. *IEEE Trans. Geoscience Remote Sensing*, 33, 896–904, 1995.

Hess, L.L., J.M. Melack, and D.S. Simonett. Radar detection of flooding beneath the forest canopy: A review. *Int. J. Remote Sensing,* 11(7) 1313–1325, 1990.

Hoffer, R.M., P.W. Mueller, and D.F. Lozano-Garcia. Multiple incidence angle shuttle imaging radar data for discriminating forest cover types. In: *Proceedings of the ACSM-ASPRS Fall Convention,* American Congress on Surveying and Mapping, American Society for Photogrammetry and Remote Sensing, Falls Church, VA, pp. 476–485, 1985a.

Hoffer, R.M., S.E. Davidson, P.W. Mueller, and D.F. Lozano-Garcia. A comparison of X- and L-band radar data for discriminating forest cover types. In: *Proceedings of the Pecora Symposium,* American Society for Photogrammetry and Remote Sensing, Falls Church, VA, pp. 439–440, 1985b.

Imhoff, M., M. Story, C. Vermillion, F. Khan, and F. Polcyn. Forest canopy characterization and vegetation penetration assessment with space-borne radar. *IEEE Trans. Geoscience Remote Sensing,* 24, 535–542, 1986.

Imhoff, M.L. Radar backscatter and biomass saturation: Ramifications for global biomass inventory. *IEEE Trans. Geoscience Remote Sensing,* 33, 511–518, 1995a.

Imhoff, M.L. A theoretical analysis of the effect off forest structure on synthetic aperture radar backscatter and the remote sensing of biomass. *IEEE Trans. Geoscience Remote Sensing,* 33, 341–352, 1995b.

Jensen, J.R., H. Lin, X. Yang, E.W. Ramsey III, B.A. Davis, and C.W. Thoemke. The measurement of mangrove characteristics in southwest Florida using SPOT multispectral data. *Geocarto Int.,* 2, 13–21, 1991.

Kasischke, E.S. and L.L. Bourgeau-Chavez. Monitoring south Florida wetlands using ERS-1 SAR imagery. *Photogrammetric Eng. Remote Sensing,* 63, 281–291, 1997.

Kasischke, E.S., L.L. Bourgeau-Chavez, N.L. Christensen, Jr., and E. Haney. Observations on the sensitivity of ERS-1 SAR image intensity to changes in aboveground biomass in young loblolly pine forests. *Int. J. Remote Sensing,* 15, 3–16, 1994.

Lewis, A.J. and D. Peterson *Wetland Mapping Using Multiple Polarized Radar Data Sets: Resource Development of the Lower Mississippi River.* American Water Resources Association, Baton Rouge, LA, 1991, pp. 17–25.

Lillesand, T.M. and R.W. Kiefer. *Remote Sensing and Image Interpretation.* John Wiley & Sons, Inc., New York, 1979.

Lin, D.S., E.F. Wood, K. Beven, and S. Saatchi. Soil moisture estimation over grass-covered areas using AIRSAR. *Int. J. Remote Sensing,* 15, 2323–2343, 1994.

Lozano-Garcia, D.F. and R.M. Hoffer. Synergistic effects of combined Landsat-TM and SIR-B data for forest resources assessment. *Int. J. Remote Sensing,* 14, 2677–2694, 1993.

Lyon, J.G. *Practical Handbook for Wetland Identification and Delineation.* Lewis Publishers, Boca Raton, FL, 1993.

Lyon, J.G. and J.F. McCarthy. Seasat imagery for detection of coastal wetlands. In: *Proceedings of the Fifteenth International Symposium on Remote Sensing of Environment*, Michigan Sea Grant Publications Office, Ann Arbor, MI, pp. 1475–1485, 1981.

MacDonald, H.C., W.P. Waite, and J.S. Demarcke. Use of Seasat satellite radar imagery for the detection of standing water beneath forest vegetation. In: *Technical Papers of the American Society of Photogrammetry Fall Technical Meeting*, American Society of Photogrammetry, Falls Church, VA, pp. RS3B1-RS3B13, 1980.

McCulloch, M.B. and W.B. Yates. The use of polarimetric synthetic aperture radar for crop mapping. *Swansea Geographer*, Dept. of Geography, University College of Swansea, Wales, 1993, pp. 97–109.

Mitsch, W.J. and J.G. Gosselink. *Wetlands.* Reinhold Co., Inc., New York, 1986.

Mouginis-Mark, P.J., L.R. Gaddis, and C. Ferrall. The Mississippi River delta: The monitoring of coastal processes with spaceborne and airborne radar and multispectral experiments. In: *International Sympo-*

sium on Remote Sensing of the Environment, Third Thematic Conference, Remote Sensing for Exploration Geology, Colorado Springs, CO, pp. 875–884, 1984.

Odum, W.E., C.C. McIvor, and T.J. Smith III. *The Ecology of the Mangroves of South Florida; A Community Profile.* Office of Biological Services, Washington, DC, 1982.

Ormsby, J.P., B.J. Blanchard, and A.J. Blanchard. Detection of lowland flooding using active microwave systems. *Photogrammetric Eng. Remote Sensing,* 51, 317–328, 1985.

Pardipuram, R., W.L. Teng, J.R. Wang, and E.T. Engman. Preliminary analysis of the sensitivity of AIR SAR images to soil moisture variations. In: *Summaries of the Fourth Annual JPL Airborne Geoscience Workshop,* Jet Propulsion Laboratory, Pasadena, CA, pp. 45–48, 1993.

Paris, J.F. and S.L. Ustin. Quantitative estimation of standing biomass from L-band multipolarization data. In: *Proceedings of the International Geoscience and Remote Sensing Symposium,* Institute of Electrical and Electronics Engineers, College Park, MD, pp. 147–150, 1990.

Pierce, L.L., K. Bergen, M.C. Dobson, and F.T. Ulaby. Quantitative remote sensing for science and applications, *Geoscience and Remote Sensing Symposium.* IEEE Press, New York, 1995.

Pierce, L.L., S.W. Running, and J. Walker. Regional scale relationships of leaf area index to specific leaf area and leaf nitrogen content. *Ecol. Appl.* 4, 313–321, 1994.

Place, J.L. *Mapping of Forested Wetland: Use of Seasat Radar Images to Complement Conventional Sources.* U.S. Department of the Interior, Geological Survey, Reston, VA, 1984.

Polatin, P.F., K. Sarabandi, and F.T. Ulaby. An iterative inversion algorithm with application to the polarimetric radar response of vegetation canopies. *IEEE Trans. Geoscienc Remote Sensing,* 32, 62–70, 1994.

Pultz, T.J., R. Leconte, L. St.-Laurent, and L. Peters. Flood mapping with airborne SAR imagery: case of the 1987 Saint-John River flood. *Can. Water Resour. J.,* 16, 173–189, 1991.

Ramsey III, E.W., R. Spell, and R.H. Day. Light attenuation and canopy reflectance as discriminators of gulf coast wetland types. In: *Proceedings of the International Symposium on Spectral Sensing Research,* U.S. Army Corps of Engineers, pp. 1176–1189, 1992.

Ramsey III, E.W., R. Spell, and R.H. Day. Measuring and monitoring the wetland response to acute stress by using remote sensing techniques. In: *Proceedings of the 25th International Symposium on Remote Sensing and Global Environmental Change,* Environmental Research Institute of Michigan, Ann Arbor, MI, pp. 1143–1155, 1993.

Ramsey III, E.W., S. Laine, D. Werle, B. Tittley, and D. Lapp. Monitoring Hurricane Andrew damage and recovery of the coastal Louisiana marsh using satellite remote sensing data. In: *Proceedings of the Coastal Zone Canada '94,* Coastal Zone Canada Association, Dartmouth, NS, pp. 1841–1852, 1994.

Ramsey III, E.W. Monitoring flooding in coastal wetlands by using radar imagery and ground-based measurements. *Int. J. Remote Sensing,* 16, 2495–2502, 1995.

Ramsey III, E.W. and J.R. Jensen. Modelling mangrove canopy reflectance by using a light interaction model and an optimization technique. In: *Wetland and Environmental Applications of GIS,* Lewis Publishers, Boca Raton, FL, 1995, pp. 61–81.

Ramsey III, E.W. and J.R. Jensen. Remote sensing of mangrove wetlands: relating canopy spectra to site-specific data. *Photogrammetric Eng. Remote Sensing,* 62, 939–948, 1996.

Ramsey III, E.W., D. Chappell, and D. Baldwin. AVHRR imagery used to identify hurricane damage in a forested wetland of Louisiana. *Photogrammetric Eng. Remote Sensing,* 63, 293–297, 1997.

Ramsey III, E.W. and S. Laine. Comparison of Landsat Thematic Mapper and high resolution photography to identify change in complex coastal wetlands. *J. Coastal Res.,* 13(2) 281–292, 1997.

Ramsey III, E.W., G.A. Nelson, S.C. Laine, and R.G. Kirkman. Generation of coastal marsh topography with radar and ground-based measurements. *J. Coastal Res.,* 13(4) 1335–1341, 1997.

Ramsey III, E.W., G.A. Nelson, and S.K. Sapkota. Classifying coastal resources by integrating optical and radar imagery and color infrared photography. *Mangroves and Salt Marshes,* in press.

Ranson, K.J., G. Sun, J.F. Weishampel, and R.G. Knox. Forest Biomass from combined ecosystem and radar backscatter modeling. *Remote Sens. Environ.,* 59, 118–133, 1997.

Rebillard, P. and D. Evans. Analysis of coregistered Landsat, Seasat and SIR-A images of varied terrain types. *Geophys. Res. Lett.,* 10(4) 277–280, 1983.

Richards, J.A., P.W. Woodgate, and A.K. Skidmore. An Explanation of enhanced radar backscattering from flooded forests. *Int. J. Remote Sens.,* 8(7) 1093–1100, 1987.

Rignot, E.J., R. Zimmermann, and J.J. van Zyl. Spaceborne applications of P band imaging radars for measuring forest biomass. *IEEE Trans. Geoscience Remote Sensing,* 33, 1162–1169, 1995.

Saatchi, S. and M. Moghaddam. Biomass distribution in boreal forest using SAR imagery. In: *Multispectral and Microwave Sensing of Forestry, Hydrology, and Natural Resources,* 1994, pp. 437–448.

Shutko, A.M. Soil/Vegetation Characteristics at Microwave Wavelength. *Understanding the Terrestrial Environment: The Role of the Earth Observations from Space.* P.M. Mather, Ed., Taylor and Francis, Washington, DC, 1992.

Shutko, A.M., A. Haldin, E. Novichikhin, G. Yazerian, G. Chukhray, E. Vorobeichik, V. Agura, S. Kalashnik, V. Sarkisjants, N. Sklonnaja, B. Logan, and E. Ramsey III. *Proceedings of the Fourth International Conference: Remote Sensing for Marine and Coastal Environments*, ERIM, Ann Arbor, MI, 1997.

Stofan, E.R., D.L. Evans, C. Schmullius, B. Holt, J.J. Plaut, and J.J. van Zyl. Overview of results of spaceborne imaging radar-C, X-band synthetic aperture radar (SIR-C/X-SAR). *IEEE Trans. Geoscience Remote Sensing,* 33(4), 817–828, 1995.

Ulaby, F.T. and M.C. Dobson. *Handbook of Radar Scattering Statistics for Terrain.* Artech House, Norwood, MA, 1989.

Ulaby, F.T. and M.A. El-Rayes. Microwave dielectric spectrum of vegetation-part ii: dual dispersion model. *IEEE Trans. Geoscience Remote Sensing,* 25, 550–557,1987.

Ulaby, F.T., A. Tavakoli, and T.B.A. Senior. Microwave propagation constant for a vegetation canopy with vertical stalks. *IEEE Trans. Geoscience Remote Sensing,* GE-25, 714–725, 1987.

Ulaby, F.T., R.K. Moore, and A.K. Fung. *Microwave Remote Sensing: Active and Passive.* Artech House, Norwood, MA, 1981.

Ulaby, F.T., R.K. Moore, and A.K. Fung. *Microwave Remote Sensing: Active and Passive, Vol. II-Radar Remote Sensing and Surface Scattering and Emission Theory.* Addison-Wesley, Reading, MA, 1982.

Ulaby, F.T. and E.A. Wilson. Microwave attenuation properties of vegetation canopies. *IEEE Trans. Geoscience Remote Sensing.* GRS-23, 746–753, 1985.

Ulaby, F.T., K. Sarabandi, K. McDonald, M. Whitt, and C. Dobson. Michigan microwave canopy scattering model. *Int. J. Remote Sensing,* 11, 1223–1253, 1990.

Untawale, A.G., S. Wafar, and T.G. Jagtap. Application of remote sensing techniques to study the distribution of mangroves along the estuaries of Goa. *Wetlands Ecology and Management*, 1980, pp. 51–68.

Ustin, S.L., C.A. Wessman, B. Curtiss, E. Kasischke, J. Way, and V. Vanderbilt. Opportunities for using the EOS imaging spectrometers and synthetic aperture radar in ecological models. *Ecology,* 72, 1934–1945, 1991.

van den Broek, A.C. and J.S. Groot. Classification and soil moisture determination of agricultural fields. *Can. J. Remote Sensing,* 3, 77–80, 1993.

van Zyl, J.J. Unsupervised classification of scattering behaviour using polarimetric data. *IEEE Trans Geoscience Remote Sensing,* 27, 36–45, 1989.

van Zyl, J.J., H.A. Zebker, and C. Elachi. Imaging radar polarization signatures: theory and observations. *Radio Sci.,* 22, 529–543, 1987.

van Zyl, J.J., H.A. Zebker, and C. Elachi. Polarimetric SAR applications. *Radar Polarimetry for Geoscience Applications*, F.T. Ulaby and C. Elachi, Eds., Artech House, Norwood, MA, 1990, pp. 315–360.

Waite, W.P. and H.C. MacDonald. 'Vegetation Penetration' with K-band imaging radars. *IEEE Trans. Geoscience Remote Sensing,* GE-9, 147–155, 1971.

Waite, W.P., H.C. MacDonald, V.H. Kaupp, and J.S. Demarcke. Wetland mapping with imaging radar. In: *Digest, International Geoscience and Remote Sensing Symposium (IGARSS '81),* IEEE, New York, 1981, pp. 794–799.

Wang, Y. and M.L. Imhoff. Simulated and observed L-HH radar backscatter from tropical mangrove forests. *Int. J. Remote Sensing,* 14, 2819–2828, 1993.

Waring, R.H., J. Way, E.R. Hunt, L. Morrissey, K.J. Ranson, J.F. Weishampel, R. Oren, and S.E. Franklin. Imaging radar for ecosystem studies. *BioScience,* 45(10), 715–723, 1995.

Way, J. and E.A. Smith. The evolution of synthetic aperture radar systems and their progression to the EOS SAR. *IEEE Trans. Geoscience Remote Sensing,* 29, 962–985, 1991.

Wedler, E. and R. Kessler. Interpretation of vegetative cover in wetlands using four-channel SAR imagery. In: *Technical Papers of the 47th Annual Meeting of the American Society of Photogrammetry,* ASPRS, Falls Church, VA, 1981, pp. 111–124.

Wegmuller, U. and C. Werner. SAR interferometric signatures of forest. *IEEE Trans. Geoscience Remote Sensing,* 35, 18–24, 1997.

Werle, D. *Radar Remote Sensing: A Training Manual Dendron Resource Surveys.* Canadian Centre for Remote Sensing, Ottawa, 1988.

Wickland, D.E. Mission to planet earth: The ecological perspective. *Ecology,* 72, 1923–1933, 1991.

Wu, S.-T. and S.A. Sader. Multipolarization SAR data for surface feature delineation and forest vegetation characterization. *IEEE Trans. Geoscience Remote Sensing.* GE-25, 67–76, 1987.

CHAPTER 14

Radar Interferometry for Environmental Change Detection

Bernard Armour, Akiko Tanaka, Hiroshi Ohkura, and Genya Saito

1.0 INTRODUCTION

Interferometry is a new technique for measuring elevation, surface motion, and surface change with unprecedented spatial detail and scale. In this chapter we review the principles of interferometric SAR (InSAR) in some detail so as to introduce this technique to the reader. This is followed by a necessarily brief survey of some applications that have been identified to date where InSAR can be a rich source of information. These applications include earthquake and volcano monitoring, environmental monitoring, topographic mapping, glacier and ice sheet measurements.

1.1 Basic Principles of Interferometric SAR (InSAR)

In the following sections we give an introduction to SAR interferometry. Interferometric SAR may be abbreviated as *InSAR* or *IFSAR*. The two are equivalent and both are used in the published literature. In the following, emphasis is placed on the principles of InSAR, while the details are for the most part excluded for the sake of clarity and brevity. The reader is referred to the references for more information on the physics and processing aspects of InSAR. One of the first papers on the use of synthetic aperture radar as an interferometer was published by Graham (1974). Zebker and Goldstein (1986) were among the first to publish the use of InSAR in an airborne system. An excellent overview of interferometry is given by Massonnet (1997), which includes a good introduction and history of InSAR.

1.2 Principles of Imaging Radar

There are several excellent references in the literature introducing the theory of imaging radar and SAR. One excellent introductory text is by Elachi (1988). A more advanced text by Curlander and McDonough (1991) gives detailed information of SAR and SAR processing. In this section we give only a brief introduction to SAR. The reader is referred to the references for more detailed information.

To understand interferometric SAR, we must first understand the basic principles of a radar. Of the many variations of radar configurations, we will focus on pulsed, coherent radars. A pulsed radar emits a series of pulses, either in bursts or in continuous streams. Each pulse consists of a sinusoidal wave of energy oscillating at microwave frequencies, and lasting for a brief duration in time. The wave of energy is *coherent* if the wave consists of only one frequency at any one time. In contrast, a wave is noncoherent if it consists of a number of superimposed waves, each with its own frequency and starting reference point (e.g., a wave may start at a wave peak, a wave trough, or at some other point in the wave cycle). Thus a pulsed, coherent, radar emits a series of coherent pulses equally spaced in time.

An imaging radar is normally mounted on an aircraft or spacecraft. The radar is sidelooking, meaning that the transmitting and receiving antenna is mounted to point at right angles to the vehicle's direction of travel. Each pulse transmitted by the radar is focused by the antenna to travel mainly in the "look" direction of the antenna, and thus transmitted pulses are backscattered (reflected) only in the areas within the antenna footprint, as shown in Figure 14.1. The radar measures the backscatter wave amplitude, phase, and range distance from the radar antenna to backscattering point on the ground. The *phase* of a reflected pulse is the fraction of a wave cycle that the *reflected* wave starting point has changed by as compared to the *transmitted* wave.

The radar transmit antenna projects onto the earth surface a wide and deep footprint. As a result, many backscattering points on the ground reflect transmitted pulses. An imaging radar must be able to resolve individual scattering points in two dimensions. High resolution is achieved in the range dimension, as shown in Figure 14.1, by using pulse compression techniques to make the pulse very short. High resolution in the azimuth dimension is achieved by synthesizing a very long antenna through taking advantage of the fact that the same scatterers on the ground are hit by successive radar pulses thousands of times, but from slightly different radar positions. This enables a synthetic antenna to be constructed using the phase history of the pulses, the known position of the radar as a function of time, and the relative motion between the radar and the scatterers on the ground. Hence the name *synthetic aperture radar* or SAR.

The resolution of a SAR is independently determined in range and azimuth. Typical airborne SAR image resolutions are 0.5 to 10 meters. Typical spaceborne SAR image resolutions are 5 to 100 meters (Way and Smith, 1991). Normally the pixel size in meters of a SAR image is equal or slightly smaller than the resolution of the image. If the image contains both backscatter amplitude and phase information, then the image is *complex,* since in the mathematical sense it has a real and imaginary part (Churchill and Brown, 1984).

1.3 SAR Image Phase and Coherence

A radar measures backscattering amplitude (brightness), phase (change in wave cycle starting point), and range distance. Given that the radar measures the backscattering of a group of scatterers in a resolution cell, the phase of a resolution cell is the result of a complicated summation of backscattered pulses from individual scatterers within one resolution cell all being combined into one measurement. The phase for each of these individual scatterers is determined by the two-way propagation distance from radar to scattering element on the ground, as well as the interaction of the microwave signal with the scatterer. Moisture content, conductivity, surface structure, and orientation all affect the surface microwave interaction, and thus the reflected phase. If the relative locations of the individual scatterers in the resolution cell change, or the phase of the reflected pulse from any one of the scatterers changes substantially with respect to the other scatterers, then the resulting phase of the scatterers will change.

This idea is illustrated in Figure 14.2. Assume that the scattering cell consists of five point scatterers. The reflected pulses for each scatterer will have an amplitude and phase. Since the

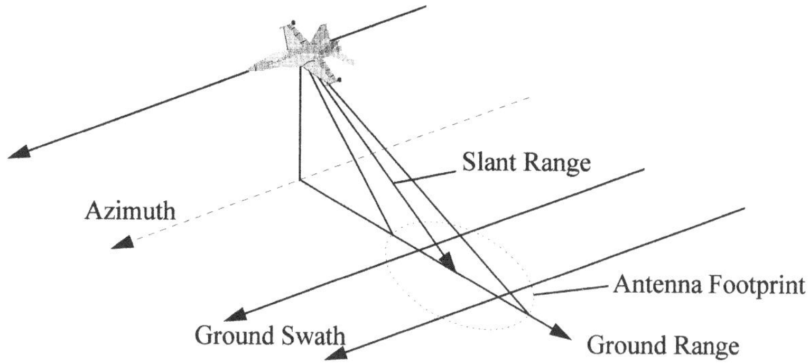

Figure 14.1. Imaging radar geometry and definitions.

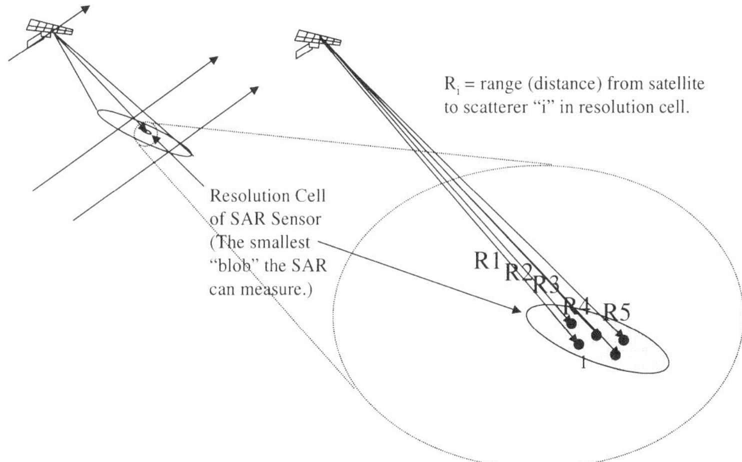

Figure 14.2. The phase of one resolution cell in a SAR image is determined by the summation of the complex amplitude and phase values due to each individual scatterer in the cell.

radar cannot resolve anything smaller than a resolution cell size, these reflected pulses all superimpose (sum) at the receive radar antenna and cannot be resolved in the signal processing that converts the reflected pulses into an image. The phase of the resolution cell as a whole is then the complex vector sum of the individual scatterer amplitude and phase values. If there is one dominant scatterer in terms of amplitude, the phase of the resolution cell is largely determined by this one scatterer. If there are many equally strong scatterers, then the phase is determined by the complex vector sum.

Now consider the case where we acquire two SAR images at different times but of the same area on the Earth's surface. From the time of the first SAR image acquisition to the time of the second acquisition if the relative positions of the scatterers within each resolution cell do not change, and if the microwave interaction with the scattering elements does not change then the phase of each resolution cell will be unchanged. If this is the case, then the two SAR images are said to be *coherent.* Now if the relative position of some of the scattering elements in a resolution cell changes by a few fractions of a radar wavelength, or if the microwave interaction with the scattering elements changes, resulting in the phase reflected changing by some fraction of a wavelength, then the resulting phase of the resolution cell will be potentially quite different. The degree of the difference will determine to what extent the coherence is lost between the two acquisitions. For SAR interferometry, we rely on the phase in two SAR images acquired of the same ground area having coherence. If the images do, then we can compare the phase difference between the two images and measure height and height change.

Two SAR images are shown in Figure 14.3. These are an ERS-1 repeat pass InSAR pair acquired three days apart in January 1994. The ground cover is predominantly frozen tundra with a small river bed running diagonally across the scene. Only the magnitude of the two SAR images is shown. The phase in this image pair is mostly correlated (i.e., the InSAR pair is coherent). The coherence image calculated by cross correlating a small sliding window of pixels across both images is shown in Figure 14.4. Coherence near 1.0 (the maximum coherence) appears as white; total loss of coherence appears as black. Low coherence areas appear in the riverbed, which is largely frozen. This is most likely due to ice growth, or melting/freezing effects. Coherence is lost in the tundra areas due to snow drifting and changes in the permafrost layer of the ground surface. Coherence is also lost in the slopes of the small mountain that face the radar. This is due to the slopes not being resolved sufficiently that the phase information for a particular resolution cell can remain coherent. This effect is commonly known as *layover* (Elachi, 1988). Though changes in coherence are a complicated process, this information does relate closely with the form and dynamics of the scene being imaged. As a result, it is the subject of active research in the remote sensing community.

1.4 Configurations for Acquiring Across-Track Interferometric SAR Image Pairs

Across-track interferometry requires that two SAR images be acquired of the same scene. For measuring topography, the two SAR receiving antennas must be located at slightly different positions in the across-track dimension. This displacement results in a small difference in incidence angle between the image acquisitions. This is shown in Figure 14.5. The two SAR images may be acquired simultaneously (single-pass interferometry) or in two separate passes of the SAR sensor (repeat-pass interferometry). Both techniques are possible with airborne and spaceborne SAR systems. Typically, airborne InSAR systems are single pass since repeat pass airborne InSAR is difficult to achieve (Gray and Farris-Manning, 1993).

Single-pass spaceborne SAR interferometry configurations including the Shuttle Radar Topographic Mission (SRTM, 1996) and the TOPSAT (Zebker et al., 1994) mission are planned or theorized for upcoming launches. Repeat-pass spaceborne interferometry in comparison to repeat-pass airborne interferometry is quite successful. Satellite orbits tend to have very regular and controlled paths that by their very nature repeat periodically. As a result, repeat pass scene revisits occur regularly. Many of these repeat passes will have orbits which have the right amount of separation for measuring topography or land deformation. The repeat pass configuration is shown in Figure 14.6.

The European Space Agency (ESA) spaceborne SARs ERS-1 (Attena, 1991) and ERS-2, the Japanese SAR JERS-1 (Nemoto et al., 1991), and the Canadian Space Agency (CSA) SAR RADARSAT (Raney et al., 1991) are all capable of acquiring repeat pass InSAR pairs of imagery with useful orbit separations. Characteristics of these missions are listed in Table 14.1. Note that a long period of time may have elapsed between the orbit repeats which may lead to a loss of coherence between the two acquired SAR images. One notable exception is the ERS-1/ERS-2 Tandem mode mission which operated intensively for just over one year in 1995/1996

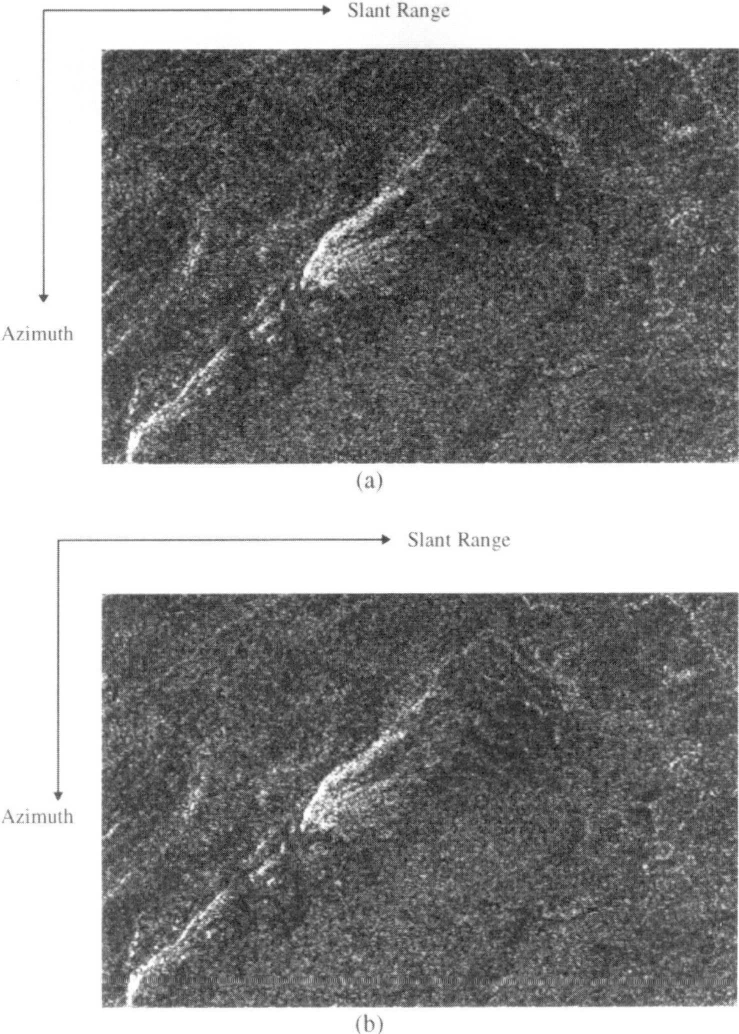

Figure 14.3. ERS-1 three-day repeat SAR image pair of an area near Schefferville, Quebec, Canada. Image (a) was acquired on January 9, 1994, and image (b) on January 12, 1994, three days later. The phase in each image is essentially random from pixel to pixel; however, the phase difference between the two images contains useful information for interferometry. Each image is approximately 3200 m in slant range by 5161 m in azimuth. (© ESA 1994. Imagery is courtesy of the Canada Centre for Remote Sensing, Natural Resources, Canada.)

mapping about 95 percent of the Earth with 1-day repeat InSAR pairs. This was achieved by configuring the orbits such that the ERS-2 satellite followed the ERS-1 satellite with only a one-day lag.

1.5 Geometry of Single-Pass InSAR

In this section we give a brief, geometry-based, explanation of single-pass interferometry which follows the conventions in Zebker et al. (1986) and Zebker et al. (1994). A basic diagram

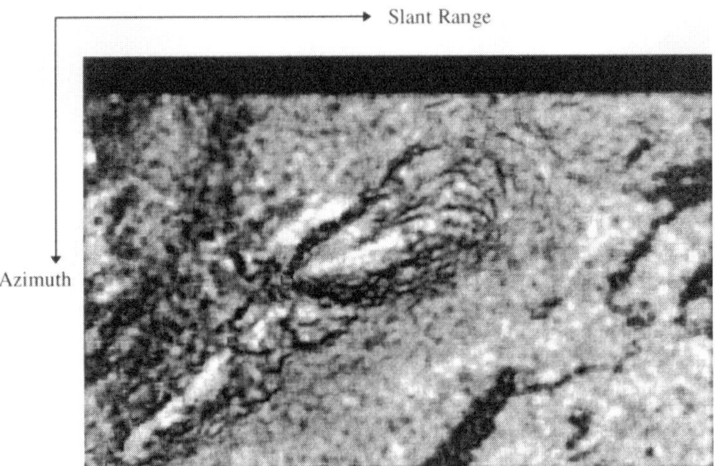

Figure 14.4. This coherence image was obtained by correlating small windows of 5 by 25 pixels in the two images in Figure 14.3. High coherence (high complex correlation) has a value of 1.0 and is represented by pure white. Low coherence (low complex correlation) has a value of 0.0 and is represented by black. Areas containing water, radar shadow and layover, or very low backscatter will typically have very low coherence. The dark shape in the lower center of the image is a lake partially frozen over. Decorrelation occurred due to ice growth over the three days between image acquisitions.

of the imaging geometry for single-pass SAR interferometry is shown in Figure 14.7. An interferogram is formed by calculating the phase difference of two precisely coregistered SAR images, where the first image is acquired through antenna A_1 and the second image is acquired through antenna A_2, as shown in Figure 14.7. The phase difference between the two SAR images is calculated by multiplying image 1 by the complex conjugate of image 2. The resulting phase in the interferogram image is proportional to the range difference δR as follows:

$$\Delta\phi = (R_1 + R_2 - 2R_1)\frac{2\pi}{\lambda}$$
$$\Delta\phi = (R_2 - R_1)\frac{2\pi}{\lambda} \tag{1}$$

where $\Delta\phi$ is the interferometric phase, and λ is the radar wavelength. The interferometric phase has inherently a 2π (360 degrees) phase ambiguity, which must be removed through a process called *phase unwrapping* (Goldstein et al., 1988). Assuming that the phase unwrapping has been performed successfully, the change in range δR can be derived from the interferometric phase as:

$$\delta R = \frac{\lambda\Delta\phi}{2\pi} \tag{2}$$

The height $z(y)$ at a point with ground range position y is determined from:

$$z(y) = h - R_1 \cos\theta \tag{3}$$

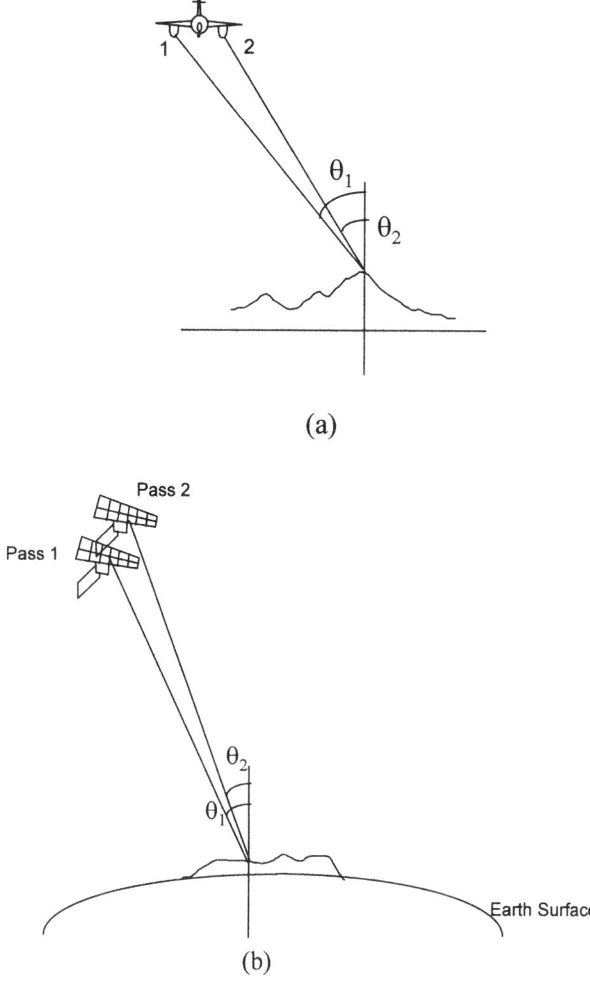

Figure 14.5. Across-track interferometry is possible from both airborne and spaceborne SAR platforms. The antenna positions for the first and second image acquisitions are slightly displaced in the across-track direction, resulting in small differences in imaging incidence angle. Vertical height is determining by differencing the phase measurements of the two SAR images and relating these to the local terrain height via accurate modeling of the imaging geometry.

where θ is the look angle of the radar. To get the local height $z(y)$, we need to know the slant range R_l very accurately. We estimate R_l from the phase of the interferogram, plus our knowledge of the baseline length B, the baseline orientation angle α, and the look angle θ, as defined in Figure 14.7.

In the case of spaceborne InSAR, the slant range is quite large, on the order of 300 to 1,500 kilometers. Because of this long slant range distance, the requirement for height errors on the order of 10 meters or less requires the platform attitude[1] to be known to within an accuracy of 0.002 degrees or better (Zebker et al., 1994). This can be difficult to determine with sufficient

[1] Platform attitude is simply the pointing direction of the SAR transmit and receive antenna.

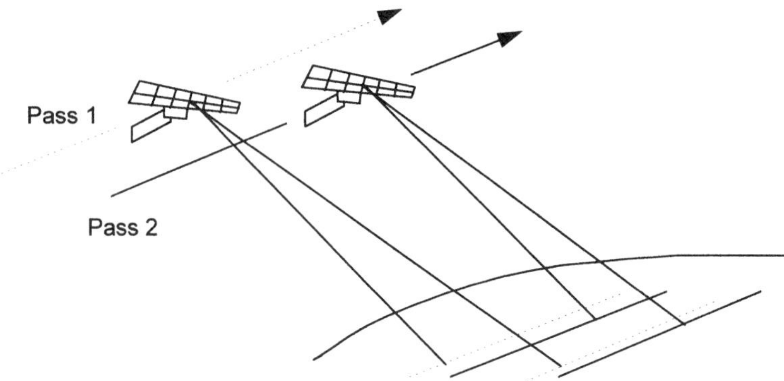

Figure 14.6. Spaceborne interferometry using repeat-pass image acquisitions. The first and second passes may be repeat orbits of the same satellite (such as ERS-1, ERS-2, JERS-1, RADARSAT), or else be an orbit of a second satellite which is following the first, such as the ERS-1/ERS-2 tandem mission.

Table 14.1. Repeat Pass Spaceborne Interferometry SAR Sensors.

Sensor	Band	Incidence Angle (deg)	Swath Width (km)	Range Resolutions (m)	Azimuth Resolution (m)	Exact Repeat Cycle (days)
ERS-1	C (5.7 cm)	23	100	20	30 (4 looks)	3, 35, 176
ERS-2	C (5.7 cm)	23	100	20	30 (4 looks)	35
JERS-1	L (23.5 cm)	38	75	18	18 (3 looks)	44
RADARSAT	C (5.7 cm)	20–59	45–500	10–100	10–100	24

precision, and ground control points may be used to "work backward" and refine the baseline modeling.

1.6 Geometry of Repeat-Pass InSAR

The geometry for repeat-pass interferometric SAR is largely the same as for the single-pass geometry case. The main difference is that both antennas A_1 and A_2 transmit and receive pulses, and this leads to an additional factor of 2 in the resulting equations. Referring to imaging geometry for single-pass interferometric SAR, antenna A_1 both transmits and receives, while antenna A_2 receives only. In practice, the equivalent antenna separation B may be doubled by alternating the transmitting antenna between A_1 and A_2 for each pulse. Single-pass InSAR is possible from both airborne and spaceborne platforms. The curvature of the Earth becomes significant in the spaceborne case.

In Figure 14.8, the equations analogous to the single pass case given in the previous section are:

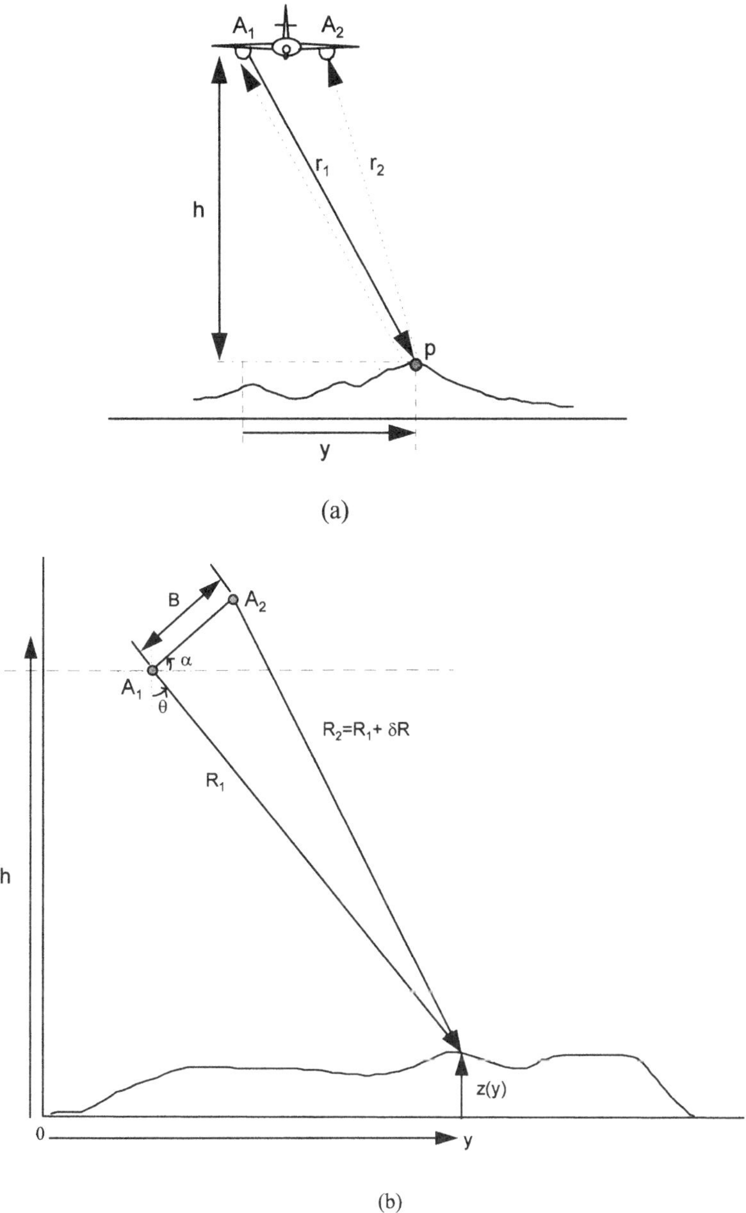

Figure 14.7. Imaging geometry for single-pass interferometric SAR. Antenna A_1 both transmits and receives, while antenna A_2 receives only. In practice, the equivalent antenna separation *B* may be doubled by alternating the transmitting antenna between A_1 and A_2 for each pulse. Single-pass InSAR is possible from both airborne and spaceborne platforms. The curvature of the Earth becomes significant in the spaceborne case.

$$\Delta\phi = (2R_2 - 2R_1)\frac{2\pi}{\lambda}$$
$$\Delta\phi = 2(R_2 - R_1)\frac{2\pi}{\lambda} \quad (4)$$

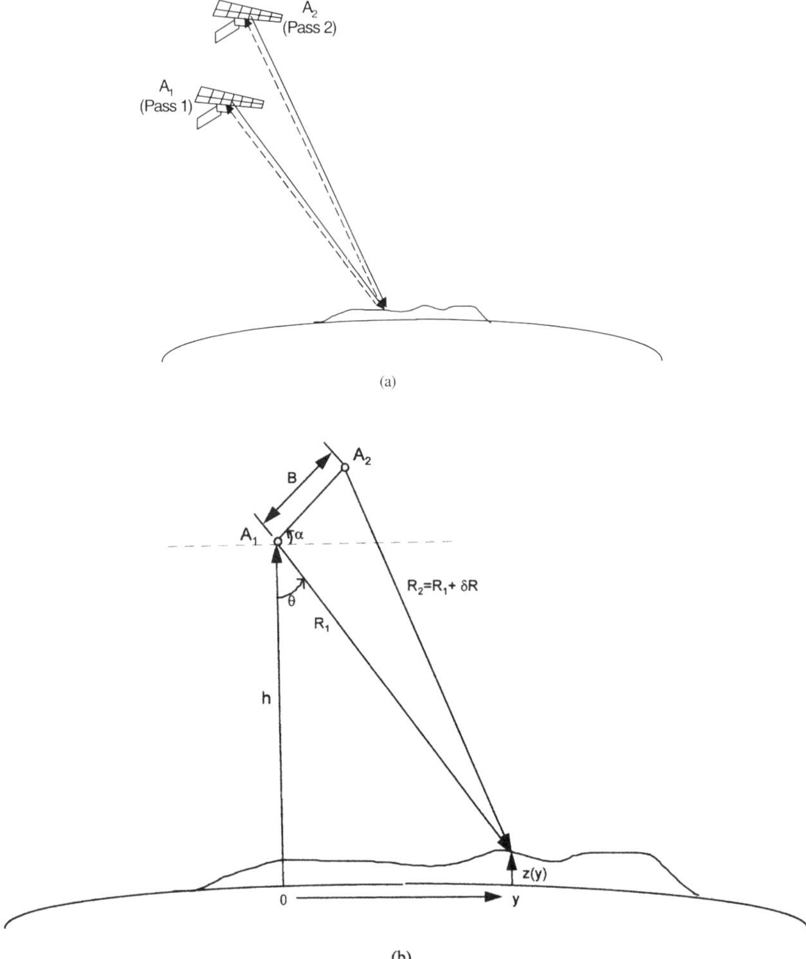

Figure 14.8. Imaging geometry for repeat-pass interferometric SAR. Antenna A_1 and antenna A_2 both transmit and receive pulses.

$$\delta R = \frac{\lambda \Delta \phi}{4\pi} \tag{5}$$

$$z(y) = h - R_1 \cos\theta \tag{6}$$

Again, the slant range R_1 is estimated from the unwrapped interferogram phase ϕ, the baseline B, and the baseline orientation α.

Knowledge of the baseline separation and orientation and how this changes with time is very important for obtaining good height accuracy. An analysis of the contributing factors to height errors for repeat-pass spaceborne InSAR is given in Li et al. (1990). The details are beyond the scope of this introduction; however, to summarize, baseline separation length accuracy on the order of millimeters is required, and baseline orientation angle accuracy must be on the order of 1 arc second. Since these requirements are not possible to achieve with current spaceborne

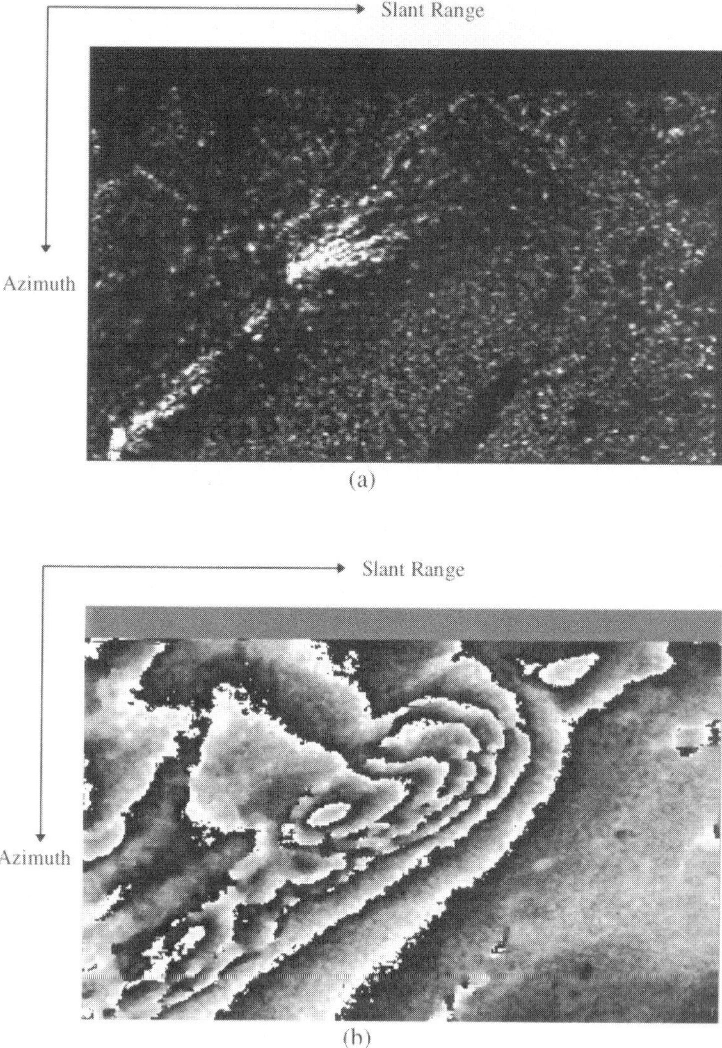

Figure 14.9. Interferogram magnitude (a) and phase (b) corresponding to the ERS-1 InSAR pair shown in Figure 14.3. The interferogram fringes in image (b) may be interpreted to first order as isocontours, meaning lines of equal height in the imaged area. This ERS-1 InSAR pair has B_{perp}=–197 m and B_{para}=–87m. The height sensitivity is $\Delta h/\Delta \phi$ =–53 m/cycle. (© ESA 1994. Imagery is courtesy of the Canada Centre for Remote Sensing, Natural Resources, Canada.)

SAR sensors, ground control points are needed to achieve optimal absolute height accuracy. In contrast, the height change from pixel to pixel is quite accurate even in the presence of significant absolute height errors.

1.7 Mapping Height with Interferometry: Height Sensitivity and Baseline

In the following, we constrain the context to repeat-pass spaceborne interferometric SAR. This is because the height sensitivity varies widely with each InSAR pair of images due to the variability of orbit repeat accuracy of today's spaceborne SAR sensor satellites. Also, the choice of repeat pass spaceborne InSAR image pairs is largely determined by the estimated height sensitivity and the expected phase coherence.

Figure 14.9 contains the interferogram corresponding to the ERS-1 InSAR image pair shown in Figure 14.3. By convention, for repeat-pass interferometry the first SAR image in an InSAR pair of images is called the *master* image, and the second SAR image, acquired in the repeat pass, is called the *slave* image. The slave image is coregistered to the master SAR image and then the master image is multiplied with the complex conjugate of the coregistered slave SAR image. The resulting interferogram may be refined by careful filtering of the image data at several stages in this process (Geudtner et al., 1994). The result is the interferogram shown in Figure 14.9. The interferogram magnitude is simply the multiplication of the two SAR image magnitudes in the InSAR pair. The interferogram phase is the phase difference between corresponding pixels in the two images. The phase contours in Figure 14.9(b) are called *fringes*. Fringes contain height and height change information, as well as artifacts related to the image acquisition and data processing.

Assuming for the moment that all the phase information corresponds only to topography, to a first order of accuracy, we can interpret the fringes as topographic isocontours. This means that a fringe of constant phase indicates ground terrain of constant height. The fringes cycle through 360 degree rotations as the ground surface height varies from one position to another across the image. The relationship between phase cycles of 360 degrees and increase/decrease in height is determined by the imaging geometry of the InSAR pair. The degree to which the height changes with each cycle of phase is the height sensitivity.

The height sensitivity of an InSAR pair is determined by the baseline, the incidence angle, the slant range distance, and the radar wavelength. Referring to Figure 14.10, the mathematical expression for height sensitivity is given as:

$$\frac{\Delta h}{\Delta \phi} = \frac{1}{\left[\frac{2}{\lambda}\frac{B_{perp}}{R_1}\frac{1}{\sin\theta_i}\right]} \quad \text{(meters/phase cycle)} \tag{7}$$

where B_{perp} is the perpendicular baseline, θ_l is the incidence angle, R_l is the slant range distance, and λ is the radar wavelength. The quantity Δh indicates the change in height as we traverse one fringe cycle of phase, $\Delta\phi$, where $\Delta\phi$ has units of fringe cycles (one cycle = one 360 degree rotation). Since the incidence angle and the slant range both change slightly from near range to far range across an image, the height sensitivity will change slightly as well. This effect is quite small and thus to a first order of accuracy for interpreting interferogram phase, we can assume that the height sensitivity across an image is constant. The height sensitivity for the interferogram in Figure 14.9 is $\Delta h/\Delta\phi = -53$ meters/cycle. This negative height sensitivity implies that for every increase in phase of one cycle, the height decreases by approximately 53 meters.

The height sensitivity is inversely proportional to the perpendicular baseline length, B_{perp}. Thus large orbit repeat separations will give high sensitivity to height ($\Delta h/\Delta\phi$ will be small), while small orbit repeat separations will give low height sensitivity ($\Delta h/\Delta\phi$ will be large). Several height sensitivities have been calculated in Table 14.2 for ERS, JERS-1, and RADARSAT spaceborne SAR satellites using typical values for each sensor. The RADARSAT sensor has three operating modes: wide (23–35.5 meter resolutions), standard (20–27.6 meter resolution),

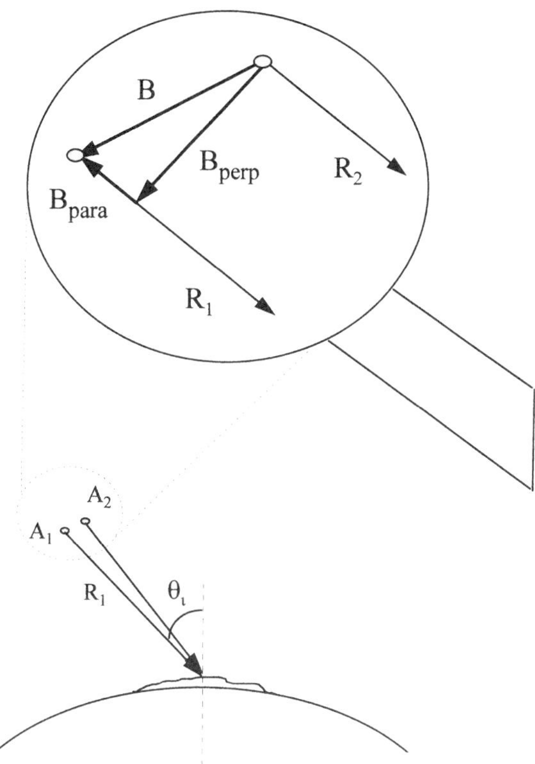

Figure 14.10. InSAR baseline and baseline components. The perpendicular baseline B_{perp}, the incidence angle θ_i, the slant range distance R_1, and the radar wavelength all determine the sensitivity of the resulting interferogram to topographic height.

Table 14.2. Height Sensitivity for Repeat Pass Spaceborne SAR Sensors for Varying Perpendicular Baselines.

Sensor	Wave-length λ (m)	Incidence Angle θ_1 (deg.)	Slant Range R (km)	Height Sensitivity $\Delta h/\Delta\phi$ m/cycle at Perpendicular Baseline, B_{perp} (m)					
				50 m	100 m	200 m	400 m	800 m	1200 m
ERS-1, ERS-2	0.056	25	860	204	102	51	25	13	8.5
JERS-1	0.235	38	715	1034	517	259	129	65	43
RADARSAT	0.056	20	820	157	78	39	20	9.8	6.5
		25	850	201	101	50	25	13	8.4
		30	880	246	123	62	31	15	10
		35	940	302	151	75	38	19	13
		40	1000	360	180	90	45	23	15
		45	1100	436	218	109	54	27	18

and fine (7.8–9.1 meter resolution). In addition, RADARSAT has an electronically steerable antenna allowing variable incidence angles. For this reason, several different scene center incidence angles are calculated for each mode. The RADARSAT standard and fine beams are particularly suitable for interferometry, especially at the larger incidence angles. This is be-

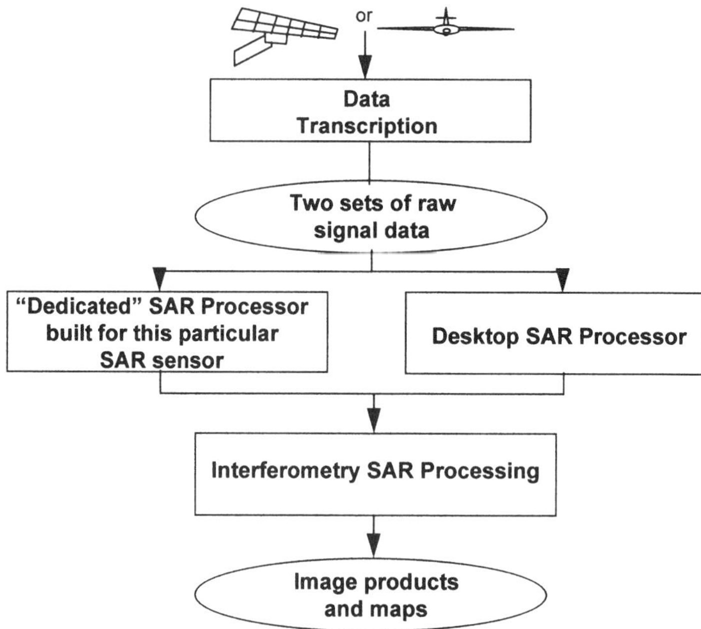

Figure 14.11. Acquired SAR images are received and transcribed into raw signal data sets. These data are SAR processed into complex SAR imagery. The SAR processing may either be done by a dedicated SAR processor at a ground station facility set up for that particular sensor, or it may be done using a SAR processor running on a desktop computer. (The dedicated SAR processors may skip the data transcription step and produce an image directly from the data produced by the sensor.) The interferometric SAR processing is then done taking the two SAR images in an InSAR pair and producing interferometric map products such as topographic maps, land deformation maps, or coherence maps.

cause of the reduced effects of layover with larger incidence angles, plus the improved ability to measure steep slopes (Vachon et al., 1996; Armour et al., 1997).

1.8 InSAR Processing

The flow of data from image acquisition to map production is shown in Figure 14.11. Typically, SAR signal data acquired by either a spaceborne SAR or an airborne SAR is recorded on high density digital tapes. These tapes are either read directly by a SAR processor or transcribed into raw signal data sets. These "raw" data sets are then processed by a SAR processor into complex imagery. The next step is to interferometrically process an InSAR pair of SAR images, converting them to interferometric map products. Several papers exist in the literature giving details of InSAR processing techniques (van der Kooij et al., 1997; Geudtner et al., 1994; Massonnet et al., 1993). In the following paragraphs, a basic outline of InSAR processing is given. The interested reader should refer to the references for additional details and information.

Figure 14.12 gives the major steps for InSAR processing. Most InSAR processing software applications perform these steps in one form or another. The first step is geometric analysis, where the parameters of the imaging geometry are calculated over the extent of the image. The next step is image coregistration, where the slave image is coregistered to the master SAR

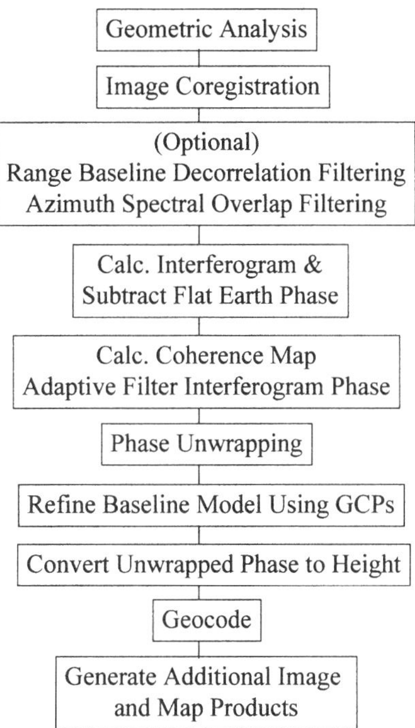

Figure 14.12. This is the basic set of InSAR processing steps that most InSAR processing software performs in one form or another. The application of filtering is not always necessary; however, especially in repeat-pass spaceborne interferometry it can significantly improve the interferogram quality. The use of ground control points is necessary when the platform position and orientation are not known to sufficient accuracy. This is normally the case with current spaceborne SAR sensors (ERS-1, ERS-2, JERS-1, RADARSAT).

image, and where the coregistration accuracy required is typically 1/5th of a pixel or about 1/8th of a resolution cell (Bamler and Just, 1993).

Image coregistration is optionally followed by range and azimuth filtering to remove parts of the image spectrum that do not contribute to information in the interferogram (Prati and Rocca, 1992). Removing these parts improves the overall coherence of the resulting interferogram. Next, the interferogram is generated and the coherence image is calculated. Coherence is simply the complex correlation between the master and the coregistered slave calculated locally over a small, say 5x5 or 10x10, pixel window in the image pair. The coherence tells us the quality of the interferometric phase. Ideally it is high (1.0) everywhere, but areas such as water, layover, and shadow will have poor coherence. Where there is some coherence, say 0.3, it is possible to improve this coherence level by averaging adjacent pixels in order to trade off phase (height) resolution versus horizontal spatial resolution in the image. The degree of filtering is determined by the coherence level. Where coherence is low, more pixels are averaged; where coherence is high, fewer or no pixels are averaged.

Once the interferogram has been enhanced, the phase is unwrapped. The process of phase unwrapping converts the cycles of phase in an interferogram to a monotonic variation in phase across the image (Goldstein et al., 1988; Pritt, 1996). Phase unwrapping is very difficult; however, good phase unwrapping algorithms do exist and this processing step has, to a large extent, been automated. Nevertheless, some manual refinement by a skilled operator is typically required to correct errors in the phase unwrapped image.

It is often necessary to refine the imaging geometry model based on ground control points. This is the case with repeat-pass spaceborne SAR since it is very difficult or impossible to know the baseline length and orientation to sufficient accuracy. The idea is to refine the InSAR pair geometry model such that the InSAR-predicted height and the ground control point heights correspond exactly, or to some least squares error criterion. This refined InSAR geometry is then used to recalculate $\Delta h/\Delta\phi$ over the whole image space.

The next step is to convert the entire phase unwrapped image, which still has phase units (either radians or cycles), into height values with units of meters. This is done by multiplying the unwrapped phase by the $\Delta h/\Delta\phi$ expression. At this point the height of the image is known to within some precision; however, the image is still in a slant range-azimuth projection. If there are large variations in height in the image, then the image will be distorted by layover and foreshortening. Since we have a precise model of the imaging geometry, plus accurate knowledge of the local heights in the imaged area, it is a simple mathematical task to correct these terrain elevation-related distortions and go on to project the height image to a common map projection. Once this is done, the height image may be called a Digital Elevation Model (DEM).

Finally, one must verify the DEM in order to determine its accuracy. This can be done by comparing the DEM with existing maps, or comparing it with ground control points acquired in the field surveys. It is possible to generate additional InSAR image products. Common ones are a geocoded coherence image, geocoded master and slave SAR images, local terrain slope images, and local incidence angle images.

Example

An example of InSAR processing is shown in color Plate 19. The master and slave SAR images are coregistered by resampling the slave image to be in register, pixel-by-pixel, with the master. Next the interferogram is calculated. The coherence image is also calculated and used to direct the phase unwrapping algorithm to some extent.

The unwrapped phase appears as an inverted height image. This is because the phase sensitivity $\Delta h/\Delta\phi = -63$ m/cycle, a negative value. Converting the unwrapped phase to height results in a more regular-looking height image, where elevations range from low (black) to high (white). Terrain correction removes foreshortening and layover effects, and geocoding transforms the image into map coordinates with North oriented directly up along image columns. Correction of the height image for height offsets and height trends may be applied at this point, provided that the orbit modeling of the satellite is accurate enough that the residual height errors are only additive offsets. This is true in this particular example. In cases where the orbit modeling is not sufficiently accurate, and where there are a sufficient number of ground control points at varying heights and positions across the image, baseline refinement modeling may be applied to improve the baseline model for the InSAR pair and obtain better accuracy height estimates across the image.

1.9 Differential Interferometric SAR

In addition to measuring topography, interferometric SAR may be used to measure surface motion, provided that the motion is coherent.[2] Coherence is maintained if the group of scatter-

[2] Remember that coherence is defined as the statistical correlation between the master and slave image magnitude and phase. As long as the phase between the two images is correlated, we can measure terrain heights and terrain horizontal and vertical position change. Our interest in differential interferometry is in land deformation: small motions of groups of scatterers.

ers within one resolution cell do not change significantly from the first pass to the second pass of the SAR sensor, and provided that the scattering mechanism of the individual scatterers does not change significantly between the two passes. If this is true, then it is possible to measure surface motion between the first pass and the second pass of the SAR sensor using repeat-pass interferometry. Since it is quite difficult to achieve repeat-pass airborne interferometry, most repeat-pass interferometry research has been carried out using InSAR pairs acquired by spaceborne SAR platforms. This is because they are particularly well suited to differential interferometry due to their regular scene revisits and well-maintained orbit repeats.

Differential interferometry is achieved by generating an interferogram which measures topography and surface motion, and then subtracting the topography, leaving only the surface motion measurement. This is illustrated in Figure 14.13. Assuming that the satellite orbits repeat exactly, there is no sensitivity to topography. This case is shown in Figure 14.13(a). Now assume that some surface motion has occurred between the image acquisition time of Pass 1 and that of Pass 2. This motion may include vertical shifts, horizontal shifts, or some combination of the two. A phase difference is measured between Passes 1 and 2 because the radar pulse must travel a slightly longer distance to the same group of scatterers on the second pass. The phase difference is simply $\Delta\phi = 2\Delta R/\lambda$ cycles, where ΔR is the change in slant range from Pass 1 to Pass 2, as shown in Figure 14.13(b), and λ is the radar wavelength. Thus one fringe cycle represents a change in slant range of $\Delta R=\lambda/2$. For C-band radars, λ=5.6 cm, and so ΔR=2.8 cm.

Next observe the imaging geometry of Figure 14.13c. In this case, the orbit repeats are not exact and thus there is a nonzero height sensitivity (i.e., $|\Delta h/\Delta\phi| > 0$ m/cycle). In addition, there is still the case of surface motion taking place between Passes 1 and 2. The resulting interferogram will therefore contain the sum of topographic phase and surface motion phase. Assuming a digital elevation model (DEM) is known a priori, it is possible to estimate the topographic phase in the absence of surface motion, and then to subtract this topographic phase contribution from the interferogram, leaving only phase due to surface motion. The a priori DEM that is used to subtract topography may be obtained through conventional elevation mapping techniques such as optical stereo, or through InSAR topographic mapping before the surface motion event, or using a three-pass InSAR technique described in Zebker et al. (1994).

Only the displacement in the slant range direction is imaged by the differential interferogram. The complete three-dimensional displacement vector cannot be measured from a single differential interferogram. Typically, differential interferograms have low coherence and as a result are difficult to phase unwrap. For this reason, the differential interferogram is often directly interpreted to infer the magnitude and direction of surface motion. Figure 14.14 shows two possible ways of interpreting the differential interferogram. If the motion is known through ground truthing, or some other means, to be predominantly horizontal or vertical, then the full displacement vector can be calculated using the incidence angle as shown in the figure. Note that if the displacement is horizontal, then we are only measuring the ground range[3] vector component of the complete horizontal two-dimensional displacement. If it is not known how much of the displacement is vertical and how much is horizontal, then all we can infer is that a certain amount of displacement occurred in the slant range direction. Some other displacement may have occurred in the orthogonal direction, but this cannot be measured in just one differential interferogram. It is possible, however, to calculate a three dimensional displacement vector by summing the displacement vectors derived from an ascending pass differential interferogram and a descending pass differential interferogram.

---

[3] The *ground range vector* is simply the slant range vector projected onto the Earth's surface.

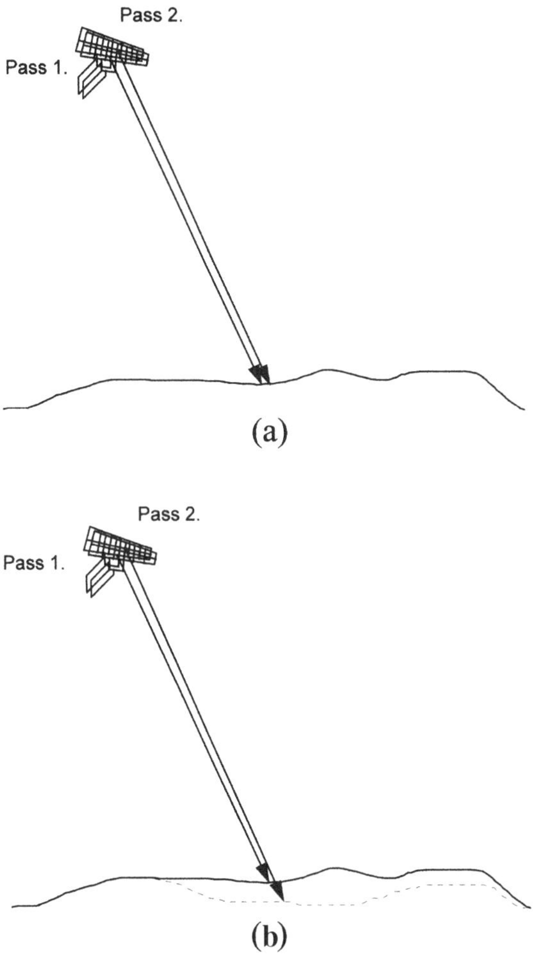

Figure 14.13. (a) The sensor orbit repeats exactly; therefore, the height sensitivity is $\Delta h/\Delta\phi$ = 0 m/cycle. (b) Now assume that the orbits still repeat exactly (i.e., $\Delta h/\Delta\phi$ = 0 m/cycle), but that ground subsidence, for example, has occurred. The surface motion between the time of Pass 1 and the time of Pass 2 will cause a phase difference to be measured of a few wavelengths. In the case of C-band radars, one wavelength is 5.6 cm. This phase difference will appear very clearly as fringes in the interferogram. Note that the surface motion measured is in the range direction only. It is *not* a three-dimensional surface motion measurement.

The processing steps for differential interferometry are almost identical to those for topographic interferometry as shown in Figure 14.12. There is an additional requirement for subtracting the topographic phase. This can be done alongside the subtraction of flat Earth phase when the interferogram is calculated. In cases where four SAR images are used as multiple InSAR pairs, images 1 and 2 can be combined to generate the a priori height data, and then images 3 and 4 can be combined as the differential InSAR pair from which the a priori height data derived from images 1 and 2 is subtracted. Finally, it is possible to combine four coherent SAR images to produce two interferograms, and then difference these interferograms, taking care to first compensate for the different InSAR geometries (differing baselines) used to acquire the interferograms (Zebker et al., 1994). The resulting phase image is the so-called double

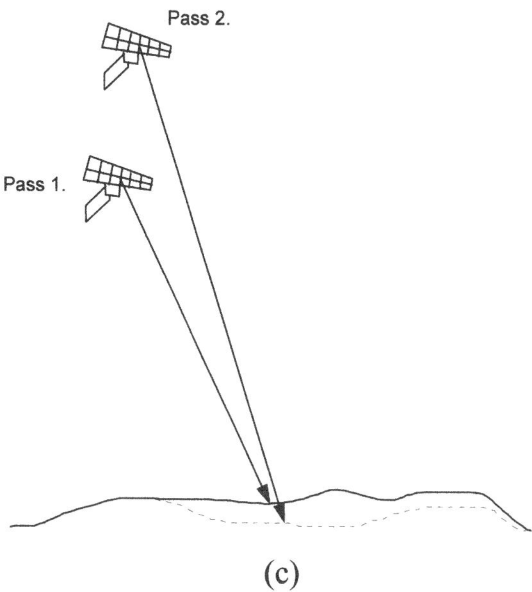

Figure 14.13. (c) In this configuration, the orbits of Pass 1 and Pass 2 do not repeat exactly, but are within a maximum separation for which topographic interferometric measurements are possible. Thus the height sensitivity is now $|\Delta h/\Delta\phi| > 0$. The interferometric measurements will now contain both topographic phase and differential phase. To recover only the differential phase, the topographic phase must be estimated and subtracted. (Please note that these figures are not drawn to scale. The ground surface motions are on the order of centimeters, while the distance from satellite to the earth is on the order of hundreds of kilometers.)

difference interferogram which is independent of topography, and therefore contains only slant range change information.

1.10 Key Requirements for Successful Repeat-Pass SAR Interferometry

There are five key requirements for successful repeat-pass SAR interferometry (Vachon et al., 1996) as follows:

- The rate of change of phase must be adequately sampled. This sets a maximum terrain slope that can be measured for a given height sensitivity. If the heights increase too quickly across an image, and if the height sensitivity is too high, then the resulting fringes in the interferogram are too closely spaced. They are literally one on top of the other in the extreme cases. The result is a loss of coherence and no height or height change information can be measured.
- The pair of SAR images must be acquired such that the phase is correlated. For this to be true, the distance between the satellite on the first (master) pass and the second (slave) pass cannot be too great, or in other words, the perpendicular baseline must not be too large. As this maximum is approached, coherence is increasingly lost. The loss of coherence in this way is called *baseline decorrelation*.
- There must be adequate coherence between the two images in an InSAR pair. Given the potentially large time intervals between orbit repeats (e.g., 1, 3, 5, 176 days for ERS, 44 days for JERS-1, 24 days for RADARSAT), the requirement for adequate coherence sets limits on the type of surface and scope of surface change that may be mapped using interferometry. The loss

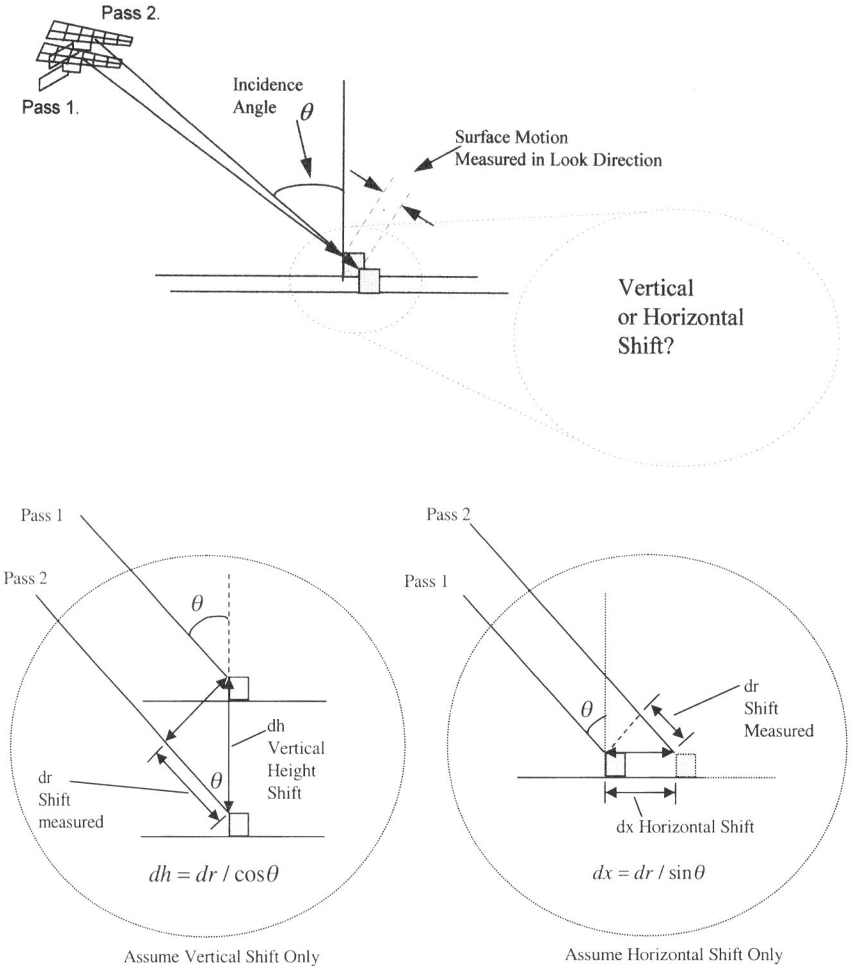

Figure 14.14. Options for interpreting the differential interferogram displacement. The displacement may be mainly vertical or horizontal, or it may be some combination. Assuming that all the displacement is either vertical or horizontal, the full displacement can be calculated as shown. Note that only the ground range component of horizontal shifts are measured. Differential interferometry is not sensitive to displacements in the along-track direction. Therefore, the measured horizontal displacement is in general *not* the total horizontal displacement.

of coherence due to surface change over time is commonly called *temporal decorrelation* (Zebker and Villasenor, 1992).

- Coherence may be lost due to weather (rainfall, snowfall, freezing or thawing, wind causing erosion), random surface movement (erosion, landslides, earthquakes, etc.), vegetation growth, etc. A few generalizations may be made with regard to obtaining good coherence. These are: (1) sparse vegetation cover is better than dense vegetation cover, (2) dry conditions are better than wet, (3) dormant vegetation or off-growing season is better than growing seasons, and (4) longer radar wavelengths are better than shorter.
- The SAR images must have adequately preserved phase information. Thus the SAR processor that converts the reflected radar pulses into complex imagery must not destroy the phase information. This requirement is for the most part a solved problem, given the ready availability of high quality SAR processors in industry.

1.11 Limitations and Problems

Temporal Decorrelation

As was discussed in the previous sections, phase coherence is one of the most difficult problems limiting the applications of repeat pass interferometry. This limitation, however, affects only repeat-pass InSAR, since there is no temporal decorrelation with single-pass SAR such as the SRTM mission (SRTM, 1996). Furthermore, temporal decorrelation is greatly reduced by using longer wavelengths. This has been demonstrated with the L-band JERS-1 SAR which has been used for mapping the land deformation due to the Kobe earthquake. Phase coherence was obtained over a two-year elapsed time between the two passes of the SAR.

The Need for Ground Control Points

Ground control points (GCPs) are used to correct height offsets, height trends, and height scaling errors in interferometrically derived digital elevation models. If very accurate sensor position and orientation is known, then GCPs are not required. This is rarely the case, however, with repeat-pass spaceborne SAR sensors operating in this decade. Errors in sensor position and orientation can be corrected with at least five GCPs. But what if you have fewer than five GCPs? With three GCPs, linear height trends and a constant height offset can be corrected in the DEM. With one GCP, then a global height offset may be corrected. With no GCPs, the DEM will have a global offset from the true absolute heights; however, the relative heights will still be largely accurate.

In the absence of GCPs, a DEM may be corrected using slope information. This is typically done where the slope or lack of slope is known with some degree of certainty. For example, coastlines may be known to have a constant elevation where the shore meets the water. If there is a height variation along the coastline, then this may be corrected by removing a linear height trend from the DEM.

Topographic maps exist today for many parts of the world. These maps have lower horizontal and vertical resolution than the DEMs that can be produced using interferometry. Neverthe less, they are useful for obtaining some height information from which slope and height offset data can be derived. Thus comparison of the InSAR DEM to existing maps is one good way of verifying and correcting the DEM in the absence of high vertical resolution GCPs.

Atmospheric Effects

Atmospheric effects are a major limiting factor for repeat-pass interferometry. Though radar does see through clouds, the propagation time of the radar pulses through air with high humidity is longer than through free space. Even clear days have atmospheric heterogeneities: small pockets of humidity which lead to different propagation times. Massonnet's publications have identified that both tropospheric and ionospheric heterogeneities and change from pass to pass can result in up to three extra fringes in the interferogram. These fringes appear to be similar to ground deformation and may be confused with true ground deformation effects. These "parasitic" fringes, however, may be identified by an informed data interpreter in some cases. One way to isolate atmospheric effects is to make multiple height or height change measurements using data acquired at different times, and then to compare the resulting measurements to separate out the atmospheric effects from the true terrain height or ground surface deformation.

2.0 EARTHQUAKE STUDIES

Each year, earthquakes cause thousands of deaths and billions of dollars in property damage worldwide (Dixon, 1994a). The loss of life and huge costs are expected to grow as populations swell and are increasingly concentrated along coast lines and mountain ranges, which often represent tectonically active plate boundary zones. Though seismic zones have been fairly well mapped worldwide, there has been little advance in our ability to predict when, where, or with what magnitude the next earthquake will strike.

Crustal deformation is a direct manifestation of the processes that lead to earthquakes. Consequently, it is one of the most useful physical measurements that can be made to improve estimates of earthquake potential. Measurement of deformation may be divided into pre-, co-, post-, and inter-seismic categories. Pre-seismic pertains to deformation of many years leading up to an earthquake. Post-seismic includes aftershocks and other visco-elastic effects lasting several years. Inter-seismic refers to the relatively long period of time (thousands of years) between seismic activity (Dixon, 1994a).

Crustal deformation in the pre-seismic and the post-seismic periods is badly undersampled by geodetic networks. Crustal deformation is spatially inhomogeneous, with most of the deformation concentrated along relatively narrow zones. The deformation rate can only be measured with adequate temporal sampling, while better spatial sampling allows short-wavelength features of the deformation field to be measured. SAR data can provide high resolution imagery of earthquake-prone areas, high resolution topography data, and a high-resolution map of co-seismic deformation generated by a large earthquake compared with other observations, but at present lacks to some degree the required temporal sampling due to large variations in orbit repeats.

Current spaceborne SAR satellites experience an occurrence of interferometric SAR image pairs unevenly in time and often one or more months apart. Thus, for immediate and continuous monitoring, an area of accumulating strain may be monitored by deploying a GPS network. Any hint that strain is accumulating faster in one area would be a valuable indicator in selecting target areas for intensive geological and geophysical study. Thus resources can be more effectively deployed.

One such GPS network has been setup and is operational in Japan (Miyazaki et al., 1996, 1997). Ground motion with millimeter to centimeter-level sensitivity is measured daily by 610 GPS receivers (as of April 1996) distributed around the country. Data is collected and disseminated in real time to the various monitoring stations. These data are useful in combination with differential InSAR for providing a calibrated deformation map of large areas. The GPS will provide three-dimensional point measurements with poor spatial sampling, while the differential InSAR provides slant range change measurements at a very dense spatial sampling.

Several successful interferometric SAR studies of coseismic deformation and post-seismic deformation have been completed and published in the literature. Massonnet et al. (1993) and Zebker et al. (1994) have extensively studied the deformation due to the Landers, California earthquake (Massonnet et al., 1993, 1994; Feigl et al., 1994, 1995; Zebker et al., 1994). Massonnet and others have studied the 1993 Eureka Valley, California earthquake. Results for the 1994 Northridge, California earthquake are given in Massonnet et al. (1996) and Murakami et al. (1996). The 1995 Hyogoken-Nanbu Earthquake, Kobe, Japan, has been studied by Tanaka and Nakano (1997), and Ohkura (1997, 1998).

2.1 Case Example: The 1995 Hyogoken-Nanbu Earthquake

After a large earthquake, SAR interferometry can be used to characterize the slip that occurred at the surface, the slip from aftershocks, and the slip distribution depth. However, interpretation of the differential interferogram requires some care, as explained in Section 1.9.

On January 17, 1995 a large earthquake struck the city of Kobe, Japan and the surrounding area. This incident resulted in over 40,000 injuries and 5,000 deaths. The surface deformation due to this earthquake was quite significant, resulting in displacements of over one meter. Using JERS-1 SAR interferometry, this deformation was successfully mapped by Tanaka and Nakano (1997) and Ohkura (1997, 1998). JERS-1 was chosen because the relatively long 23.5 cm L-band wavelength permits long time periods between acquisitions while still achieving useful phase coherence.

Two JERS-1 SAR images were identified and combined interferometrically. The images were acquired October 10, 1993 and March 22, 1995. An external Digital Elevation Model (DEM) from the Geographical Survey Institute, Japan, was used to subtract the topographic phase contribution from the interferogram (see Section 1.9). The result is shown in Plate 20. This image is in a slant range/azimuth projection. Each fringe cycle from red to red bands, represents a 11.8 cm change in slant range distance from the satellite radar antenna to the image pixel location on the ground. The long diagonal fringes at the top-right of the interferogram indicates that significant deformation reaching at least 59 cm in the slant range direction occurred in this area. The curved fringes on the island peninsula show that significant ground deformation reaching at least 106 cm in the slant range direction also occurred in this area.

To explain the observed fringes quantitatively, an elastic dislocation model (Okada, 1985) is used to model the co-seismic surface displacement field (Tanaka and Nakano, 1997). It is specified by nine parameters describing the rectangular patches of rupture on the fault planes: three centroid coordinates X (34.61°N), Y (135.04°E), h (8 km); strike (203°) and dip (86°); along strike length (40 km) and down-dip width (10 km); and the magnitude (2.1 m) and direction (167°) of the slip vector. The results are shown in Figure 14.15. An almost vertical fault plane is modelled running from the northwest coast of Awaji Island to the southwest coast of the mainland near the city of Kobe. Note that the slant range component of the deformation, as shown in Figure 14.15(b) is markedly different in appearance as compared to the total displacement.

Comparing the model results in Figure 14.15(b) with those shown in color Plate 20 indicates that the results from the fairly simple elastic dislocation model largely follow the measured land deformation. The location of the fringes are however slightly displaced, and the magnitude of deformation is somewhat different. If we proceeded by trial and error increasing progressively the complexity of the fault model to fit quantitatively the details of the interferogram information, we would increase the probability that it is closer to reality although the problem has no unique solution. The inversion techniques could be used, but these less easily incorporate widely different data sets such as GPS data and differential interferograms.

2.2 Use of Differential InSAR for Monitoring of Seismic Activity

One of the most important scientific objectives for geologists is to understand the earthquake cycle, including local peculiarities that arise because of different tectonic environments. Since earthquakes do not recur with sufficient frequency to acquire the necessary data in one location, earthquakes must be studied worldwide. Patterns may then be recognized and associated with broad sets of geological and tectonic circumstances, which may apply to other similar areas. Spaceborne SAR interferometry provides (Dixon, 1994a): (1) an ability to identify subtle surface features not easily recognized from the ground; (2) an ability to identify nonhomogeneous surface deformation; and (3) fast and easy access to measurements of remote areas. There is the

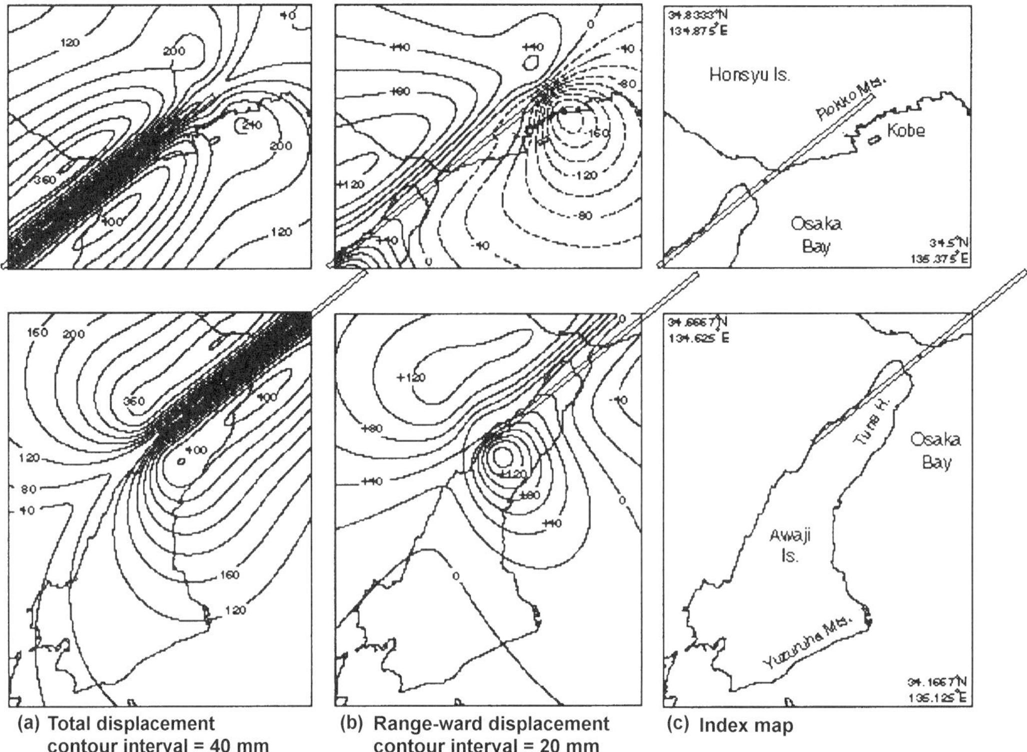

Figure 14.15. (Tanaka and Nakano, 1997) The (a) total, and (b) slant range-ward displacements in millimeters due to the 1995 Hyogoken-Nanbu earthquake calculated by the elastic dislocation model for infinite half space (Okada, 1985). The long diagonal rectangle in the center of the scene indicates the edge of the fault plane used in the model. Diagram (c) indicates the index map of the studied area.

potential to detect blind-thrust structures which are difficult to recognize from surface observations [e.g., the 1994 Northridge, California earthquake (Dixon, 1994a)]. SAR interferometry may in fact be the only way to study coseismic and postseismic activity in remote areas of the world where geodetic measurements are not available.

Though limited results for the detection of secular strain in tectonic regions (preseismic crustal deformation and the associated accumulation of strain) have been achieved using SAR interferometry ERS-1 shows promise in arid regions where the deformation is rapid and concentrated in relatively narrow zones. In contrast, the JERS-1 L-band SAR has shown very good coherence over long time periods of 17 months or more. It is therefore quite likely that preseismic deformation may be measurable with JERS-1.

An envisaged operational use of repeat-pass spaceborne InSAR for earthquake monitoring would be one where a large number of sites are routinely monitored, but with low temporal sampling of up to 3 to 12 months. Revisit time would be in the same season and attempt to coincide with typically dry season conditions. Any areas that are identified as having significant detected deformation would be monitored at a more rapid sampling rate, using every available interferometric pair. If GPS stations are deployed, then surface motion rate of change would also be used to indicate areas to be monitored from space using interferometry.

3.0 VOLCANO MONITORING

Though many volcanoes experience periods of unrest and deformation, it appears well accepted that many of the world's volcanoes that do erupt experience significant preeruption surface deformation (Newhall and Dzurisin, 1988; Thatcher, 1990; Shimada et al., 1990). It is likely, then, that precise monitoring of the world's active volcanoes could lead to accurate predictions of volcanic eruptions (Dixon, 1994b). Repeat-pass differential SAR interferometry provides a safe means of monitoring ground deformation of volcanoes. The potential of SAR interferometry to precisely monitor deformation of potentially hazardous volcanoes around the globe has the potential to warn of impending eruptions, and thus the potential to save lives and property.

Surface deformation may be measured by using leveling, tiltmeters, trilateration, and GPS. SAR interferometry, however, provides a denser spatial sampling, but with slower temporal sampling given current spaceborne SAR orbits. Nevertheless, surface deformation, densely sampled over a large area, and the time evolution of surface deformation, may indicate the nature of the subsurface processes. Source depth may then be estimated using simple models such as an underground point source model of deformation (Dixon, 1994b).

One successful study of surface deformation due to deflation of a volcano was the work by Massonnet et al. (1995) in their study of Mount Etna, Sicily. In this work, they combined ERS-1 C-band SAR pairs acquired in 1992 and 1993 with as long as 13 months between the acquisitions to produce differential interferograms mapping the gradual subsidence of the ground surface as the magma drained away beneath the mountain. The sensitivity of the C-band differential interferometry is impressive, since one fringe in the interferogram represents 28 mm of slant range change. Over a 13-month interval, a change of about 110 mm was measured.

Several problems have arisen in recent years relating to the use of SAR interferometry for the detection of land deformation. Atmospheric effects may show fringe patterns similar to volcanic deformation patterns. These effects can be mitigated by using multiple interferograms of the same area to discriminate deformation due only to atmospheric effects. Poor quality topographic data may also lead to false deformation patterns. An up-to-date reference topographic map may be generated using SAR interferometry, SAR stereo (though to much less accuracy as compared to SAR interferometry), or optical stereo techniques. Alternatively, three-pass interferometry may be directly used to calculate a differential interferogram (Zebker, 1994).

A critical question is whether or not SAR images obtained at different times from heavily vegetated or snow-covered volcanoes can be correlated, thus allowing differential interferograms to be calculated. Results to date indicate that this can be done in arid and semiarid areas over months to years with C-band wavelength data (Massonnet et al., 1995). Many active volcanoes, however, are covered by snow and ice at least seasonally, or else have significant vegetation cover. It is still unknown to what extent multiple-pass interferometry may be done in these areas. Thus the rate of surface decorrelation as a function of volcanic surface cover and wavelength is an important question to be answered.

An example of interferometry in the area of the Mt. Pinatubo volcano in the Philippines is shown in Figure 14.16 and Figure 14.17. This mountain surface cover consists of barren surface cover on the west side and dense vegetation on the east side. The data sets were acquired after the major eruption of the volcano in June 1991. The scene was imaged using ERS-1/ERS-2 Tandem Mode with a very short 24-hour time elapse between image acquisitions. Both ascending and descending InSAR pairs were acquired and processed. Phase coherence could not be obtained on the east side of the mountain due to the changes in scattering behavior of the vegetation cover after this 24-hour period. This is in spite of selecting image acquisitions dur-

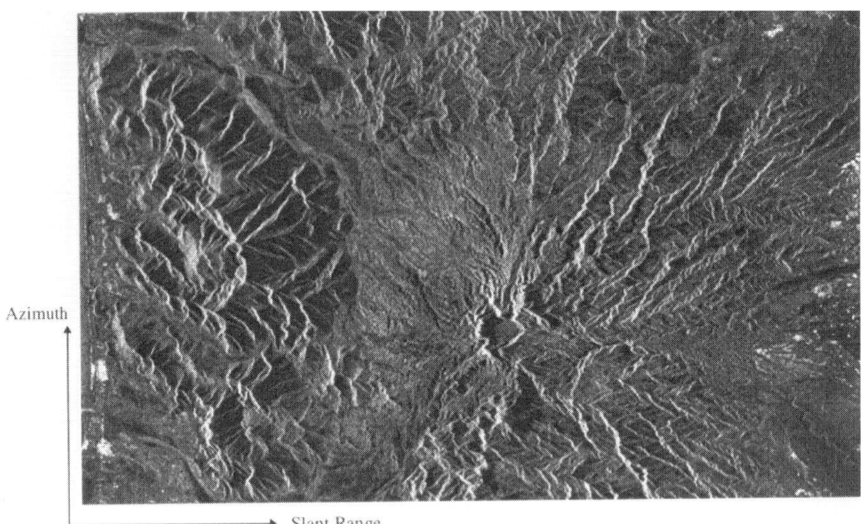

Figure 14.16. Master SAR Image of Mt. Pinatubo in the Philippines. This image was acquired ERS-1 on April 21, 1996. (© ESA 1996.)

ing periods other than the rainy season so that the effects of precipitation on the radar backscattering are at a minimum. Useful coherence is, however, obtained on the west side of the volcano. This area was largely devegetated due to pyroclastic flows and lahars. As a result, the surface is mostly barren and thus retains coherence very well over the 24-hour period between passes of ERS-1 and ERS-2.

Surface decorrelation may be overcome by differencing DEMs obtained by interferometry in cases where the surface deformation exceeds the vertical resolution of the DEM (typically 5 to 20 meters). To avoid deformation phase being confused with topographic phase in the interferogram, the imagery should be acquired simultaneously if possible, or else using tandem mode configurations such as ERS-1/ERS-2 during 1995 and 1996. Another alternative is the deployment of radar corner reflectors; however, this does not give the desired dense spatial sampling for which SAR interferometry is used in the first place.

Thus, use of SAR interferometry for volcano eruption prediction is still being researched, and in particular the decorrelation question is still being explored.

4.0 ENVIRONMENTAL MONITORING

4.1 Topographic Mapping

Spaceborne interferometric SAR has the ability to measure terrain elevation with height accuracies on the order of a few meters and horizontal resolution as small as 10 meters. This three-dimensional resolution capability combined with the ability to map large areas up to 100 kilometers with a few pairs of images makes spaceborne InSAR a cost-effective method for producing up-to-date and accurate topographic maps. A good example of RADARSAT fine beam mode-2 repeat pass interferometry for topographic mapping is shown in Figure 14.18. This scene is approximately 36 by 54 kilometers, and has a horizontal resolution of 10x16 meters to 20x32 meters, depending on the locally-adaptive spatial filtering applied. The verti-

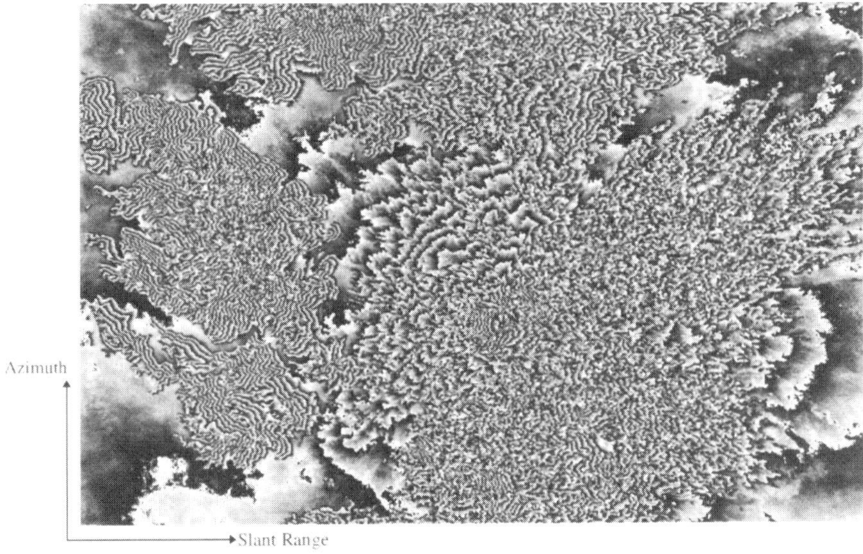

(a)

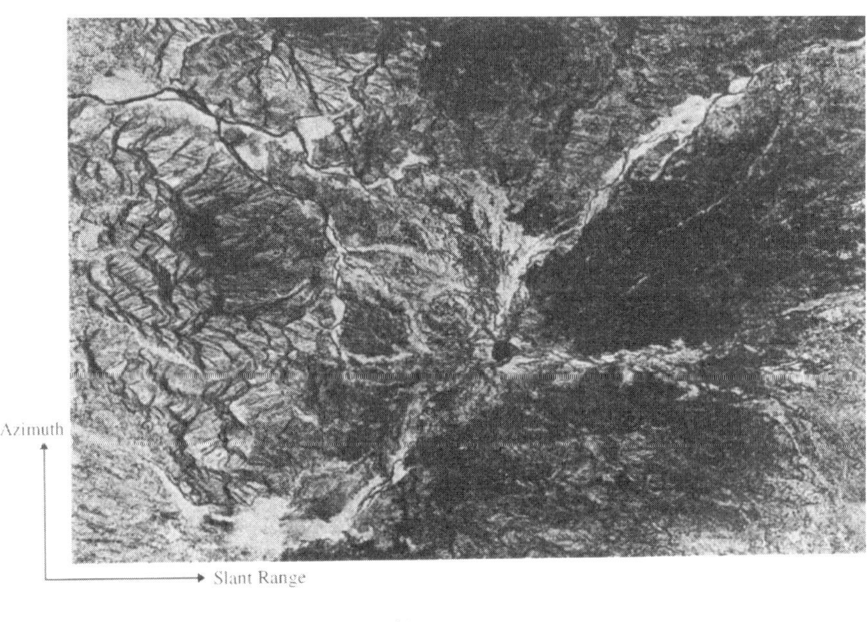

(b)

Figure 14.17. (a) Interferogram phase and (b) coherence image corresponding to the SAR image in Figure 14.16. The slave image was acquired by the ERS-2 sensor one day after the master image was acquired. Observe that even after a 24-hour elapsed period, there are large areas which have low coherence (darker areas), and therefore no interferometric information.

cal height RMS error was measured using a small number of "quality assurance" ground control points and found to be about seven meters.

Sources of repeat-pass InSAR data include ERS-1, ERS-2, JERS-1, and RADARSAT. Most notably, the ERS-1/2 Tandem mode mission over approximately one year, from June 1995 to

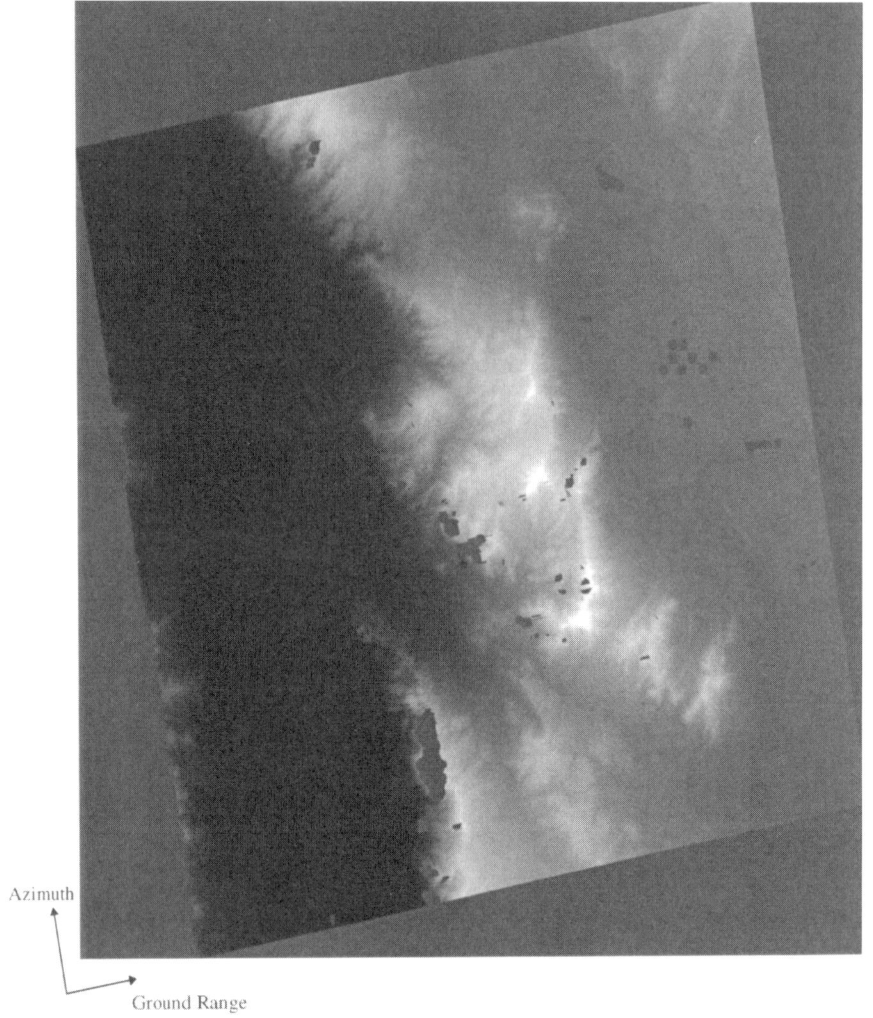

Figure 14.18. DEM using two RADARSAT Fine Beam 2 repeat-pass spaceborne InSAR pair images. The area is Death Valley, California. The height noise is in the range of 1 to 2 meters. The spatial resolution is in the range of 10 to 32 meters, depending on the local adaptive filtering applied. The scene is approximately 36 km wide by 54 km length. (Original SAR imagery is © CSA 1996. Signal data are courtesy of RADARSAT International.)

August 1996, imaged over 90 percent of the Earth's continents with 1-day repeat-pass acquisitions. This was accomplished by arranging the ERS-2 satellite to follow the ERS-1 satellite orbit track with a delay of 24 hours. The result is that most of the data acquired will have good coherence and be useful for interferometry. The exception is over dense rain forest regions where very low coherence is still experienced.

As is the case with all repeat-pass acquisitions from space, atmospheric heterogeneities will contribute additional fringes in the interferogram, which in turn will cause errors in the measured heights. This problem can be reduced by combining multiple InSAR pairs acquired under different atmospheric conditions into one high-confidence height estimate. Another problem is the inevitable gaps and holes that appear in the interferogram, and thus the derived height model. This is due to the effects of layover, shadow, or low backscatter return in areas of the image. Layover and shadow areas can be filled in by acquiring both ascending and descending InSAR pairs of the same image area. An example with JERS-1 repeat-pass interferometry of Mt. Fuji, Japan is given in Figure 14.19. Here, the foreshortening of the mountain slopes on the right-hand side of the peak contribute to small areas of low phase coherence. The low backscatter areas (the dark areas in Figure 14.19a) to the left and right of the peak also results in areas of low coherence due to the poor radar signal to noise ratio in these areas.

4.2 Measurement of Erosion and Landslides

Erosion tends to cause major changes to the surface structure imaged by the radar sensor. As a result, local erosion events such as landslides will entirely destroy phase coherence and differential interferometry is not possible to map the before and after effects due to landslides or other major erosion events (Dixon, 1994c). It is, however, possible to measure the small shifts of large sloping areas which may lead up to a major landslide. This is possible if a large area on the order of hundreds of meters is displaced a small distance as a single coherent entity. This motion may be measured using differential interferometry and related back to the slope stability in that area.

Small landslide detection has been achieved using ERS1/ERS-2 Tandem Mode data acquired of two sites in southern France (Fruneau et al., 1997). This study indicate that InSAR can measure displacements of small areas of landslide with respect to the surrounding stable zone. These results have been verified using laser measurements. If phase coherence is lost due to substantial erosion changing the character of the backscattered phase, then it may still be possible to measure height change, provided that the changes are larger than the absolute height accuracy of the InSAR-derived DEMs. Since repeat-pass spaceborne InSAR will generate height accuracies on the order of five to twenty meters root-mean-square (RMS) in favorable imaging conditions, the height change due to erosion must be substantially larger. The exception is when multiple InSAR pairs of the same area are combined to make one high-accuracy DEM before the change event, and then again after. If the processing is done well, then height accuracies up to two meters RMS can be obtained (Carrasco et al., 1997).

The ability of interferometry to measure relative heights in a scene with very good accuracy of a few meters, implies that local slopes can be mapped accurately. These data may then be used to study the mechanisms of topographic erosion (Dixon, 1994c).

4.3 Tectonic Motion and Ocean Tidal Loading

Tectonic motion measurements using repeat-pass spaceborne differential interferometry has some promise for areas experiencing rapid motion at a rate of a few centimeters per year (Peltzer et al., 1997; Werner et al., 1997). Care in the use of repeat-pass differential InSAR must always be applied, since the atmospheric heterogeneities will contribute extra fringes in the interferogram that are due to changes in the atmospheric water vapor content, and not the ground surface deformation. For this reason, multiple acquisitions forming multiple InSAR pairs are required. The measurements can be combined using a "majority rules" approach to determining the correct level of deformation. In general, long wavelengths (e.g., 23.5 cm, L-band) are better than short (e.g., 5.6 cm, C-band), since temporal decorrelation due to vegetation growth, soil moisture changes, etc. are much less with longer wavelengths. The tradeoff, however, is less sensitivity to deformation, since the sensitivity is equal to exactly half the

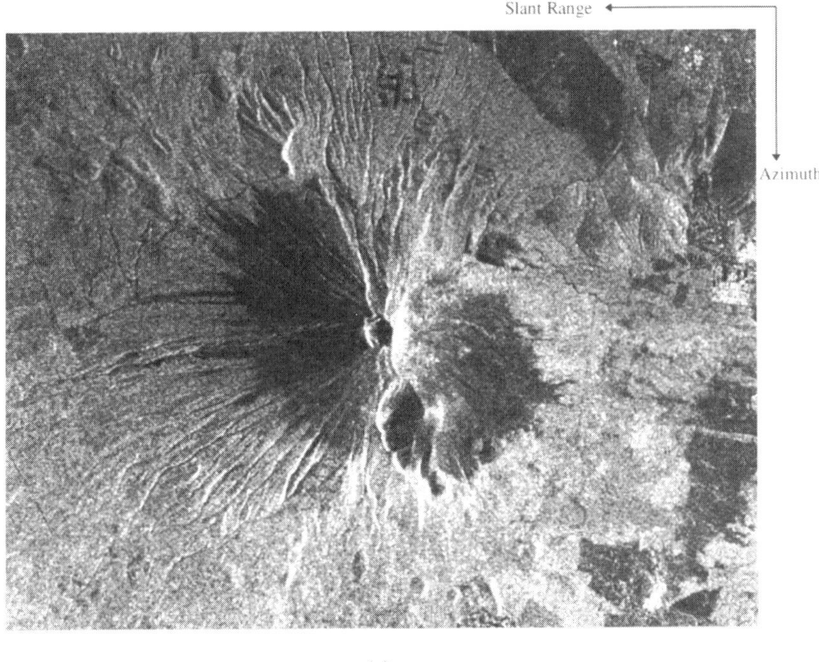

(a)

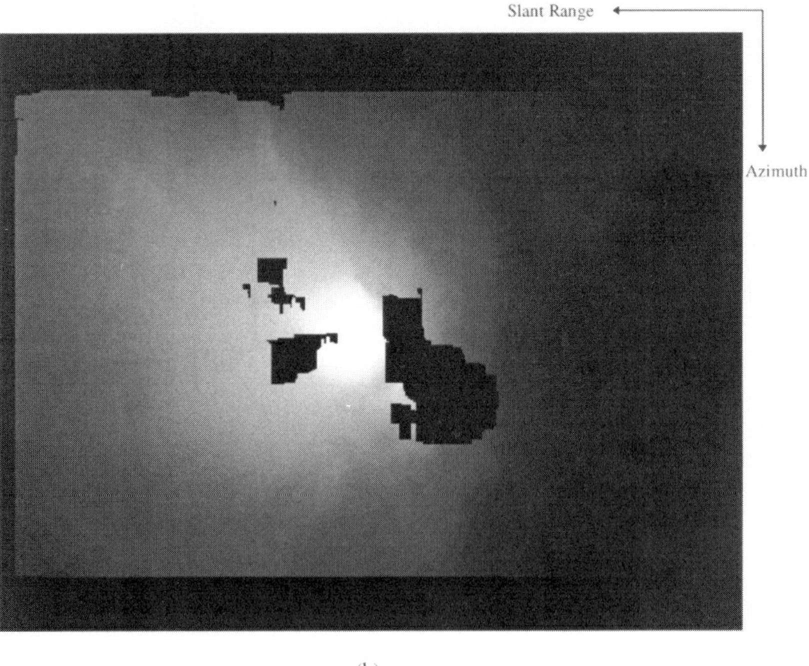

(b)

Figure 14.19. Example of a height image derived from a JERS-1 InSAR pair of Mt. Fuji, Japan. The SAR backscatter image is shown in (a), and (b) the corresponding height image calculated using interferometry. The gaps in the height image due to the foreshortened area and areas of low signal strength in the SAR backscatter image. Special processing may improve this result, but at the expense of both horizontal and vertical resolution. (MITI/NASDA retains ownership of data.)

wavelength of the radar. One approach to improve coherence is to produce a time series of differential interferograms, making a series of time steps of deformation. This is effective provided that each time step is long enough for a measurable amount of deformation to have occurred.

Ocean tidal loading effects result in up to centimeter-level crustal displacement. These motions are on the order of the sensitivity of differential interferometry. Results demonstrating the successful measurement of ocean-tide loading are difficult to obtain due to phase decorrelation and atmospheric effects in repeat-pass acquisitions. Nevertheless, the possibility exists that this technique could be used with L-band SAR interferometry, and additionally corner reflectors could be used to make point measurements.

4.4 Subsidence

Subsidence is a common occurrence in cities and populated areas due to water, natural gas, or oil extraction. Also, mining activities result in significant levels of subsidence which have a physical and economic impact on the surrounding regions. Examples of subsidence due to oil extraction have been published recently showing that it is possible to measure the time progression of subsidence in an area surrounding an oil field (van der Kooij, 1997). This information is useful in understanding the effect of oil extraction on the surface dynamics as well as providing a means of monitoring the deformation rate and adjusting oil extraction activities accordingly. The deformation due to oil extraction is quite localized and centered around each oil well. Thus it is quite easy to differentiate atmospheric effects from surface subsidence. In addition, given the rapid deformation in easily accessible areas, GPS measurements are easily obtained to cross reference against and verify the differential interferometry measurements.

Subsidence due to water extraction and natural gas extraction should also be measurable using differential InSAR, provided that adequate phase coherence is obtained and the effects of atmospheric heterogeneities are controlled. In particular, subsidence of urban areas is a major problem in several cities due to water extraction including Houston, Texas, and Mexico City. Urban areas will tend to have better coherence over long periods of time (months to years) as compared to vegetated areas. Atmospheric effects may be controlled by combining multiple differential interferograms and by obtaining meteorological data for the time of image acquisitions.

5.0 GLACIERS AND ICE SHEETS

The cryosphere consists of ice sheets, ice streams, floating ice shelves, and mountain glaciers. The cryosphere plays a critical role in the Earth's climate, oceans, and atmosphere system. Global warming, for example, could result in ice melting in the Antarctic or Greenland ice sheets and lead to a rise in sea level. For glacier and ice sheet research, topographic data are required. Furthermore, measurement of the rate of change of the elevation data is important for the estimation of how the ice mass is changing (Dixon, 1994d).

SAR interferometry is one technique for mapping remote polar areas with high vertical accuracy provided that the necessary ground control points are available, and adequate coherence can be achieved. The height accuracy requirements are, however, very demanding at less than 1 meter absolute height error (Dixon, 1994d). This is just beyond the capabilities of current spaceborne repeat-pass SAR sensors. Differential InSAR may be used to measure comparatively rapid motion in the ice sheets. It can also be used to differentiate floating ice shelves from grounded ice shelves, and measure ice flow rates (Goldstein et al., 1993).

Mountain glaciers are suspected to play a major role in sea-level change. In particular, mountain glacier wastage is a major factor in sea-level rise. Thus, there is a requirement to map and monitor a large number of mountain glaciers around the world in order to assess the contribution of mountain glacier melting to sea-level rise (Dixon, 1994d). Spaceborne SAR interferometry can meet this requirement; however, improved polar coverage is required.

At present only RADARSAT can image almost the entire North pole, and two short Antarctic phases are being conducted for RADARSAT to map the South Polar regions. The baselines, however, converge toward zero for latitudes approaching the poles. As a result, a large number of orbit repeats will have too small a baseline for accurate topographic maps to be derived using interferometric techniques. On the other hand, small baselines are ideal for surface motion measurements using differential InSAR.

Finally, it remains to be determined what the height accuracy will be over areas with significant snow and ice cover, and how ice motion (differential phase) will be distinguished from topographic phase in the interferogram. Furthermore, how does the height accuracy change with different snow types and ice types? What effect does the radar wavelength have on height accuracy? What is the expected level of coherence for mountain glaciers and ice sheets in the various climates and temperature-related conditions? The answer to these questions is the subject of current research.

6.0 FUTURE TECHNOLOGICAL DEVELOPMENTS

There is clearly a requirement for a spaceborne SAR mission that is specifically designed for single-pass SAR interferometry. Though the ERS-1/ERS-2 Tandem Mode mission was designed for interferometry, it is still a repeat-pass approach, and suffers the problems that repeat-pass interferometry experiences for topographic mapping: temporal decorrelation and atmospheric heterogeneities.

One possible mission scenario has been designed by JPL called TOPSAT (Zebker et al., 1994). This mission concept is to use two L band SAR satellites to image the Earth simultaneously and derive topographic height and height change data. Another mission planned for launch in May 2000 is the Shuttle Radar Topographic Mission (SRTM) (SRTM, 1996). This short mission is planned to use a SCANSAR radar and two antennas to acquire C-band single-pass InSAR data of the entire Earth land mass from latitudes of –57 degrees South to +60 degrees North. The sensor is intended to be flown onboard the NASA Space Shuttle and use a 60-meter-long boom to separate the second antenna from the first. High accuracy GPS and attitude measurements are to be used to obtain height accuracies with errors less than 16 meters, without the requirement for ground control points. Should this mission go ahead, a large percentage of the Earth's land surface will be mapped to unprecedented height accuracies.

It is clear from the volume of research in labs around the world that interferometric SAR applications will continue to be explored and better understood. For example, the use of coherence data to map flooding and post-flood damage assessment is an emerging new application for spaceborne SAR interferometry. We stress that interferometry provides an *additional* source of information for understanding the Earth's processes and mechanisms. Interferometry is not an answer in itself in many cases, but provides a rich source of data which, when used with care, can be combined with other measurements to provide solutions to problems in natural hazard monitoring and prediction, climate change, and environmental change.

ACKNOWLEDGMENTS

The authors wish to thank Dr. Paris Vachon, Canada Centre for Remote Sensing, Natural Resources, Canada, for his review of the section *Basic Principles of Interferometric SAR*, and his contribution to the *Glaciers and Ice Sheets* section.

REFERENCES

Armour, B., J. Ehrismann, M. van der Kooij, and S. Sato. New software technology for repeat pass spaceborne interferometric SAR (InSAR) processing. *Proceedings of the International Symposium Geomatics in the ERA of RADARSAT, Ottawa, Canada,* Paper No. 188, 1997.

Armour, B., P.W. Vachon, and P. Farris-Manning. Suitability of RADARSAT's orbit for repeat pass SAR interferometry. *Proceedings of the International Symposium Geomatics in the ERA of RADARSAT, Ottawa, Canada,* Paper No. 182, 1997.

Attena, E.P.W. The Active Microwave Instrument on Board the ERS-1 Satellite. *Proceedings of the IEEE.* 79, 791–799, 1991.

Bamler, R. and D. Just. Phase statistics and decorrelation in SAR interferograms. *Proceedings of IGARSS '93.* 1, 980–984, 1993.

Carrasco, D., J. Diaz, and A. Broquetas. Ascending-descending orbit combination SAR interferometry assessment. *3rd ERS Symposium 'Space at the Service of our Environment,'* 1997.

Churchill, R.V. and J.W. Brown. *Complex Variables and Application*, McGraw-Hill, New York, 1984.

Curlander, J.C. and R.N. McDonough. *Synthetic Aperture Radar Systems and Signal Processing*, John Wiley & Sons, New York, 1991.

Dixon, T.H., Ed. Earthquake Studies, In: *SAR Interferometry and Surface change Detection.* T.H. Dixon, Ed., Boulder, CO, February 3–4, 1994a.

Dixon, T.H., Ed., Global Volcanic Monitoring, In: *SAR Interferometry and Surface Change Detection,* T.H. Dixon, Ed., Boulder, CO, February 3–4, 1994b.

Dixon, T.H., Ed., Topography, Tectonics, and Erosion, In: *SAR Interferometry and Surface Change Detection,* T.H. Dixon, Ed., Boulder, CO, February 3–4, 1994c.

Dixon, T.H., Ed., Glaciers and Ice Sheets, In: *SAR Interferometry and Surface Change Detection,* T.H. Dixon, Ed., Boulder, CO, February 3–4, 1994d.

Elachi, C. *Spaceborne Radar Remote Sensing: Applications and Techniques.* IEEE Press, New York, 1988.

Feigl, K.L., G. Peltzer, and K.W. Hudnut. Slip distribution during the June 28 Landers-Big Bear earthquake sequence estimated from coseismic radar interferograms. *J. Geophys. Res.,* 1995.

Feigl, K.L., A. Sergent, and D. Jacq. Estimation of an earthquake focal mechanism from a satellite radar interferogram: Application to the December 4, 1992 Landers aftershock. *Geophys. Res. Lett.*, 1994.

Fruneau, B., C. Delacourt, J. Achache, and C. Carnec. Landslide monitoring in south of France with tandem data. *3rd ERS Symposium 'Space at the Service of our Environment,'* 1997.

Geudtner, D., M. Schwabisch, and R. Winter. SAR-interferometry with ERS-1 data. Proceedings of PIERS '94, 1994.

Goldstein, R.M., E. Englhardt, B. Kamb, and R.M. Frolich. Satellite radar interferometry for monitoring ice sheet motion: application to an antarctic ice Stream. *Science*, 262, 1525–1530, 1993.

Goldstein, R.M., H.A. Zebker, and C. Werner. Satellite radar interferometry: Two-dimensional phase unwrapping. *Radioscience*. 23, 713–720, 1988.

Graham, L.C. Synthetic interferometer radar for topographic mapping. *Proceedings of the IEEE.* 62, 763–768, 1974.

Gray, A.L. and P.J. Farris-Manning. Repeat-pass interferometry with airborne synthetic aperture radar. *Geoscience and Remote Sensing*, 180–191, 1993.

Inanaga, A., T. Yamanokuchi, S. Tanaka, and M. Ono. JERS-1/SAR interferograms of the Great Hanshin earthquake. *Int. J. Remote Sensing,* 18, 1649–1656, 1997.

Li, F.K. and R.M. Goldstein. Studies of multibaseline spaceborne interferometric synthetic aperture radars. *IEEE Trans. Geoscience Remote Sensing.* 28, 88–97, 1990.

Massonnet, D., K. Feigl, M. Rossi, and F. Adragna. Radar interferometric mapping of deformation in the year after the Landers earthquake. *Nature*, 369, 227–230, 1994.

Massonnet, D., K.L. Feigl, H. Vadon, and M. Rossi. Coseismic deformation field of the M=6.7 Northridge, California earthquake January 17 1994 recorded by two radar satellites using interferometry. *Geophys. Res. Lett.,* 23, 969–972, 1996.

Massonnet, D., M. Rossi, C. Carmona, F. Adragna, G. Peltzer, K. Feigl, and T. Rabaute. The displacement field of the Landers earthquake mapped by radar interferometry. *Nature,* 364, 138–142, 1993.

Massonnet, D., P. Briole, and A. Arnaud. Deflation of Mount Etna monitored by spaceborne radar interferometry. *Nature,* 375, 567–570, 1995.

Massonnet, D. Satellite radar interferometry. *Scientific American,* 276(2) , 46–53, 1997.

Miyazaki, S., H. Tsuji, Y. Hatanaka, Y. Abe, A. Yoshimura, K. Kameda, K. Kibayashi, H. Morishita, and J. Iimura. Establishment of the nationwide GPS array (GRAPES) and its initial results on the crustal deformation of Japan. *Bulletin of Geographical Survey Institute,* 42, 27–41, 1996.

Miyazaki, S., T. Saito, M. Sasaki, Y. Hatanaka, and Y. Iimura. Explanation of GIS's nationwide GPS array. *Bulletin of Geographical Survey Institute,* 43, 23–34, 1997.

Murakami, M., M. Tosita, S. Fujiwara, T. Saito, and S. Masaharu. Coseismic crustal deformation of 1994 Northridge, California, earthquake detected by interferometric JERS–1 synthetic aperture radar. *J. Geophys. Res.,* 101, 8605–8614, 1996.

Nemoto, Y., H. Nishino, M. Ono, H. Mizutamari, K. Nishikawa, and K. Tanaka. Japanese earth resources satellite-1 synthetic aperture radar. *Proceedings of the IEEE.* 79, 800–809, 1991.

Newhall, C.G. and D. Dzurisin. Historical unrest at large calderas of the world. *U.S. Geol. Surv. Bull., 1855*, 1988.

Ohkura, H. Applications of SAR data to monitoring of earthquake disasters. *Advances in Space Research*, 19, 1429–1436, 1997.

Ohkura, H. Applications of SAR data to earth surface changes and displacement. *Advances in Space Research*, 21, 485–492, 1998.

Okada, Y. Surface deformation due to shear and tensile faults in a half-space. *Bull. Seism. Soc. Am.,* 75, 1134–1154, 1985.

Peltzer, G., F. Rogez, P. Rosen, and K. Hudn. Crustal deformation in Southern California using SAR interferometry. *3rd ERS Symposium 'Space at the Service of our Environment,'* 1997.

Prati, C. and F. Rocca. Range Resolution Enhancement with Multiple SAR Surveys Combination. *Proceedings of IGARSS '92.* 1576–1578, 1992.

Pritt, M.D. Phase unwrapping by means of multigrid techniques for interferometric SAR. *IEEE Trans. Geoscience Remote Sensing.* 34, 728–738, 1996.

Raney, R.K., A.P. Luscombe, E.J. Langham, and S. Ahmed. RADARSAT. *Proceedings of the IEEE.* 79, 839–849, 1991.

Shimada, S., Y. Fujinawa, S. Sekiguchi, S. Ohmi, T. Eguchi, and Y. Okada. Detection of a volcanic fracture using GPS. *Nature*, 343, 631–633, 1990.

SRTM, Future Topographic Radar Mission Will Map 80% of the Earth, #96512 Public Information Office, Jet Propulsion Laboratory, 1996.

Tanaka, A. and T. Nakano. Some problems on displacement field mapped by SAR interferometry. *J. Seismol. Soc. Japan*, 50, 88–99, 1997.

Thacher, W. Precursors to eruption. *Nature,* 343, 550–591, 1990.

Vachon, P.W., D. Geudtner, A.L. Gray, and R. Tousi. ERS-1 Synthetic Aperture Radar Repeat-Pass Interferometry Studies, Implications for RADARSAT, *Can. J. Remote Sensing,* 21, 441–454, 1996.

van der Kooij, M. Use of ERS SAR data to measure land subsidence caused by oil exploitation. *3rd ERS Symposium 'Space at the Service of our Environment,'* 1997.

Way, J.B. and E.A. Smith. The evolution of synthetic aperture radar systems and their progression to the EOS SAR. *IEEE Trans. Geoscience Remote Sensing,* 29, 962–985, 1991.

Werner, L., P. Rosen, S. Hensley, E. Fielding, E. Chapin, S. Buckley, and P. Vincent. Detection of aseismic creep along the San Andreas fault near Parkefield, California with ERS-1 radar interferometry. *3rd ERS Symposium 'Space at the Service of our Environment,'* 1997.

Zebker, H.A. and J. Villasenor. Decorrelation in Interferometric Radar Echoes. *IEEE Trans. Geoscience Remote Sensing,* 30, 950–959, 1992.

Zebker, H.A., P. Rosen, R.M. Goldstein, A. Gabriel, and C. Werner. On the derivation of coseismic displacement fields using differential radar interferometry: the Lander earthquake. *J. Geophys. Res.* 99, 19,617–19,634, 1994.

Zebker, H.A. and R.M. Goldstein. Topographic mapping from interferometric synthetic aperture radar observations. *J. Geophys. Res.,* 91, 4993–4999, 1986.

Zebker, H.A., T.G. Farr, R.P. Salazar, and T.H. Dixon. Mapping the world's topography using radar interferometry: the TOPSAT mission. *Proceedings of the IEEE,* 82, 1774–1786, 1994.

CHAPTER 15

Sampling Systems for Change Detection Accuracy Assessment

Gregory S. Biging, David R. Colby, and Russell G. Congalton

1.0 INTRODUCTION

Assessing the accuracy of a one-point-in-time (OPIT) thematic map derived from remotely sensed data is a complex, but achievable task. Assessing the accuracy of a change map is significantly more difficult and requires extensive consideration and planning. Figure 15.1 presents an overview of the many factors that can contribute to the error in any change detection project. Failure to deal effectively with these factors dooms the project to certain failure.

In this chapter we present a suite of statistical sampling designs and analysis techniques which can be utilized alone, or in combination, in assessing the accuracy of a change detection map. While there are many similarities between the statistical techniques that we employ in accuracy assessment of change detection, there are also significant differences from those used for assessing the accuracy of an OPIT thematic map. These differences and similarities will be discussed in more detail later in this chapter. For an overview of state-of-the art for OPIT accuracy assessment see Congalton (1991), Congalton and Green (1998), and Stehman and Czaplewski (1998).

There are a number of statistics that users of remote sensing products are potentially concerned with (Stehman, 1995). Stehman (1997) examines various accuracy measures and their interpretation. These statistics are derived from the sample error matrix (Table 15.1a and Figure 15.2). We utilize Stehman's (1997) notation. The accuracy measures include the overall accuracy measure (P_c), which is the proportion of correctly classified polygons (pixels) for which there are reference data, the observed proportions of polygons (pixels) $p_{ij,}$ classified as category i and in reference category j (for i=1,...,q and j=1,...,q), the row marginals ($P_{Ui}= p_{ii}/p_{i+}$: the User's accuracy for class i), the column marginals ($P_{Aj}= p_{jj}/p_{+j}$: the Producer's accuracy for class j), and the kappa coefficient of agreement (κ) which adjusts P_c for chance agreement. The p_{ij} are computed from the sample data and serve as estimates of the accuracy parameters. Story and Congalton (1986) document the error matrix and provide useful interpretations of both producer's and user's accuracies (Table 15.1a and Figure 15.2).

Change maps are an increasingly important product to a variety of professions for monitoring environmental and land cover/use change. We focus on how to determine the accuracy of these maps. Knowledge of the accuracy of a change map is a fundamental requisite in using the map to make appropriate environmental policy and management decisions. All the cost and logistic considerations that are present for accuracy assessment of an OPIT thematic map are compounded for a change map (Figure 15.1).

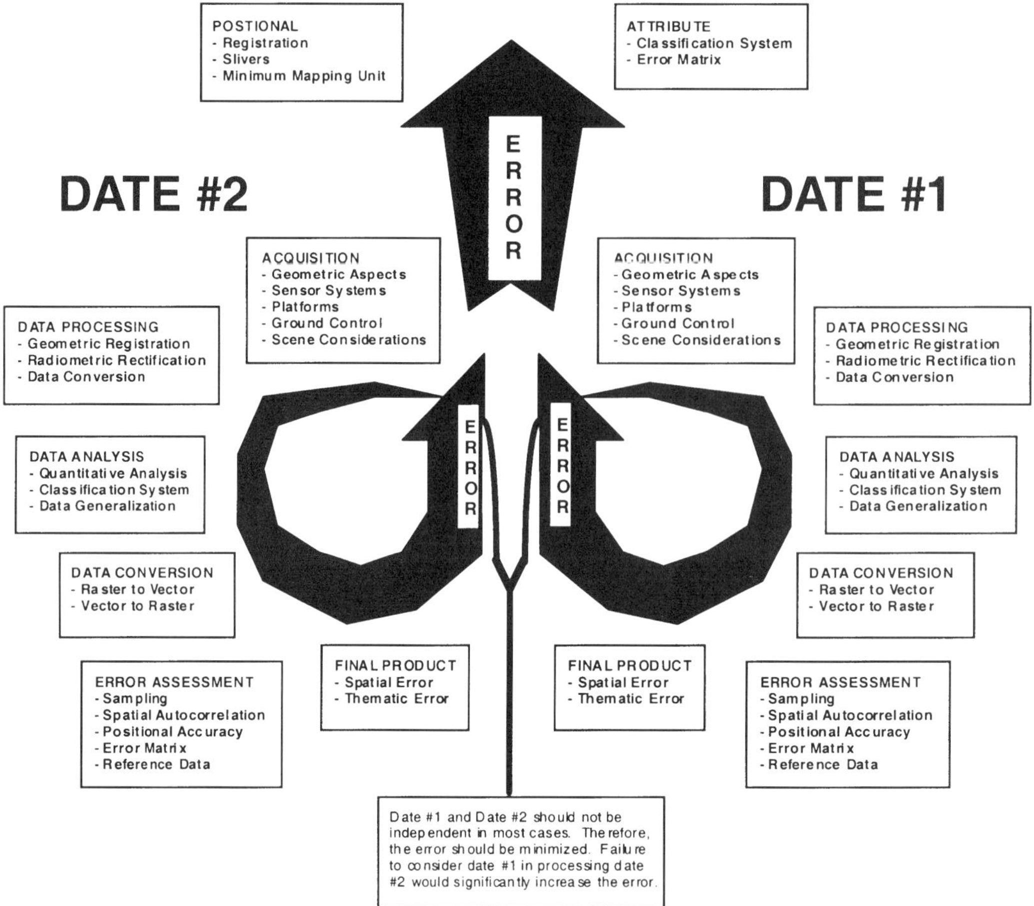

Figure 15.1. Sources of error using remotely sensed data for change detection mapping (modification of Lunetta et al., 1991).

In this chapter we are primarily concerned with assessing the accuracy of landscape change detection products at various scales. We generally assume that we are assessing the accuracy of a change map for a land cover/land use (LCLU) based classification such as the USGS Land Cover/Land Use Classification system (Anderson et al., 1976) or the NOAA Coastal Change Analysis Program (C-CAP) scheme (Dobson et al., 1995). Later in the chapter, however, we also consider approaches for cases in which the change variable is not a class, but instead is a continuous variable such as biomass or leaf area index (LAI).

We primarily discuss the methods for assessing the accuracy of a polygon change map. Alternatively, a pixel based assessment could be pursued. While there are some statistical and conceptual advantages to conducting a pixel based accuracy assessment, the current state-of-the-art does not allow us to consistently locate a pixel with sufficient precision for accuracy assessment. Even with geometric rectification algorithms, a 30 meter pixel can currently only be georectified with a RMS error of $\leq \pm 0.5$ pixel (Khorram et al., 1998).

Polygon and pixel based maps lend themselves to somewhat different statistical methods for accuracy assessment of change detection. Although we concentrate on the polygon based mapping methods, we make observations concerning the pixel based methods as well. Throughout

Table 15.1a. An OPIT Population Error Matrix. (P_{ij} is the proportion of area in the *LCLU* class i and reference *LCLU* class j. P_c is the overall proportion of area correctly classified.)

Reference Data

CLASSIFIED DATA	1	2	...	q	Row Proportion	User's Accuracy
1	p_{11}	p_{12}	...	p_{1q}	$p_{1+}=\sum_{j=1}^{q} p_{1j}$	$P_{U1}=\frac{p_{11}}{p_{1+}}$
2	p_{21}	p_{22}	...	p_{2q}	$p_{2+}=\sum_{j=1}^{q} p_{2j}$	$P_{U2}=\frac{p_{22}}{p_{2+}}$
.	.	.	...	.	.	.
.	.	.	...	.	.	.
.	.	.	...	.	.	.
q	p_{q1}	p_{+2}	...	p_{qq}	$p_{q+}=\sum_{j=1}^{q} p_{qj}$	$P_{Uq}=\frac{p_{qq}}{p_{q+}}$
Column proportion	$p_{+1}=\sum_{i=1}^{q} p_{i1}$	$p_{+2}=\sum_{i=1}^{q} p_{i2}$	...	$p_{+q}=\sum_{i=1}^{q} p_{iq}$	$P_C=\sum_{j=1}^{q} p_{jj}$ overall accuracy	
Producer's accuracy	$P_{A1}=\frac{p_{11}}{p_{+1}}$	$P_{A2}=\frac{p_{22}}{p_{+2}}$	...	$P_{Aq}=\frac{p_{qq}}{p_{+q}}$		

Table 15.1b. A Comparison Between a Single Date (*OPIT*) Error Matrix and the Equivalent Change Detection Error Matrix.

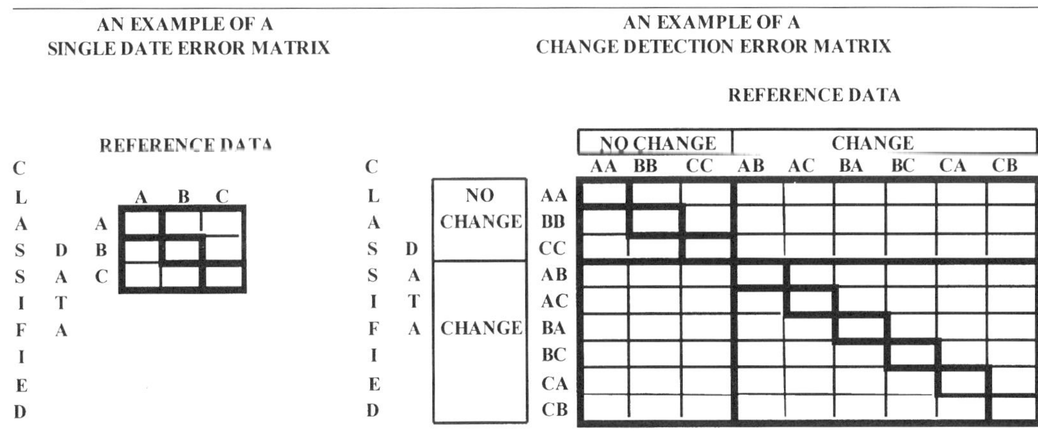

this chapter we denote "polygons (pixels)" to mean that the discussion applies to both the polygon and pixel classification model. In general, these approaches are not interchangeable. Many ecological and LCLU classification schemes provide for a minimum mapping unit (MMU) often 5–10 acres in size. This means that within a class we expect inclusions of other elements whose size is below the MMU. Thus a polygon need not be pure to be classified as a given type. For example, a forest polygon may include a small meadow, or rock outcropping. For a polygon-based mapping exercise it does not make sense to conduct a pixel-based accuracy assessment because it would tend to underestimate map accuracy. And, if the pixels cannot be accurately

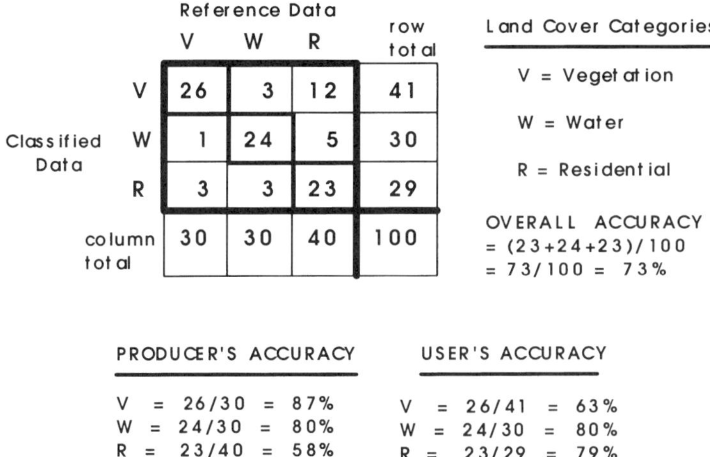

Figure 15.2. A typical one-point-in-time (*OPIT*) sample error matrix from a simple random sample (SRS) design.

located, the underestimation of accuracy would be even more pronounced. Conversely, if in the less common case that the MMU *is* the pixel, then generalizing to polygons tends to inflate the estimated accuracy of the remote sensing product.

Although some classifications allow mixed categories, the basic paradigm that we assume is that each location on the map belongs to one category only. However, techniques for classifying mixed pixels include fuzzy classification in which pixels may have partial membership in several classes (e.g., Wang, 1990; Jensen, 1996) and using spectral unmixing models to estimate the proportions of classes contained within a pixel [see Gong et al. (1994)].

There are many ancillary issues surrounding accuracy assessment that will not be addressed in this chapter, but which are important in their own right. These include validation of historical data, the effect of the spatial scale of the remote sensing system used to assess a landscape, merging data of different spectral resolutions, using standardized field forms, errors in photointerpretation, and boundary location error of polygons. For an overview and discussion of some of these issues refer to Khorram et al. (1998).

Finally, we should note that ideally for a LCLU change analysis one would select and document the LCLU at a substantial number of randomly or systematically[1] selected locations at the beginning of the study. These sites would be revisited at the end of the time interval and their LCLU again documented. These data then would provide the essential reference database for rigorously assessing the accuracy of estimates of change from one category to another, based upon remote sensing. Unfortunately, this logical scenario is generally not feasible, as a hypothetical example will illustrate.

Suppose the change analysis is to be performed on a scene encompassing one million pixels. Next, suppose that cost considerations are ignored and one percent of the pixels (10,000) are

[1] A disadvantage of systematic sampling is that it does not, for example, allow unbiased estimation of the variance of overall accuracy $V(\hat{p}_c)$. In addition, the formula for $V(\hat{p}_c)$ under systematic sampling includes a term for correlation between pairs of units in the sample. Autocorrelation can be persistent. For example, in examining difference images (classified - reference) in agricultural, rangeland, and forest environments, Congalton (1988) reported significant autocorrelation from 30 pixels away. Clearly, the sampling interval should be carefully selected to minimize $V(\hat{p}_c)$.

selected at random or systematically at the beginning of the study to serve as reference sites. Next, suppose that five percent of the one million pixels (50,000) do in fact change over the time interval of interest. Now, let's ask the question: How many of the 10,000 reference pixels can be expected to have been among those 50,000 that have changed? If we assume that the changes occur at random across the landscape, then the answer is 500 pixels. Thus, after a prodigious, costly effort "to do things right," we would have only 500 data for use in rigorously assessing the accuracy of our estimates of change from remotely sensed data. If we had only 10 LCLU classes, we would have 100 potential from-to change classes, and thus on average only five reference sites per class! And since it is difficult to imagine what need would justify the expenditure required to visit the 10,000 reference pixels twice, one shouldn't be surprised that the "ideal" change analysis has not been accomplished, nor is it likely to be. Thus our necessary resort here to other approaches to accuracy assessment that entail sampling elements of the change map.

2.0 SAMPLING METHODS

A sampling strategy consists of both a sampling design and an estimator (Schreuder et al., 1993; Stehman, 1996b). A sampling design is a method for selecting sampling units from a population, whereas an estimator is a mathematical formula for computing an estimate of a statistical parameter of interest from the data in the sample. It is the hallmark of modern statistical survey sampling to utilize probability sampling designs to collect data in a scientifically valid process. With probability sampling each sample unit (polygon or pixel) in the population has a nonzero known probability of selection and units are selected by a random process until the desired sample size is reached. This paradigm allows construction of confidence intervals for the parameters of interest and elimination of bias inherent in nonprobability sampling.

Statistical designs are usually chosen on the basis of efficiency, which means that we select estimators having lower expected variances than those expected for simple random sampling (SRS) for a given level of sampling effort. The ratio of the expected variance from a more complicated design to that expected for SRS with the same sample size is called the design effect (Kish, 1965). The design effect is used in sample size calculations and in evaluating the efficiency of more complex designs (Cochran, 1977).

2.1 Change Detection

Sampling to determine the accuracy of a change map is inherently different from sampling for accuracy assessment of an OPIT thematic map because the change categories usually represent a very small portion of the change image, or thematic map. As such they can be considered rare classes. While there can be classes of the OPIT thematic map which have low or even rare frequency, a change map often has many (change) classes with low or rare frequencies of occurrence.

Traditional sampling techniques often prove to be expensive and inefficient for change detection accuracy assessment. One reason for this is the substantial cost of establishing "permanent" sample locations that can be revisited through time to provide reference documentation of change. In addition, the optimal allocation of samples for each of the change categories would be different than the optimal allocation of samples to the original thematic classes, and additional sample points would be needed over the original base of "permanent" sample points. A second reason is that the change polygons may cover only a small portion of the original image and would not be well detected with, say, simple random sampling or systematic sampling unless the sampling intensity was very high. On the other hand, recent developments in sampling theory offer real promise for their useful application in accuracy assessment of change detection maps.

2.2 The Change Detection Error Matrix

Table 15.1b presents a comparison between an OPIT error matrix and the corresponding error matrix for change detection (Macleod et al., 1995; Macleod and Congalton, 1998). The table shows an OPIT error matrix for three vegetation/land cover categories (A, B, and C). The matrix is of dimension 3 × 3. One axis represents the three categories as derived from the remotely sensed classification and the other axis shows the three categories identified from the reference data. The major diagonal of this matrix indicates correct classification. In other words, when the classification indicates the category was A and the reference data agrees that it is A, then the [A, A] cell in the matrix is tallied. The same logic follows for the other categories B and C. Off-diagonal elements in the matrix indicate the different possibilities of error.

Table 15.1b also shows a change detection error matrix generated for the same three vegetation/land cover categories (A, B, and C). Note, however, that the matrix is no longer of dimension 3 × 3 but rather 9 × 9. This is because we are no longer looking at a single classification but rather a change between two different classifications generated at different times. In an OPIT error matrix there is one row and column for each map category. However, the dimensions of a change detection error matrix are the square of the number of categories. The question of interest in change detection is, "What category was this polygon (pixel) at time 1 and what is it at time 2?" The answer has nine possible outcomes for each dimension of the matrix (A at time 1 and A at time 2, A at time 1 and B at time 2, A at time 1 and C at time 2, ..., C at time 1 and C at time 2), all of which are indicated in the error matrix. It is then important to note what the remotely sensed data said about the change and compare it to what the reference data indicate. This comparison uses the exact same logic as for the OPIT error matrix, it is just complicated by the two time periods (i.e., the change). The major diagonal indicates correct classification while the off-diagonal elements indicate misclassification errors or errors of commission or omission. Commission error occurs when detected change is false; i.e., "false positive." Omission error occurs when change is undetected; i.e., "false negative."

Before designing sampling strategies for accuracy assessment it is appropriate to consider the conceptual framework of the types of error that can arise in a post-classification comparison of change detection. The general strategy is documented in Khorram et al. (1998). As in Table 15.1b, a hypothetical classification of land covers contains simply three classes A, B, and C. The first letter signifies the category (reference or classified) of the first (base) period (T_b), and the second letter signifies the category of the second period ($T_{b\pm1}$). Under this notation an entry CB classified and BA reference means that the classification category was C at time T_b and B at time $T_{b\pm1}$, whereas the reference data indicated that the category was B at time T_b and A at time $T_{b\pm1}$.

Table 15.2 identifies the sources of error in the classified change stratum and the classified no change stratum. We adopt the terminology of Khorram et al. (1998) so that the numbers in the various cells in the change detection matrix are defined as in Table 15.3. The major diagonal, (upper left to lower right), of the matrix has nine cells outlined in bold. The first three cells of the major diagonal represent the case ([1]) when there was no change in the land cover class and the polygons (pixels) were correctly classified at both times. The lower six cells of the major diagonal represent the case ([2]) when there was change and it was correctly classified. Four types of errors can occur and are labeled as [3], [4], [5], and [6]. This type of assessment

Table 15.2. Change Detection Matrix with the Sources of Error Rearranged in Logical Sequence.

			Reference No Change			Reference Change					
	T_b		A	B	C	A	A	B	B	C	C
		$T_{b\pm1}$	A	B	C	B	C	A	C	A	B
Classified No Change	A	A	1	3	3	5	5	5	5	5	5
	B	B	3	1	3	5	5	5	5	5	5
	C	C	3	3	1	5	5	5	5	5	5
Classified Change	A	B	4	4	4	2	6	6	6	6	6
	A	C	4	4	4	6	2	6	6	6	6
	B	A	4	4	4	6	6	2	6	6	6
	B	C	4	4	4	6	6	6	2	6	6
	C	A	4	4	4	6	6	6	6	2	6
	C	B	4	4	4	6	6	6	6	6	2

Table 15.3. A Summary of the Various Possible Combinations of Change/No Change Represented by the Change Detection Error Matrix.

Symbol	Actual Status	Classification Status	Error type	Occurrences	
[1]	true no change	correctly classified no change class	no error	3	N
[3]	true no change	incorrectly classified no change class	classif. error	6	N (N-1)
[5]	true change	incorrectly classified no change class	false neg.	18	N^2 (N-1)
[2]	true change	correctly classified change class	no error	6	N (N-1)
[4]	true no change	incorrectly classified change class	false pos.	18	N^2 (N-1)
[6]	true change	incorrectly classified change class	classif. error	30	$N^2 (N-1)^2$-N(N-1)
			total	81	N^4

is possible only for polygons (pixels) for which suitable reference data are available for both points in time.

The classified no change stratum comprises categories [1], [3], and [5], whereas the classified change stratum comprises categories [2], [4], and [6]. In general, if the classification algorithm allows for N (in this example, 3) LCLU classes, there will be N^4 (in this example, 81) different possible outcomes, of which N represent correct measurement of no change (category [1]) in land cover and N(N-1) (in this example, 6) represent correct measurement of change (category [2]). Of the remaining possible outcomes, N^3-N^2 (in this example, 18) are errors of omission (false negatives) contained in category [5], N^3-N^2 (18) are errors of commission (false positives) contained in categories [4], and the remaining N^2 $(N-1)^2$ (36), consisting of categories [3] and [6] are cases of misclassification errors at either or both times. The matrix thus makes it easy to consider the various possible land cover realities and corresponding land cover assignments for a sample at the two times.

The classified change stratum will typically be considerably smaller in area than the classified no change stratum and that gives us some advantages for sampling. For the classified change stratum we need to estimate at a minimum three components of change denoted as [2], [4], and [6] in Table 15.3, up to a maximum of [N^4-N^3] specific "from-to" categories. It should be pointed out that not all of the from-to categories are possible. For example, upland forests cannot change to wetlands, nor can clear-cut forests change to dense forests within, say, a decadal time frame. For this reason we normally will only need to estimate a subset of the possible from-to categories. However, in varied and heterogeneous landscapes this subset can be a relatively large fraction of the original set of possibilities.

The real advantage of partitioning a change map, into a change and a no change stratum is that we can conduct different kinds and intensities of sampling of the two strata. In some cases we may be more interested in the accuracy of estimates of overall change than in the accuracies of estimated frequencies of particular from-to classes.

2.3 Inclusion Probabilities for Polygons and Pixels Under Stratification

Accuracy statistics need to be adjusted to reflect the different sampling intensities between the strata. The adjustments are the inverse of the inclusion probabilities of the polygons or pixels. No adjustment is needed when we take a SRS of n pixels from N pixels in a thematic map. In this case the estimator for the cell probabilities is $\hat{p}=(n_{ij}/n)$ where n_{ij} denotes the number of reference sample pixels in reference class *j* classified into map category *i*. However, with stratified random sampling adjusting for different inclusion probabilities (Stehman and Czaplewski, 1998) yields: $\hat{p}=(n_{ij}/n_{i+})(N_{i+}/N)$, with n_{i+} and N_{i+} defined as the sample and population sizes for pixels in stratum *i*. Stehman (1996b) shows that the overall accuracy statistic with a stratified design is given as:

$$\hat{p}_c = \sum_{i=1}^{q} (n_{ii}/n_{i+})(N_{i+}/N) = (1/N)\sum_{i=1}^{q} (N_{i+}/n_{i+})(n_{ii})$$

If pixels are of equal size, then p_{ij} is unchanged using either pixel counts or areas. This is not true with polygons. Polygon counts yield different p_{ij} values than obtained using polygon areas. In this chapter we will be assuming that the primary interest is in defining p_{ij} as the proportion of area in LCLU class *i* and reference class *j*. Under SRS the cell proportions will then be estimated as: $\hat{p}_{ij}=(a_{ij}/a)$ where a_{ij} is the sample polygon area in LCLU class *i* and reference class *j*, and *a* is the area total of sampled polygons. With stratified sampling the estimate of the cell proportions is: $\hat{p}_{ij}=(a_{ij}/a_{i+})(A_{i+}/A)$ where A_{i+} is the area total in LCLU class *i* and A is the total area of the thematic map. It also follows that the overall accuracy statistic in this case is:

$$\hat{p}_c = \sum_{i=1}^{q} (a_{ii}/a_{i+})(A_{i+}/A) = (1/A)\sum_{i=1}^{q} (A_{i+}/a_{i+})(a_{ii})$$

2.4 Sample Size Determination for Estimating Overall Accuracy (P_c) or User's Accuracy Within Stratum*i*(P_{Ui})

We can use the binomial sample size formula (below) to calculate the sample size needed to estimate overall accuracy (P_c) to a specified level of precision. These same formulas can be used to calculate the sample sizes needed for estimating User's accuracy for each stratum ($P_{Ui} = p_{ii}/p_{i+}$), when simple random sampling is used within strata (Stehman, 1996b).

For the binomial distribution the probability that a sample of n units contains m units which are correctly classified is (Cochran, 1977, sec. 4.4):

$$pr(m) = \frac{p^m \cdot q^{n-m} \cdot n!}{m! \cdot (n-m)!}$$

Let $\hat{p} = m/n$, $\hat{q} = 1 - \hat{p}$. The variance of the estimated proportion of correctly classified units ($\hat{p}$) under SRS without replacement is (Murthy, 1967):

$$v(\hat{p}) = \frac{p \cdot q}{n-1} \cdot \left(\frac{N-n}{N}\right) = (1-f) \cdot \frac{p \cdot q}{n-1}$$

The variance function is greatest when the population is equally divided between the two classes and is symmetrical about this point. The standard deviation of $\hat{p}$ changes relatively little between $0.3 \leq p \leq 0.7$. However, at $p < 0.3$ or $p > 0.7$ the sampling effort needed decreases rapidly for a fixed level of error tolerance.

In the formula for sample size below, $\alpha = \Pr(|\hat{p} - P| \geq \delta)$ is the risk that the actual error is larger than δ, which is called the error tolerance. We assume that $\hat{p}$ is normally distributed. If the precision of the estimate has probability $1-\alpha = 0.95$ then the sample size is given as (Cochran, 1977, p. 75):

$$n \cong \frac{t^2 \cdot p \cdot q}{\delta^2}$$

Where t is the $1-\alpha/2$ quantile of the standard normal distribution. If we choose the error tolerance to be $\delta = 0.06$ take $t \cong 2$, and if $p = 0.75$ = the overall proportion correctly classified, then we need $n = 208$ samples (see Table 15.4) to estimate p within a bracket ± 0.06. When $p = 0.5$ then $n = 278$ samples are needed to meet the desired precision levels (see Table 15.4). Even though this seems counterintuitive when the population is evenly split, the level of sampling effort required to achieve a specified level of precision is much higher than when the population moves toward a dominant and rare composition. This is, as we have seen, because the variance of $\hat{p}$ is a maximum when $p = 0.50$.

It is often better to specify varying levels of precision for the different possible values of p than to try to achieve one absolute tolerance limit. For example, we might want to estimate $0.3 \leq p \leq 0.7$ within a tolerance $\delta = 0.10$; $0.7 < p \leq 0.9$ within an error tolerance $\delta = 0.06$, and for $p > 0.9$ within an error tolerance $\delta = 0.02$. The choice of error tolerances (δ) depends upon the relative importance of obtaining highly precise numbers, given the importance of the ecological area under investigation and the cost of obtaining the samples. In this last scenario, with varying error tolerances, we would need n = 100, 208, and 475 for $p = 0.5$, 0.75, and 0.95, respectively (see Table 15.4). So in this example, by employing changing error tolerances we needed fewer samples for $p = 0.5$ than when $p > 0.5$. This is the reverse of what occurs when we utilize one fixed error tolerance for all levels of p. It should be evident that the required error tolerances have a pronounced effect on the required sample size, as is implied by the above sample size formula.

Table 15.4. Sample Sizes for Multinomial and Binomial Sampling for Various Combinations of Numbers of Categories and Proportions of the Category in the Population for Several Confidence Interval Half Widths (δ) at α=0.10.

		Multinomial Sampling	Multinomial Sampling	Binomial Sampling
		k = 54 categories χ^2 = 9.69	k = 3 categories χ^2 = 4.53	k = 2 categories
δ	p	n = sample size	n = sample size	n = sample size
0.015	0.50	10767	5033	4444
0.015	0.75	8075	3775	3333
0.015	0.90	3876	1812	1600
0.015	0.95	2046	956	844
0.015	0.97	1253	586	517
0.02	0.50	6056	2831	2500
0.02	0.75	4542	2123	1875
0.02	0.90	2180	1019	900
0.02	0.95	1151	538	475
0.02	0.97	705	330	291
0.06	0.50	673	315	278
0.06	0.75	505	236	208
0.06	0.90	242	113	100
0.06	0.95	128	60	53
0.06	0.97	78	37	32
0.10	0.50	242	113	100
0.10	0.75	182	85	75
0.10	0.90	87	41	36
0.10	0.95	46	22	19
0.10	0.97	28	13	12

2.5 Simple Random Sampling for the Accuracy of Thematic Classes

In the previous section we focused on estimating the sample size needed to estimate the summary parameter (P_c) or User's accuracy within stratum *i* (P_{Ui}) to a given level of precision. In our experience we find that it is common to want to estimate the accuracy of many, if not all, the "from-to" components of the classified no change stratum and the classified change stratum. As previously noted, we normally work from a subset because not all the from-to components can occur. This corresponds to estimating the accuracy of many of the cells (P_{ij}) of the accuracy assessment matrix (Table 15.1a, Table 15.1b) which are defined as the proportion of area in the classified LCLU class *i* and reference LCLU class *j*.

This is a difficult proposition for several reasons. As will be seen in Section 3.1, the required sample size can be relatively large. And finding sufficient reference samples can be difficult. This is because with change detection we expect true change (change in the reference class) to be relatively rare, or sparse on the landscape. For example, in Dobson and Bright's (1994) study of the Chesapeake Bay, approximately three percent of the area had experienced change over a five year period. This means that each of the change categories was scarce enough to be considered a rare population since it was less than three percent of the total. While there is no formal definition of what frequency of occurrence constitutes a rare population, generally a frequency below 10 percent can be considered rare (Kalton and Anderson, 1986). This implies that we may encounter difficulties in locating the reference (true) change classes.

In OPIT accuracy assessment we calculate sample size to achieve a specified level of precision for the accuracy of the various classes, but we generally don't have to search hard to find the categories of interest. We may simply choose a subset of the classified polygons (pixels) with a sampling design such as SRS for comparison to the reference data. However, reference change classes tend to be rare, so they will not generally be found with small sample sizes under SRS. Thus, we need different techniques for "finding" the true change polygons or pixels. We use sampling techniques for rare events to uncover the true change polygons (pixels) which reside in both the classified change and no change strata. In Section 3.1 we consider how much to sample, to detect and estimate the accuracy of the individual (or groups of) change and no-change classes, and in Section 3.2 we discuss how to estimate the overall proportion of area correctly classified using disproportionate sampling.

Let Figure 15.3 represent the simple case where we are evaluating the User's accuracy (P_{Ui}) of one of the change map classes. Assume that we have conducted a complete inventory to determine where true change occurs. In Figure 15.3 true change pixels are depicted in black and true no change pixels are depicted in white. If we employ an equal probability design (systematic or SRS) with a sample size of n=20 pixels on this 16x29 pixel grid to estimate the User's accuracy, then the odds of finding very many of the rare (true change) pixels is quite low (see Figure 15.4 and Table 15.5). With this sample size the probability of not detecting any of the change pixels at all would be 0.59 (Table 15.5). And the odds of finding 1 of the 12 rare cases is about 30 percent, and finding two is about eight percent—assuming there are no periodicities in the pattern of occurrence of change pixels. The probability of uncovering all 12 of the change pixels with a sample of size 20 is under 1/1000th of one percent. In Figure 15.4 the circles represent the cases where the sample detected a true change pixel. So while such a design would lead to an unbiased estimate of the User's accuracy (P_{Ui}) (i.e., 12/464), this is small compensation for the fact that our chance of detecting any change pixels with a single sample of 20 is about 40 percent. In fact, the sample size would need to be 27 before our probability of detecting one or more change pixels would exceed 0.50. So, for this example we would need to sample about 5.8 percent (27/464) of the entire stratum to stand an even chance of detecting any true change at all.

This means that if we used traditional sampling techniques (e.g., SRS or systematic sampling) for estimating the User's accuracy (P_{Ui}), or the overall accuracy (P_c) using "reasonable" sample sizes, we are likely to detect only a small portion of the rare cases (true change). For example, in the Dobson and Bright study (1994) one hundred 500-meter circles were selected to assess the accuracy of their change map of the Chesapeake Bay region. Of those 100 circles, only eight contained any change. When we fail to uncover enough of these rather rare change reference classes, the precision of our estimate of User's accuracy will be quite low. The easiest conceptual way to overcome this problem is to vastly increase the sample size because as sample size increases, the detection rate increases. Unfortunately, costs of the sampling exercise escalate as well, often beyond the resources of the interested group(s). An alternative to simple random sampling for estimating the accuracy of rare classes is to use a stratified design, which is discussed in Sections 3.1 and 3.2.

3.0 INCREASING PRECISION

Designing an efficient sampling system is one of the most challenging tasks for a survey statistician, especially when some of the categories are rare (Kalton and Anderson, 1986). A

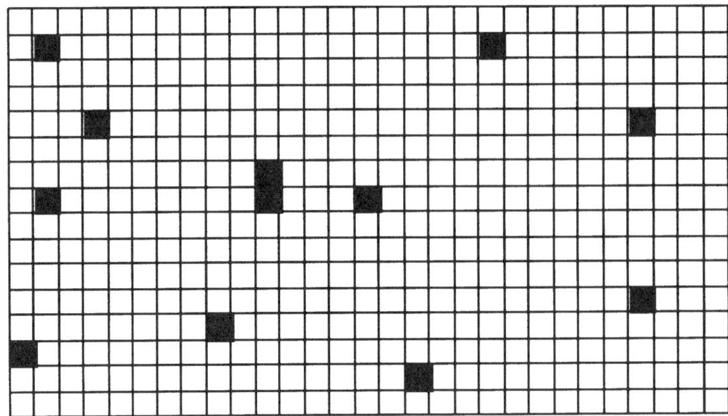

Figure 15.3. Reference map showing the locations of true change and no change pixels from a change stratum of the thematic map.

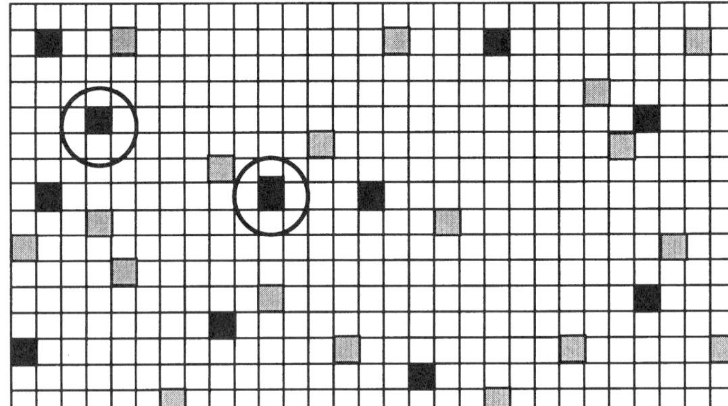

Figure 15.4. Results of sampling to determine the user's accuracy (P_{u_i}) from a change stratum i of the thematic map.

Table 15.5. Probabilities of Selecting 0, 1, 2 or More Change Pixels in a Simple Random Sample of 20 Sample Units from a Population of 464.

Number of Change Pixels Detected	Probability
0	0.59212
1	0.31440
2	0.07930
3	0.01263
4	0.00143
5	0.00012
6 or more	0.00001

number of methods have been devised to increase sampling efficiency, which are discussed in this section. These methods are applicable for either accuracy assessment measures (parameters) such as P_c or P_{Ui}, and/or for the individual cell probabilities (p_{ij}) of the error matrix.

3.1 Stratified Random Sampling

Stratified random sampling is widely used because it allows reporting of statistics by strata and in many cases it is more efficient than SRS. Stratification is also a useful tool in concentrating sampling effort for rare cases. Without stratification, simple random or systematic sampling uncover very few rare cases, and the precision of the accuracy measures for these classes will suffer accordingly.

In stratified sampling, the population of N pixels or polygons is divided into *k* subpopulations of sampling units N_1, N_2, ..., N_k termed strata. Sampling efficiency is improved if the strata are more homogeneous than the population as a whole. Raj (1968) says that all prior knowledge, personal intuition, and judgment can be used to bring about similarity in the strata. According to Cochran (1977) if done intelligently, stratification almost always is more efficient than SRS. The use of two strata: classified change and classified no change should serve this purpose well, owing to the fact that we expect there to be substantive differences in the accuracy between the classified change stratum and its complement. Another possible stratification includes target features discussed later.

Sample Sizes for Estimating Error Matrix Cell Probabilities in Individual Stratum

We can utilize multinomial sampling methods to determine the sample size needed to estimate the cell probabilities within each row of the error matrix. The sample size needed to estimate all the cell probabilities for a stratum, as we will show, is larger than needed for estimating User's accuracy for a stratum. Because the sample sizes inferred from multinomial sampling can be large and because there may not be interest in all cells within a row, Stehman (1996a) argues that it is often best to put the sampling effort into estimating User's accuracy. While this may be the case for a given project, in our experience we find that it is common to want to estimate the frequency of many, if not all, the "from-to" components of the classified no change stratum and the classified change stratum. Thus, we discuss it here. As previously noted, we normally work from a subset because not all the from-to components can occur.

The multinomial distribution is a discrete distribution in which N elements can be classified into *k* classes. We consider N to be the number of pixels or polygons in the remotely sensed scene. Under simple random sampling if p_i is the proportion of the population in the ith category (i=1, ..., k) and we draw a sample of size n, the probability of obtaining n_1 occurrences of class *1*, up to n_k occurrences of class k is given by:

$$p(n_1,\ldots,n_k) = \frac{n!}{\left[n_1! \ldots n_k!\right]} \cdot p_1^{n_1} \cdots p_k^{n_k}$$

Tortora (1978) calculates the approximate large sample equations for the simultaneous confidence limits of the multinomial distribution. Using Tortora's method, the sample size n_i for a category i (i=1, ..., k) is given by:

$$n_i = \chi^2_{(1,1-\alpha/k)} \cdot \frac{p_i \cdot (1-p_i)}{\delta_i^2}$$

Where $\chi^2_{(1,1-\alpha/k)}$ is the upper $(\alpha/k)\times 100^{th}$ percentile of the distribution χ^2 with one degree of freedom, δ_i is the half width of the desired confidence interval (error tolerance) for the ith category, and k is the number of categories to be estimated.

In general, it is necessary to make k calculations (i=1, ... , k) to determine the sample size. We choose the largest n_i as the desired sample size. When δ_i is the same for all categories, the only required calculation is for the p_i closest to 1/2. This provides for a conservative estimate of sample size since the largest sample size is required exactly at p=0.5.

The sample size calculated with this formula is appropriate for SRS without stratification. However, it can also be applied to a stratified design, provided we make this calculation for each of the individual strata. Consider the error matrix in Table 15.2. In this matrix, the classified change stratum has 54 individual classes, whereas each row has nine elements. If we calculate a separate n for each row (map class) of the classified change stratum (there are six rows), we arrive at a different sample size than if we calculate the sample size based on the entire 54 classes found in this stratum. (Note that these are really two different designs.) Within a row, the cell frequencies can be considerably higher than when calculated for the stratum as a whole, and this results in higher sample sizes for a fixed alpha and delta. Because the individual cell frequencies on a stratum basis are lower than calculated on a row basis we would probably want to decrease delta, which would tend to bring us to more equivalent sample sizes determined on a row or stratum basis. Nonetheless, these two approaches will not, in general, yield identical results.

In Table 15.4 it can be seen that 242 samples are required to estimate a category with a true p_i=0.5 and a confidence interval half-width (δ) equal to 0.10 when there are 54 categories (i.e., the number of categories in the classified change stratum for this example). The sample size reduces to 113 when there are three summary categories (categories [2], [4], and [6]). If we are willing to simply estimate the proportion of correctly and incorrectly classified pixels, we can recalculate the sample size needed using the binomial distribution, which in this case is 100 samples. Thus, reducing from three to two categories only produces minor reductions in required sample size. The greatest gain is from reducing 54 to three categories under the multinomial sampling model. Nonetheless, we expect that the normal case will be that we want to estimate all feasible elements of the Change Detection Error Matrix. Sample sizes for different values of p and δ (half-width of the confidence interval) are given in Table 15.4 for α=0.10 for multinomial and binomial models.

It is clear that the sample size needed to achieve a given level of precision increases as the number of categories (elements of the matrix) being estimated increases. However, the increase in sample size is not proportional to the increase in k, the number of categories for which accuracy assessment statistics are required, because k enters the calculation as the divisor of alpha.

3.2 Disproportionate Stratified Sampling

Often information is available indicating that rare population members (change pixels or polygons) may be concentrated in a stratum or strata. If the mapping procedure was done well, it is reasonable to assume that the classified change stratum would be home to large numbers of the rare population (true change). We might also have expert knowledge which leads us to expect that true change is concentrated in locations such as urban centers, fringes of urban centers, or in rural forested areas in specific locales. Or change may be related to other spatial/geographic variables such as steep terrain or riparian zones. Let's call these areas target features. If so, it may be possible to designate a spatial buffer around these target features where we suspect change activity would have a high probability of taking place. We term the target features and buffers to be target areas. Some target areas can be expected to occur in the classified change stratum, but others can occur in the complement (classified no change stratum). The target areas from the complement can be augmented with the classified change stratum to form a stratum from which to sample for the rare population. We term this the augmented classified change stratum. Alternatively we could treat the change stratum and the target areas from the complement as separate strata, both with expected large concentrations of the rare population.

Under stratified sampling the allocation of initial sampling units to estimate a parameter such as P_c or P_{Ui} can follow any number of allocation rules such as equal, proportional, optimal, or Neyman (Raj, 1968) with the selection of the sample units for separate strata made independently. With N-proportional allocation (proportional to stratum size) the sampling intensities are constant across strata and all subgroups of the population are sampled proportionally to their size. When replacement sampling is used with proportional allocation it is guaranteed to be more efficient than SRS. And when costs of sampling across strata are equal, then optimal (disproportionate) allocation is superior to proportional allocation (Cochran, 1977), as discussed below.

Disproportionate sampling is a design that increases the sampling fraction in the stratum in which the rare population is concentrated. In our application, disproportionate sampling uses a higher sampling fraction for the augmented classified change stratum than would be indicated using proportional allocation for stratified sampling. In this sense an optimum allocation scheme can be considered disproportionate sampling. Even though disproportionate sampling can potentially confer substantial efficiencies for the augmented classified change stratum, it comes at the expense of reduced efficiency in the complement. Obviously, for a fixed sampling effort increasing sample size in one stratum requires decreased sampling effort in another, with resultant decreases in precision. However, disproportionate allocation will also likely favor the precision of accuracy estimates for the combined strata.

Khorram et al. (1998) advocate disproportionate sampling as an efficient means of conducting accuracy assessment of change detection. Allocation of sampling effort to the strata can be accomplished to reflect the varying likelihoods of change, so that most of the effort is expended on polygons (pixels) most likely to have changed. Substantial gains in efficiency, as chronicled below, can be expected from such an approach, because most of the sampling effort will be expended where the changes in land cover are most likely. However, efficiency gains from optimal allocation come at the cost of increasing complexity in estimation (Stehman, 1996b).

First we examine the gains in efficiency of stratified sampling with proportional allocation as compared to SRS. Then we compare the gains in efficiency of disproportionate sampling over proportional allocation. The variance of $\hat{p}$ in *SRS* is pq/n where p is the population proportion, $q = 1-p$ and n is the total sample size; and under stratified sampling with proportional allocation the approximate variance of $\hat{p}$ is:

$$\frac{1-f}{n-1} \cdot \sum w_h p_h q_h$$

(see Cochran, 1977) where w_h is the stratum weight, p_h is the User's accuracy in the hth stratum and $w_h = (1-p_h)$.

The design effect is the ratio of variance under a more complex design (stratified sampling with proportional allocation) to variance under SRS. However, it is more convenient to take the inverse of this ratio. Defined in this way, as we shall do throughout the chapter, the design effect is the gain in precision from using the more complex design (Stehman, 1996a).

The design effect for stratified sampling (over simple random sampling) is given in Table 15.6 for specified values of overall thematic map accuracy (P_c). We consider overall map accuracies between 70–85 percent; change stratum User's accuracies (P_{U1}) between 30–60 percent and percent of total area (W_1) within the augmented classified change stratum between 5–25 percent. These results show in general that sampling efficiencies increase with: increases in the overall thematic classification accuracy, increases in percent area in the augmented classified change stratum, and decreases in the User's accuracy (P_{U1}) of the augmented classified change stratum. Some fairly substantial efficiency gains (well in excess of 25 percent) can be realized in some conditions by using stratified random sampling. Obviously, sampling efficiency implies that fewer samples can be taken to achieve a given precision of the estimator. This reduces the total cost of the sample survey. Stratification also has the benefit of increasing the odds of uncovering true change pixels (polygons) because the proportion of true change pixels (polygons) within the augmented classified change stratum should increase with effective stratification. There is a trade-off, however. Greater sampling efficiencies result as the difference between the overall thematic map accuracy and the accuracy of the augmented classified change stratum increases. Working counter to this is that as the accuracy of the augmented classified change stratum decreases, the odds of finding true change pixels (polygons) decreases.

Cochran (1977) shows that the largest gains in efficiency in stratified sampling relative to SRS occur when strata proportions strongly differ. This can be seen in our results in Table 15.6. Because we expect the normal case to be that the accuracy of classification for change classes will be lower than for nonchange classes, we expect there will be a substantial enough differential in the accuracy of classification of these various strata to achieve meaningful efficiencies from stratification. Of course, the analyst should never intentionally seek to degrade the classification accuracies of the change stratum to boost this effect.

Next we compare the gains in efficiency of disproportionate sampling (Neyman allocation) over proportional allocation. Neyman allocation assumes equal costs of sampling in each stratum so that total cost (C) $C=c_0+cn$ where c_0 is overhead cost, c is the cost of sampling per unit, and n is the sample size. In this case the optimum allocation of units which minimizes variance for a fixed cost reduces to optimum allocation for fixed sample size (Cochran, 1977). The actual allocation formula for sampling effort is:

$$\frac{n_h}{n} = \frac{N_h \cdot S_h}{\sum N_h \cdot S_h}$$

Where n is the total sample size, n_h is the sample size of the hth stratum, N_h is the size of the hth stratum, and $S_h = p_h q_h / n_h$ which is the standard deviation of the variable of interest ($\hat{p}_h$) in the hth stratum. In general, the gains in precision from optimum allocation over proportional allocation are not as large as the gains in precision from proportional allocation over simple random sampling.

Table 15.7a gives the design effect for Neyman allocation (equal cost) computed for the overall accuracy measure (P_c) and User's accuracy (P_{U1}). Efficiency ratios of 1.04 or better are

Table 15.6. The Design Effect (Relative Precision: RP) for Stratified Random Sampling with Proportional Allocation Compared with Simple Random Sampling (*SRS*) for Estimating the Overall Accuracy Measure (P_c) or User's Accuracy (P_{U1}). (RP values greater than 1.10 are highlighted in bold.)

		Values of W_1					
P_C=70.0%	P_{U1}	5	10	15	20	25	Definitions
	60%	1.00	1.01	1.01	1.01	1.02	P_c proportion of area correctly
	55%	1.01	1.01	1.02	1.03	1.04	classified
	50%	1.01	1.02	1.03	1.05	1.07	
	45%	1.02	1.03	1.06	1.08	**1.11**	
	40%	1.02	1.05	1.08	**1.12**	**1.17**	W_1 percent of area in stratum 1
	35%	1.03	1.07	**1.11**	**1.17**	**1.24**	
	30%	1.04	1.09	**1.16**	**1.24**	**1.34**	P_{U1} User's accuracy in stratum 1
P_C =75.0%	P_{U1}	5	10	15	20	25	
	60%	1.01	1.01	1.02	1.03	1.04	P_{U2} User's accuracy in stratum 2
	55%	1.01	1.02	1.04	1.06	1.08	
	50%	1.02	1.04	1.06	1.09	**1.13**	
	45%	1.03	1.06	1.09	**1.14**	**1.19**	stratum 1
	40%	1.04	1.08	**1.13**	**1.20**	**1.28**	augmented classified change
	35%	1.05	1.10	**1.18**	**1.27**	**1.40**	stratum
	30%	1.06	**1.14**	**1.24**	**1.37**	**1.56**	
P_C=80.0%	P_{U1}	5	10	15	20	25	stratum 2
	60%	1.01	1.03	1.05	1.07	1.09	the complement of stratum 1
	55%	1.02	1.05	1.07	**1.11**	**1.15**	
	50%	1.03	1.07	**1.11**	**1.16**	**1.23**	
	45%	1.04	1.09	**1.16**	**1.24**	**1.34**	
	40%	1.06	**1.13**	**1.21**	**1.33**	**1.50**	
	35%	1.07	**1.16**	**1.29**	**1.46**	**1.73**	
	30%	1.09	**1.21**	**1.38**	**1.64**	**2.09**	
P_C=85.0%	P_{U1}	5	10	15	20	25	
	60%	1.03	1.06	1.09	**1.14**	**1.20**	
	55%	1.04	1.09	**1.14**	**1.21**	**1.31**	* Note that slightly less than the
	50%	1.05	**1.12**	**1.20**	**1.32**	**1.47**	nominal P_C=85.0% value is
	45%	1.07	**1.16**	**1.28**	**1.46**	**1.72**	achievable for this combination
	40%	1.09	**1.21**	**1.39**	**1.66**	**2.13**	of W_1 and P_{U1} values. In this
	35%	**1.12**	**1.28**	**1.53**	**1.96**	**2.24***	case RP is computed assuming
	30%	**1.14**	**1.36**	**1.72**	**2.46**	**2.43***	100% accuracy in stratum 2.

highlighted in bold in Table 15.7a. These results show that there are additional efficiencies to be gained by optimal allocation, but they are more modest than those gained from proportional allocation versus SRS. Additional substantive gains occur for the higher overall thematic map accuracies (P_c: 80–85 percent), when the percent of area in stratum 1 (W_1), the augmented classified change stratum, is between 15–25 percent and for the lowest User's accuracies.

Table 15.7a also provides selected values for K_0, which is the ratio of the optimum sampling rate in stratum one (the augmented classified change stratum) to that in stratum two (the complement of stratum 1). For example, when P=80 percent, P_{U1}=45 percent and the percent of area in stratum 1 (W_1) is 15 percent, then K_0=1.44. This means that the ratio of the sampling rate in the change stratum to the complement is 1.44. This compares to a ratio of 0.18 (0.15/0.85) ex-

Table 15.7a. The Design Effect (Relative Precision: RP) for Neyman (Optimal) Allocation Relative to Proportional Allocation for Estimating the Overall Accuracy Measure (P_c) or User's Accuracy (P_{U1}). (RP values greater than 1.03 are highlighted in bold.)

			Values of W_1				
P_C=70.0%	P_{U1}		5	10	15	20	25
	60%	RP	1.00	1.00	1.00	1.00	1.00
	55%	RP	1.00	1.00	1.00	1.00	1.00
		K_0	1.09	1.10	1.12	1.13	1.15
	50%	RP	1.00	1.00	1.00	1.00	1.01
	45%	RP	1.00	1.00	1.00	1.00	1.01
		K_0	1.10	1.12	1.14	1.17	1.21
		Z_0	10.48	5.28	3.55	2.69	2.19
	40%	RP	1.00	1.00	1.00	1.00	1.01
	35%	RP	1.00	1.00	1.00	1.00	1.01
		K_0	1.06	1.09	1.12	1.17	1.23
	30%	RP	1.00	1.00	1.00	1.00	1.01
P_C=75.0%	P_{U1}		5	10	15	20	25
	60%	RP	1.00	1.00	1.00	1.01	1.01
	55%	RP	1.00	1.00	1.01	1.01	1.01
		K_0	1.17	1.19	1.21	1.24	1.29
	50%	RP	1.00	1.00	1.01	1.01	1.02
	45%	RP	1.00	1.00	1.01	1.01	1.02
		K_0	1.17	1.21	1.25	1.31	1.39
		Z_0	10.80	5.47	3.70	2.83	2.33
	40%	RP	1.00	1.00	1.01	1.02	1.03
	35%	RP	1.00	1.00	1.01	1.02	**1.04**
		K_0	1.14	1.18	1.24	1.34	1.49
	30%	RP	1.00	1.00	1.01	1.02	**1.04**
P_C=80.0%	P_{U1}		5	10	15	20	25
	60%	RP	1.00	1.01	1.01	1.02	1.03
	55%	RP	1.00	1.01	1.02	1.03	**1.04**
		K_0	1.28	1.32	1.37	1.44	1.55
	50%	RP	1.00	1.01	1.02	1.03	**1.06**
	45%	RP	1.00	1.01	1.02	**1.04**	**1.08**
		K_0	1.29	1.35	1.44	1.57	1.80
		Z_0	11.27	5.75	3.94	3.06	2.57
	40%	RP	1.00	1.01	1.02	**1.05**	**1.11**
	35%	RP	1.00	1.01	1.02	**1.06**	**1.16**
		K_0	1.25	1.34	1.46	1.69	2.19
	30%	RP	1.00	1.01	1.02	**1.07**	**1.23**
P_C=85.0%	P_{U1}		5	10	15	20	25
	60%	RP	1.01	1.02	**1.04**	**1.07**	**1.11**
	55%	RP	1.01	1.02	**1.05**	**1.09**	**1.18**
		K_0	1.46	1.55	1.68	1.89	2.28
	50%	RP	1.01	1.03	**1.06**	**1.12**	**1.29**
	45%	RP	1.01	1.03	**1.07**	**1.17**	**1.53**
		K_0	1.48	1.62	1.84	2.28	3.89
		Z_0	11.95	6.18	4.32	3.48	3.18
	40%	RP	1.01	1.03	**1.08**	**1.23**	**4.00**
	35%	RP	1.01	1.03	**1.09**	**1.34**	-
		K_0	1.45	1.63	**1.98**	**3.06**	-
	30%	RP	1.01	1.03	**1.10**	**1.59**	-

Definitions	
P_c	proportion of area correctly classified
W_1	percent of area in stratum 1
P_{U1}	User's accuracy in stratum 1
P_{U2}	User's accuracy in stratum 2
stratum 1	augmented classified change stratum
stratum 2	the complement of stratum 1
K_0	ratio of samples in stratum 1 to stratum 2 for Neyman allocation
Z_0	ratio of samples in stratum 1 under Neyman allocation relative to proportional allocation

pected under proportional allocation. This means that using Neyman allocation (which is based on variance) we put 3.94 times as many samples into the rare stratum as would be required with proportional allocation. We term this last ratio to be Z_0 and provide selected values in Table 15.7a. In this table it can be seen that relative to proportional allocation, Z_0 doubles the sam-

Table 15.7b. The Design Effect (Relative Precision: RP) for Neyman (Optimal) Allocation Versus Simple Random Sampling (*SRS*) for Estimating the Overall Accuracy Measure (P_c) or User's Accuracy (P_{U1}). (RP values greater than 1.10 are highlighted in bold.)

		Values of W_1					
P_C=70.0%	P_{U1}	5	10	15	20	25	Definitions
	60%	1.00	1.01	1.01	1.01	1.02	P_c proportion of area correctly
	55%	1.01	1.01	1.02	1.03	1.04	classified
	50%	1.01	1.02	1.04	1.05	1.07	
	45%	1.02	1.04	1.06	1.09	**1.12**	W_1 percent of area in stratum 1
	40%	1.02	1.05	1.08	**1.13**	**1.18**	
	35%	1.03	1.07	**1.12**	**1.18**	**1.25**	
	30%	1.04	1.09	**1.16**	**1.24**	**1.35**	P_{U1} User's accuracy in stratum 1
P_C=75.0%	P_{U1}	5	10	15	20	25	
	60%	1.01	1.02	1.03	1.04	1.05	P_{U2} User's accuracy in stratum 2
	55%	1.01	1.03	1.04	1.07	1.09	
	50%	1.02	1.04	1.07	1.10	**1.15**	stratum 1
	45%	1.03	1.06	1.10	**1.15**	**1.22**	augmented classified change
	40%	1.04	1.08	**1.14**	**1.21**	**1.32**	stratum
	35%	1.05	**1.11**	**1.19**	**1.29**	**1.45**	
	30%	1.06	**1.14**	**1.24**	**1.39**	**1.63**	stratum 2
P_C=80.0%	P_{U1}	5	10	15	20	25	the complement of stratum 1
	60%	1.02	1.04	1.06	1.09	**1.12**	
	55%	1.02	1.05	1.09	**1.14**	**1.20**	
	50%	1.03	1.08	**1.13**	**1.20**	**1.31**	
	45%	1.05	1.10	**1.18**	**1.29**	**1.45**	
	40%	1.06	**1.14**	**1.24**	**1.40**	**1.67**	
	35%	1.07	**1.17**	**1.32**	**1.55**	**2.00**	
	30%	1.09	**1.22**	**1.41**	**1.75**	**2.58**	
P_C=85.0%	P_{U1}	5	10	15	20	25	
	60%	1.03	1.08	**1.14**	**1.21**	**1.33**	
	55%	1.05	**1.11**	**1.20**	**1.32**	**1.54**	
	50%	1.06	**1.15**	**1.28**	**1.48**	**1.89**	
	45%	1.08	**1.20**	**1.38**	**1.70**	**2.63**	
	40%	1.10	**1.25**	**1.50**	**2.04**	**8.50**	
	35%	**1.13**	**1.32**	**1.67**	**2.63**	-	
	30%	**1.15**	**1.40**	**1.90**	**3.91**	-	

pling fraction in stratum 1 when it is 25 percent of the area, and approximately quadruples it when the percent area in stratum 1 is 15 percent. Clearly, this increases the precision of the estimate of User's accuracy in the augmented classified change stratum and also results in uncovering more of the rare reference classes than would be found with SRS.

The effect of Neyman allocation relative to SRS is given in Table 15.7b. This table shows the benefit of stratification *and* optimal allocation relative to SRS. The same major trends as appear in Table 15.6 are evident in Table 15.7b owing to the dominance of stratification in increasing the design effect.

Optimum allocation of sampling effort among strata makes provision for incorporating differential costs into the process. While we have thus far assumed that costs of measuring a sampling unit will be equivalent from one stratum to another (Neyman allocation), that need not be so. In some cases assessing polygons (pixels) in a change stratum might be less expensive to measure (observe) on average than in a no-change stratum by virtue of the fact that some types of changes of land cover are obvious. Examples of obvious change include recent clear-cutting of mature forest, construction of a new shopping center in what was formerly a meadow, or the filling of a lake behind a recently built dam, etc. The costs of verifying these sorts of changes are accordingly low relative to more subtle types of change and changes that require sophisticated on-the-ground measurements to confirm. In general, our approach is to increase our sampling effort in strata in which the costs are low, and reduce our sampling effort in strata in which the location and measurement of sample units are especially expensive (Cochran 1977).

3.3 Double Sampling with Ratio Estimation

Double sampling is a useful design which can be used when there exists auxiliary information (x) which is reasonably well correlated with the variable of interest (y). A beneficial feature of double sampling is that auxiliary information is required for only a subsample of the population, not for the entire population. Double sampling can be applied with stratification as well.

With double sampling, to contain costs, generally the main sample is somewhat reduced so that information on the covariate can be collected and analyzed. In the first large sample of size n', only x_i is measured, and in the second sample of size n, both x_i and y_i are measured. This design is beneficial when the increase in precision from the ratio estimates offsets the loss in precision due to the decrease in the sample size of the main sample (Cochran, 1977).

This sampling design can be used to increase efficiency for estimating the overall accuracy statistic ($\hat{p}_c$) or User's accuracy for stratum i ($\hat{p}_{U1}$). This can be seen by defining: y_i to be the proportion correctly classified in the second sample, and x_i to be the proportion of correctly classified with auxiliary data (refer to Section 3.5). Under the polygon model y_i and x_i are the proportions of area correctly classified in the second sample and with the auxiliary data, respectively. This method can analogously be used for improving efficiency in estimating individual cell probabilities (p_{ij}).

So far in this chapter we have been discussing LCLU change detection whose classes are discrete or categorical. Even though the underlying metric for some of the LCLU variates may be continuous, the classes formed from these underlying continuous variates are discrete. However, there are other change variables which are clearly continuous in nature that we may desire to track for change detection. For example, one may be interested in changes in leaf area index (LAI), the normalized difference vegetation index (*NDVI*), or biomass. For the case when the y variable is continuous, we can use double sampling with ratio or regression estimation at times T_b and $T_{b\pm1}$ (discussed below) to improve our sampling efficiency. Change is estimated as the difference in the values estimated at these two points in time.

Using LAI as an example, we know that it is expensive and difficult to measure this variate at two points in time for collecting change reference data which we will then compare with classified data. Double sampling used at two points in time can increase sampling efficiency. To do so at each point in time, we take a large sample and measure some other x variate such as tree diameter or height which is well correlated with *LAI*. Then on a much smaller sample we measure both *LAI* and the x variate.

Consider the following setup. In the first phase, n' units are selected with replacement in a SRS design and the mean of the x observations ($\bar{x}'$) forms an estimate of $\bar{X}$. In the second phase, a random subsample without replacement of size n is taken from the first sample, and the means of the y and x variables are calculated. Then a double sampling ratio estimator (*dR*) (Kish, 1965; Raj, 1968; Schreuder et al., 1993) is:

$$\bar{y}_{dR} = \frac{\bar{y}}{\bar{x}} \cdot \bar{x}' = \hat{R} \cdot \bar{x}'$$

However, $\bar{y}_{dR}$ is a biased estimator. The variance and bias of this estimator is given in the above references. If the second sample is chosen independently of the first, then the variance formula has one fewer term than when the second sample is a random subsample of the first sample. However, the differences between these two variances are small when the first phase sample is much larger than the second phase sample (Kish, 1965).

Murthy (1967) shows that the condition for the ratio estimator to be more efficient than SRS (i.e., the design effect) under both with replacement and without replacement sampling is when

$$\rho > \frac{1}{2} \cdot \frac{c(x)}{c(y)}$$

where c(x) and c(y) are the population coefficients of variation in x and y, respectively, and ρ is the correlation coefficient between the x and y variables. When the coefficient of variation is approximately equal for x and y, then ratio estimation is more efficient than SRS for $\rho > 0.5$. Murthy assumes that it is normally the case that c(x) will be less than c(y) so that the correlation coefficient required for efficient sampling can be less than one-half.

3.4 Double Sampling with Regression Estimation

As for double sampling with ratio estimation, double sampling with regression can be used to increase efficiency for estimating overall accuracy (P_c), User's accuracy for stratum i (P_{Ui}), or for improving efficiency in estimating individual cell probabilities (p_{ij}) with suitably defined x and y variables.

Similar to ratio estimators, double sampling linear regression (*dlr*) estimators use auxiliary information x. However, the adjustment is more complicated because we need to estimate the slope parameter (β) for the relationship between y and x using linear regression. Assume that the sampling is SRS without replacement at the first phase, and the second phase sample is drawn with SRS without replacement from the first phase sample. Then:

$$\bar{y}_{dlr} = \bar{y} + b \cdot (\bar{x}' - \bar{x})$$

where b is the least squares estimator.

Assuming that n is sufficiently large so that b well approximates β, then the large-sample variance can be derived (see Raj, 1968). The design effect is the gain in precision from using the more complex design. This design effect ratio is: $[1-\rho^2(1-n/n')]^{-1}$. Double sampling for regression is always more efficient than taking a direct sample of size n for y (i.e., without taking the auxiliary information). Stehman (1996a) points out that auxiliary information has a cost to it and double sampling with linear regression only becomes practical if ρ^2 is sufficiently high. For example, if ρ is 0.30 and n/n'=0.25, then the design effect is 7.2 percent (i.e., $[1-\rho^2(1-n/n')]^{-1} = 1.072$) with a resulting reduction in standard error of 3.6 percent. When ρ is increased to 0.5 or 0.7 with the same ratio of second to first phase units, then the design effect is 23.1 percent and 58.1 percent, respectively, with a resulting reduction in standard error of

10.9 percent and 25.7 percent. In general, real gains are achieved whenever $\rho \geq 0.5$, and this holds across a wide range of values for the ratio of n/n'. It is also interesting to note that based upon simulation results, Stehman (1996a) found that the first phase sample sizes needed to make var($\bar{y}_{dlr}$) nearly as precise as the general regression estimator (using auxiliary data for all N pixels in the population) are relatively small.

Murthy (1967) derives the optimum values of n' and n to minimize variance for a fixed cost C where the cost of sampling is of the form $C = c_0 + n'c_1 + n\,c_2$ (c_0 is overhead cost, c_1 and c_2 are the unit costs for acquiring information on x and y, respectively). Raj (1968) derives the relationship specifying when double sampling with a linear regression estimator is better than taking a direct sample of y for a specified cost. Assuming no overhead cost, this occurs when $\rho^2 > 4\,c_1\,c_2\,/\,(c_1 + c_2)^2$. If the cost of sampling for x and y are equal, then double sampling is never better than taking a direct sample of y.

3.5 Double Sampling with Regression Estimation for Proportions

As above, this sampling design can be used to increase efficiency for estimating the overall accuracy statistic (P_c) or User's accuracy for stratum i (P_{Ui}), or for improving efficiency in estimating individual cell probabilities (p_{ij}) with suitably defined x and y variables. It can be used with a pixel or polygon paradigm.

Stehman (1996a) shows how to use double sampling with regression to improve the estimate of P_c, the proportion of pixels correctly classified by thematic map. This is one of the key parameters of interest in accuracy assessment. The estimator still is of the general form listed above since p is a parameter over the interval [0,1]. If $\hat{p}_y$ is defined as the sample proportion of pixels correctly classified using the reference data in the second phase sample, $\hat{p}_x$ is the sample proportion of pixels correctly classified by the auxiliary data in the second phase sample, $\hat{p}'_x$ is the sample proportion of pixels labeled as correctly classified using the auxiliary data for the first phase sample, and b is the regression estimator then:

$$\hat{p}_{dlr} = \hat{p}_y + b \cdot (\hat{p}'_x - \hat{p}_x)$$

Cover maps and data from other remote sensing platforms including videography or aerial photography are potential data sources for auxiliary data useful for classifying pixels into LCLU classes (Stehman, 1996a). Any other spatial data such as digital elevation models, soil maps, and weather maps can also be used to form a prediction of LCLU, or the variate itself can be used as an auxiliary variate. Stehman shows that to obtain a five percent reduction in standard error with double sampling with regression estimation for proportions, a 90 percent accurate target thematic map needs a 67 percent accurate auxiliary map, and a 70 percent accurate target map needs a 55 percent accurate auxiliary map.

These types of auxiliary map accuracies are achievable in many instances. For example, Biging et al. (1991) examined the accuracy of forest photointerpretation as compared to reference data derived from extensive field survey. They found that photointerpretation accuracies for species composition were at least 75 percent and size class accuracies were near 75 percent. Crown closure was much more difficult to photointerpret, and the accuracies averaged around 40 percent. For this example, the auxiliary data for species and size would be more than sufficient to achieve good sampling efficiencies.

This approach is most natural for pixel based comparisons. However, it is possible to utilize this procedure with polygons. In Figure 15.5 there is a polygon from the change map and a soils

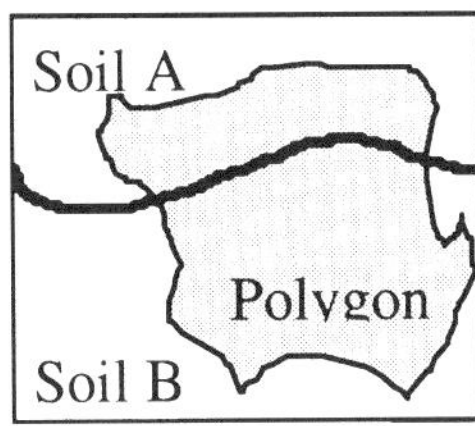

Figure 15.5. An example of using auxiliary data with polygon data.

map is overlaid. The soils map constitutes the auxiliary information. There are two soils types intersecting the polygon. To use the auxiliary information for the polygon we would need to derive a quantitative measure which is correlated with the LCLU type for the polygon. We can use a majority rule which assigns the soil type with the greatest area, in this example soil type B, or when more detailed information is available we can use another metric such as a weighted soil fertility where the weight is area of the polygon in a certain soil type multiplied by the organic carbon to a fixed depth (kg/m^2), for example. Then the weighted mean for a polygon is computed and used as x_i in calculating $\bar{x}'$.

3.6 Regression

Regression uses auxiliary data as in double sampling to increase precision. However, the difference is that a regression estimator requires that auxiliary information be available for the population (which may be the entire thematic map). When the relationship between y and x is linear, but not through the origin, then a regression estimator is more efficient than a ratio estimator. The general form of this estimator (Cochran, 1977) is:

$$\bar{y}_{lr} = \bar{y} + b(\bar{X} - \bar{x})$$

Where the subscript lr denotes linear regression, b is the slope of the line relating y to x, $\bar{X}$ is the population mean of the auxiliary variable which is presumed known, and $\bar{x}$ is the sample mean of the auxiliary variable.

The design effect for the regression estimator is $[1-\rho^2]^{-1}$ where ρ is the population correlation between y and x. So to achieve a 25 percent gain in precision from using the regression estimator, we need $\rho \cong 0.45$.

As for double sampling with ratio estimation, this sampling design can be used to increase efficiency for estimating overall accuracy (P_c), User's accuracy for stratum i (P_{Ui}), or for improving efficiency in estimating individual cell probabilities (p_{ij}). It can be used with a pixel or polygon paradigm. We expect that this approach, while efficient, will often be supplanted by double sampling with ratio or regression estimation which is nearly as efficient.

3.7 Adaptive Sampling

Adaptive sampling (Thompson, 1990) is a sequential statistical method in which it is possible to adaptively increase sampling effort in the vicinity of observed values that satisfy a condition of interest. Because many populations naturally cluster, or are clustered due to factors such as dispersal patterns, or patchiness of the environment (Cormack, 1988), adaptive designs can be superior to conventional designs as measured by increased power and lower expected sample size (Siegmund, 1985). Adaptive designs can increase efficiency when the location and shape of aggregations cannot be predicted before a survey. It is assumed that the clustering occurs over an area larger than the sampling unit. Conversely, if the patterns of occurrences are known beforehand, then conventional stratification can be combined with adaptive sampling to increase precision by allocating more sampling effort to the strata with the highest variability (Thompson, 1991).

This sampling design can be used to increase efficiency for estimating the overall accuracy statistic (P_c), User's accuracy for stratum i (P_{Ui}), or for improving efficiency in estimating individual cell probabilities (p_{ij}). It may be applied with polygons, but the details remain to be worked out. Let's say that we are estimating the overall accuracy of our simple change/no change example found in Figure 15.4 and discussed in Section 2.5. First a random sample of pixels is chosen for observation, as indicated by the gray pixels. If any of the sampled pixels from the predicted change map belong to a true (reference) change category, then additional pixels (top, bottom, left, and right) within a specified neighborhood of this pixel are added to the sample. If any of these supplemental pixels are determined to be change pixels, then the process is repeated. In Figure 15.6, the adaptive sampling strategy is shown for the eight pixels with a bold border. When the sample identified a change pixel (rare case represented as a black pixel within the group of eight pixels), then four additional pixels are added to the sample (top, bottom, left, and right). From this, another change pixel is encountered in the top pixel which is represented with a dot pattern. Then three additional pixels (top, left, and right) are established, but no other additional change pixels are uncovered so the adaptive procedure is stopped.

Classical sample estimators such as the sample mean used in this design are biased. However, Thompson (1990) derives unbiased estimators of the mean and variance using both a modified Hansen-Hurwitz estimator (Hansen and Hurwitz, 1943) and a modified Horvitz-Thompson estimator (Horvitz and Thompson, 1952). Both types were improved upon by the Rao-Blackwell theorem. Because the adaptive sampling procedure (the estimator together with the design) is design unbiased, the results do not depend on the type of population being sampled. However, the efficiency of the adaptive design relative to simple random sampling does depend on the type of the population to be investigated. Thompson (1990) shows that if the initial sample is simple random without replacement, the adaptive strategy using the Hansen-Hurwitz estimator has lower variance than its SRS counterpart if the within-network[2] variance of the population is relatively high. See Thompson (1990) for the exact conditions under which this occurs.

3.8 Stratified Adaptive Sampling

In the context of accuracy assessment for change detection, standard adaptive sampling can be useful when we have no prior knowledge about the location of change pixels which we can use in the survey design. However, when we have information about the likelihood of change occurring on the landscape, then stratified adaptive sampling or other types of sampling such as disproportionate sampling described above may be preferred. A stratified adaptive sampling

[2] The union of all the units that are observed as a result of the initial selection of unit i is termed a cluster. There is a subcollection of units within a cluster, called a network, which has the property that when any unit in the network is selected it would lead to inclusion in the sample of every other unit in the network. The subcollection of units which contain the change classes in Figure 15.5 form a network.

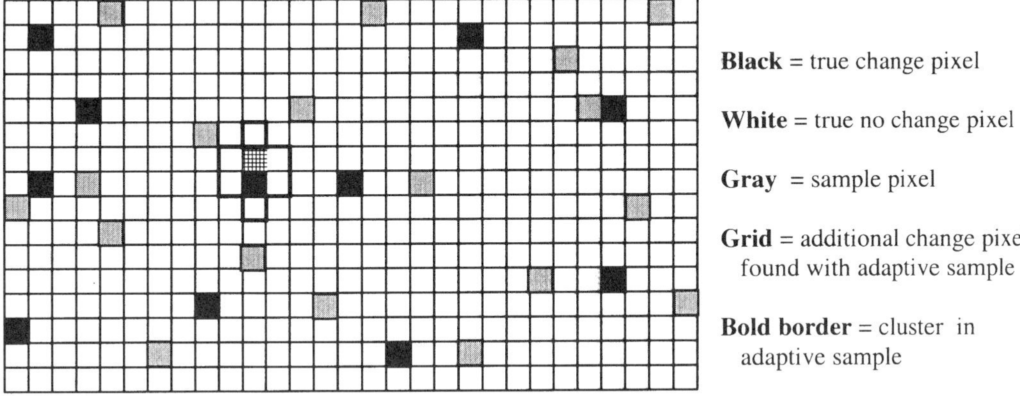

Figure 15.6. An example of using adaptive sampling to increase precision of the overall or User's accuracy statistics.

design can be used to increase efficiency for estimating the overall accuracy statistic (P_c), User's accuracy for stratum *i* (P_{Ui}), or for improving efficiency in estimating individual cell probabilities (p_{ij}). It may be applied with polygons as previously stated, but the details remain to be worked out.

Stratified adaptive sampling combines stratified sampling with adaptive sampling (Thompson, 1991). In this design, an initial without replacement sample of units is selected from the population using stratified random sampling. The allocation of initial sampling units can follow any number of allocation rules such as equal, proportional, optimal, or Neyman (Raj, 1968), with the selection of the sample units for separate strata made independently.

As with the standard adaptive procedure, whenever an observed sampled pixel belongs to a change category then additional pixels within a specified neighborhood of this pixel are added to the sample. If any of these supplemental pixels are determined to be change pixels, then the process is repeated. The conventional stratified sample mean estimator is biased under the adaptive design. Thompson (1991) derives unbiased estimators of the mean and variance using the techniques discussed under *Adaptive Sampling*. Efficiency gains for these estimators ranged from 13 percent to over 400 percent for the spatially aggregated Poisson cluster process simulated by Thompson (1991).

4.0 SUMMARY

Remote sensing data are invaluable and largely unsubstitutable for studying landscape level land use-land cover change. However, change maps developed from these data can be expected to contain significant errors because of the classification scheme employed, classification and rectification errors, remote sensor system characteristics, environmental conditions, and the change detection algorithms (Khorram et al., 1998). Therefore, credible interpretation of a change detection map is contingent upon an accuracy assessment to uncover the types and frequency of errors inherent in it.

Accuracy assessment of a change detection map is much more difficult than assessing the accuracy of an OPIT thematic map because the number of change classes is the square of the number of thematic map classes, and also because many of the "from-to" classes are infrequent or rare. High costs of establishing sufficient permanent reference sites means that assessment of the accuracy will generally entail probability sampling of pixels or polygons within the change map. The change detection error matrix, like its far simpler cousin, the OPIT error matrix, offers both conceptual and practical advantages for both representing the "potential" results, and for summarizing the "actual" results of a change analysis. Depending upon the classification system used, it may be possible to eliminate some change classes as virtually impossible and hence recognize any occurrences as incontrovertible errors.

Because assessing the accuracy of a change detection map is complex and challenging it is critical that any effort be well thought out and adequately planned. Simply using SRS in the interest of simplicity will certainly lead to much effort being consumed with very little information being collected on the accuracy of the change class. Significant advantages can be gained if one or more of the strategies outlined in this chapter are considered and implemented.

A number of alternative sampling designs and strategies of allocation of sampling effort are available for accuracy assessment. The sampling strategies discussed in this chapter all offer increases in precision over SRS. However, it is difficult to make general recommendations because the advantages and disadvantages of each approach depend crucially upon characteristics of the environmental change taking place and whether or not auxiliary variables are available for incorporation into statistical estimators. On the other hand, stratification of the population being sampled into strata having inherently greater or lesser likelihoods of change should consistently increase efficiency and reduce costs. Use of urban expansion models and other types of models to extrapolate past spatial-temporal trends of change into the spatial-temporal universe of interest are particularly promising in this regard.

Once stratification is completed, a number of sampling strategies can be used. These include disproportionate sampling, adaptive sampling, and double sampling. These, of course, vary in their difficulty of implementation and the simpler approaches may be more practical, if less efficient. We tend to favor disproportionate sampling because it should be relatively easy to apply and because it takes the highest proportion of samples from the strata for which it is most likely to find true change reference classes. Because disproportionate sampling uses a higher sampling fraction for the augmented classified change stratum than would be indicated using proportional allocation for stratified sampling, more true change pixels (polygons) should be sampled with the net result of increasing the precision of the estimates of User's accuracy. However, disproportionate allocation will also likely favor the precision of accuracy estimates for the combined strata.

Finally, we note that LCLU change is a continuous process in space and time. Boundaries separating classes may be easy to delineate on the one hand or require considerable subjective judgment on the other. Similarly, plant successional processes can blur transitions from one class to another. Consequently, some types of change can be confirmed with a glimpse of the site in question, while other types of change may require expensive measurements. Lastly, we must recognize the fact that a change detection map represents an attempt to capture environmental dynamics with a static product. Thus the value of any particular change map unfortunately begins to degrade as soon as it is produced, as does its assessed accuracy.

ACKNOWLEDGMENTS

The authors would like to thank Stephen V. Stehman for his technical review of this chapter and for the many helpful suggestions leading to substantive improvements in this work. The authors are also indebted to NOAA. The authors participated from 1992–1996 in a NOAA panel charged with recommending the remote sensing and sampling technology needed for detecting land use change in shoreline terrestrial and aquatic systems throughout the coastal

U.S. Through these workshops we developed the basis for many of the ideas found in this chapter. The authors also would like to recognize the support of this work through McIntire Stennis projects 6154MS (Biging) and MS-33 (Congalton).

REFERENCES

Anderson J., E. Hardy, J. Roach, and R. Wither. A land use and land cover classification system for use with remote sensor data. U.S. Geological Survey Professional Paper 964, 1976.

Biging, G., R. Congalton, and E. Murphy. A comparison of photointerpretation and ground measurements of forest structures, *ASPRS* Tech. Pap. 3:148–157, 1991.

Cochran, W. *Sampling Techniques*. John Wiley & Sons, New York, 1977.

Congalton, R. Using spatial autocorrelation analysis to explore errors in maps generated from remotely sensed data. *Photogrammetric Eng. Remote Sensing*. 54, No. 5, 587–592, 1988.

Congalton, R. A review of assessing the accuracy of classification of remotely sensed data. *Remote Sens. Environ*. 37:35–45, 1991.

Congalton, R. and K. Green. *Assessing the Accuracy of Remotely Sensed Data: Principles and Practices*, CRC/Lewis Press, Boca Raton, FL, in press, 1998.

Cormack, R. Statistical challenges in the environmental sciences: A personal view. *J. R. Statist. Soc.* Series A, 151, 301–310, 1988.

Dobson, J. and E. Bright. Large-area change analysis: The Coastwatch change analysis project (C-CAP). Proceedings of Peccary 12. Sioux Falls, SD. August 24–26, 1993. American Society for Photo. and Remote Sensing, Bethesda, MD, 1994, pp. 73–81.

Dobson, J., R. Ferguson, D. Field, L. Wood, K. Haddad, H. Iredale, J. Jensen, V. Klemas, R. Orth, and J. Thomas. NOAA Coastal Change Analysis Program (C-CAP): Guidance for Regional Implementation, NOAA, NMFS #123, 1995.

Gong, P., J. Miller, and M. Spanner. Forest canopy closure from classification and spectral unmixing of scene components—Multisensor evaluation of an open canopy. *IEEE Trans. Geoscience Remote Sensing* 32(5): 1067–1079, 1994.

Hansen, M. and W. Hurwitz. On the theory of sampling from finite populations. *Annals Math. Stat.* 14:333–362, 1943.

Horvitz, D. and D. Thompson. A generalization of sampling without replacement from a finite universe. *JASA*. 47:663–685, 1952.

Jensen, J. *Introductory Digital Image Processing. A Remote Sensing Perspective*. Prentice-Hall, Inc., Simon & Schuster, Upper Saddle River, NJ, 1996.

Kalton, G. and D.W. Anderson. Sampling rare populations. *J. R. Statist. Soc. A*. 149, Part 1, 65–82, 1986.

Kish, L. *Survey Sampling*. John Wiley & Sons, New York, 1965.

Khorram, S., G. Biging, D. Colby, R. Congalton, J. Dobson, R. Ferguson, M. Goodchild, J. Jensen, and T. Mace. Accuracy assessment of remote sensing derived change detection. *Photogrammetric Eng. Remote Sensing, Monograph*, in press, 1998.

Lunetta, R., R. Congalton, L. Fenstermaker, J. Jensen, K. McGwire, and L. Tinney. Remote sensing and geographic information system data integration: error sources and research issues. *Photogrammetric Eng. Remote Sensing*. 57(6):677–687, 1991.

Macleod, R., R. Congalton, and F. Short. Using quantitative accuracy assessment techniques to compare various change detection algorithms for monitoring eelgrass distributions in Great Bay, NH generated from Landsat TM data. Proceedings of the Sixty First Annual Meeting of the American Society of Photogrammetry and Remote Sensing, Charlotte, NC, Volume 3, pp. 876–885, 1995.

Macleod, R. and R. Congalton. A quantitative comparison of change detection algorithms for monitoring eelgrass from remotely sensed data *Photogrammetric Eng. Remote Sensing*. 64(3), in press, 1998.

Murthy, M. *Sampling Theory and Methods*. Statistical Publishing, Calcutta, India, 1967.

Raj, D. *Sampling Theory*. McGraw-Hill, New York, 1968.

Schreuder, H., T. Gregoire, and G. Wood. *Sampling Methods for Multiresource Forest Inventory.* John Wiley & Sons, New York, 1993.

Siegmund, D. *Sequential Analysis; Tests and Confidence Intervals.* Springer-Verlag, New York, 1985.

Stehman, S. Thematic map accuracy assessment from the perspective of finite population sampling. *Int. J. Remote Sensing.* 16(3):589–593, 1995.

Stehman, S. Use of auxiliary data to improve the precision of estimators of thematic map accuracy. *Remote Sens. Environ.* 58:169–176, 1996a.

Stehman, S. Sampling design and analysis issues for thematic map accuracy assessment. ASPRS/ACSM Proceedings. Baltimore, MD, Volume 1, pp. 372–380, 1996b.

Stehman, S. Selecting and interpreting measures of thematic classification accuracy *Remote Sens. Environ.* 62:77–89, 1997.

Stehman, S. and R. Czaplewski. Design and analysis for thematic map accuracy assessment: fundamental principles. *Remote Sens. Environ.,* in press, 1998.

Story, M. and R. Congalton. Accuracy assessment: a user's perspective. *Photogrammetric Eng. Remote Sensing.* 52(3):397–399, 1986.

Thompson, S. Adaptive cluster sampling. *JASA.* 85:1050–1059, 1990.

Thompson, S. Stratified adaptive cluster sampling. *Biometrika* 78(2):389–397, 1991.

Tortora, R. A note on sample size estimation for multinomial populations. *The American Statistician.* 32(3):100–102, 1978.

Wang, F. Improving remote sensing image analysis through fuzzy information representation. *Photogrammetric Eng. Remote Sensing.* 56(8):1163–1169, 1990.

Index

R

S